Biomechanics

Biomechanics

J. Sahoo
Professor
Department of Orthopaedics, IMS & SUM Hospital
Centre for Biomechanical Sciences
Siksha O Anusandhan University
Bhubaneswar, Odisha, India

R. K. Saxena
(Ex. Professor Biomedical, IIT Delhi)
Professor of Eminence
Department of Bioengineering
Shobhit University, Modipuram
Meerut, India

H. C. Das
Professor
Department of Mechanical Engineering
Institute of Technical Education & Research
Siksha O Anusandhan University
Bhubaneswar, Odisha, India

A. K. Dash
Associated Professor
Department of Mechanical Engineering
Institute of Technical Education & Research
Siksha O Anusandhan University
Bhubaneswar, Odisha, India

I.K. International Publishing House Pvt. Ltd.
NEW DELHI • BANGALORE

Published by
I.K. International Publishing House Pvt. Ltd.
S-25, Green Park Extension
Uphaar Cinema Market
New Delhi–110 016 (India)
E-mail:info@ikinternational.com
Website: www.ikbooks.com

ISBN: 978-93-82332-86-2

Published by Krishan Makhijani for I.K. International Publishing House Pvt. Ltd., S-25, Green Park Extension, Uphaar Cinema Market, New Delhi–110 016 and Printed by Rekha Printers Pvt. Ltd., Okhla Industrial Area, Phase II, New Delhi–110 020.

Scientific Advisory Committee

Preface

Biomechanics is an interdisciplinary area where the knowledge and methods of mechanics are applied to the structure and function of the living human system. It integrates biological systems by means of mechanical principles for the improvement of performances, prevention and alleviation of injury by qualitative and quantitative analysis. Biomechanics of fluids provides answer to complex biological transport problems, mass transfer and heat transfer. Mechanical and structural properties of biomaterials is necessary for successful medical implant designs. Mechanical behaviour of materials, i.e., elastic deformation, viscoelasticity and time-dependent behaviour, multiaxial loading and complex stress states, yielding and failure theories, and fracture mechanics. Orthopaedics, cardiovascular devices, dentistry and soft tissue implants are essential for proper and quick healing and rehabilitation. Application of mechanical engineering principles in mathematical and finite element modeling for designing methodologies and analysis has helped the diagnosis and treatment of various diseases and disabilities. Rehabilitation biomechanics addresses the impact of disability and the effectiveness of rehabilitation therapies to enhance human performance. Biomechanics has helped in the improvement of sports performance, techniques and prevention of injuries.

The vision of this book is to disseminate recent developments in the field of Biomechanical Sciences to the students, researchers, teachers, sports scientists, physiotherapists and orthopaedic surgeons, so they should able to enhance their skills and methodologies for research and development critically and optimally. Also, the objective of the book is to generate a dedicated team of researchers and teachers in Biomechanics to enhance the progress in the basic knowledge, research and facilities, so the resources can be pooled together and duplication of research work can be avoided.

Research articles reflecting almost all areas of biomechanical sciences have been received in NCBS-2014; proves the enthusiasm and dedication of scientists and researchers for biomechanics in India. Also, it has provided platform to young scientists for the evaluation of their work and guidelines for the futuristic research work and to develop improved methodologies.

In preparation of the book efforts were made to provide reader's flawless presentation of research work of our biomechanics scientists, however any omission or error is sincerely regretted. We express our heartfelt thanks to all delegates for their brilliant contributions for the progress of biomechanics.

We, the editors are grateful to Dr. M. R. Nayak, President of SOA University, Bhubaneswar for his support and encouragement. We are thankful to Prof. R. P. Mohanty, Vice Chancellor, SOA University for his constant guidance and advice.

Editors

Contents

1

Determination of Axial and Transverse Stiffness of Joshi External Stabilizing System by Experimental Method

Rajeev Kumar[1], R. C. Gupta[1], V. P. Sharma[2] and Sanjay Mishra[3]

[1]*Dept. of Mechanical Engineering, Institute of Engineering & Technology, Lucknow, UP*
[2]*Dept. of Physical Medicine and Rehabilitation, CSM Medical University, Lucknow, UP*
[3]*Department of Science & Technology, New Mehrauli Road, New Delhi–110066*

Email: sanjaykr.mishra@nic.in

ABSTRACT

The aim of the present study was to experimentally investigate the axial and transverse stiffness of Joshi External Stabilizing System for two types of specimen with identical geometrical configuration but different sizes of k-wires used. Two types of the JESS were configured in a material testing laboratory based on the measurements taken on a 62-year-old male patient with a metaphyseal tibial fracture. For the first specimen, k-wires of size 2 mm were used and for the second, k-wires of 3 mm size were used. Four samples of each specimen were configured (total 8 specimens). The metaphyseal tibial fracture was created by providing a 15 mm gap in the JESS construct. The specimens were tested experimentally under axial compression and under 3-point transverse in an INSTRON Universal Testing Machine. The average axial stiffness of the JESS with 2 mm size k-wires was 32.51 N/mm while the stiffness of the fixator with 3 mm k-wires was 45.81 N/mm. Further, the three point transverse stiffness of fixator in asterior-posterior plane with 2 mm wire was 554.49 N-mm/degree and with 3 mm k-wire was 868.08 N-mm/degree. These stiffness values were then compared with the stiffness of Illizarov external fixator reported in literature. The study provides the experimental data of axial and transverse stiffness of a specific type of JESS which will be useful for surgeons to have an accurate estimation of mechanical properties of the JESS compared to other stabilisation devices for fracture treatment. Further, by using 3 mm k-wire instead of 2 mm k-wire, the axial and transverse stiffness of the JESS are increased by 41% and 55% respectively.

Keywords: JESS; tibial fracture; axial stiffness; transverse stiffness.

1.1 INTRODUCTION

Joshi External Stabilising System (JESS) is a bone stabilising device used in the Indian subcontinent for the last 30 years [1]. The JESS was indigenously designed, fabricated and used by Dr B B Joshi in late seventies for the hand surgery however, due to its versatile nature and ease of applicability it is now being used in the treatment of variety of musculoskeletal disorders. Due to its simple design, reliance on readily available materials, versatility, light weight, easy manoeuvrability and low cost it has been used in the treatment of post burn contractures of the hand and wrist [2], interphalangeal joint contractures in leprosy [3], intra-articular distal radial fractures [4], management of idiopathic clubfoot [5], hand trauma and its sequels [6], calcaneal fractures [7] and CTEV [8, 9]. Many experimental procedures have been employed to study the behaviour of external fixation devices. Yilmaz et al. [10] conducted an experimental study to determine the stiffness characteristics of standard and hybrid Ilizarov circular fixators. Hasenboehler et al. [11] conducted an experimental study to compare the fatigue behaviour of Ilizarov frame with that of a tibial interlocking nail in a comminutedtibial fracture model. Bronson et al. [12] conducted a multi-variable study of stability of external circular fixation to evaluate how the manipulation of parameters of fixation and components of circular frame could improve and maintain optimal stability of bone fragments. It was concluded that contribution of each component to overall stability is dependent upon the mode of loading. Stein et al. [13] performed a biomechanical study on hybrid ring tubular external fixator to measure and compare the mechanical properties in transverse of the four-ring and three ring/one tube hybrid fixator.

Schrøder et al. [14] performed experimental investigations of four different configurations of the Hoffmann external fixation system using 4 mm steel pins to assess the mechanical properties. The study conducted the laboratory evaluation of mechanical properties of various unilateral frame configurations and compared it with the Hoffmann-Vidal frame. Doyle et al. [15] investigated the effect of same segment pin pretensioning on rigidity of a simple fixator system. Based on the observations, it was recommended that if same fragment pins are to be prestressed when an external fixator is applied, it should be done by distraction. Gardner et al. [16] evaluated the mechanical performance of the Pinless and Centrafix fixators for rapid application to tibial fractures. Experimental studies were conducted to measure the stiffness, maximum service loads and fatigue strengths of the fixators. Kumar et al. [17] investigated the axial compressive stiffness of JESS by experimental and by finite element method.

The JESS device was not patented by Dr BB Joshi, therefore, it is manufactured and marketed by many companies in India with common "JESS" name but each manufacturer has slightly different versions of the JESS. JESS is in extensive use by medical fraternity in India for more than twenty years. However, there arevery few studies in the literature describing its mechanical properties. As a result, estimation of its mechanical performance solely lies with the experience of the orthopaedic practitioner. Therefore, aim of the present study was to conduct an in-vitro mechanical test and to elucidate the mechanical properties of a selected JESS device which is used in the treatment of tibial fractures.

1.2 MATERIALS AND METHODS

A typical JESS frame for treatment of tibial fracture mounted on a cadaver tibia bone is shown in Fig. 1.1 and its laboratory specimen is shown in Fig. 1.2.

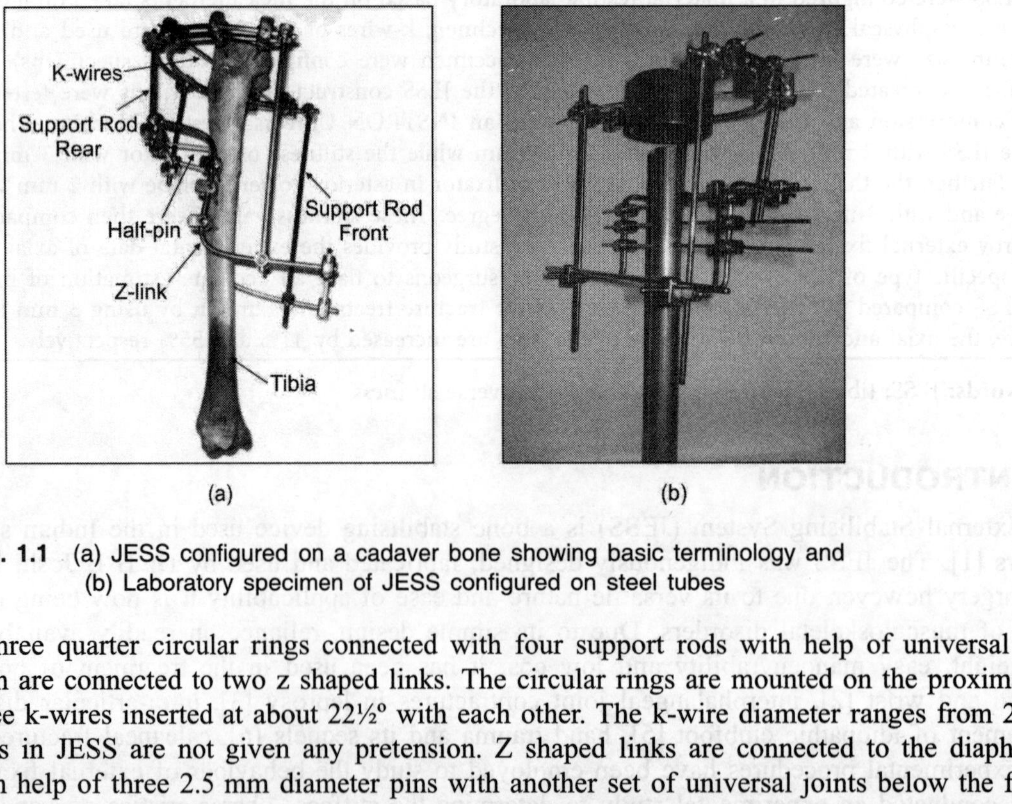

Fig. 1.1 (a) JESS configured on a cadaver bone showing basic terminology and (b) Laboratory specimen of JESS configured on steel tubes

It consists of two three quarter circular rings connected with four support rods with help of universal joints. The support rods in turn are connected to two z shaped links. The circular rings are mounted on the proximal tibia region with help of three k-wires inserted at about 22½° with each other. The k-wire diameter ranges from 2.0 mm to 3.0 mm. The k-wires in JESS are not given any pretension. Z shaped links are connected to the diaphysis in medio-lateral plane with help of three 2.5 mm diameter pins with another set of universal joints below the fracture fragment. In addition, one half pin of diameter 2 mm is inserted from posterior to provide further fragment stability.

In the present study, we have prepared two types of specimens. The dimensions of the JESS device were measured from a 62 year old male patient undergoing treatment for a metaphysical facture using JESS.All the specimens were prepared as per those measured data.In both kind of specimens, all the parameters were kept same except that first type of specimen used k-wires of diameter 2.0 mm while for the second specimen we used k-wires of size 3 mm. Table 1.1 lists the standard sizes of the different components used in making the JESS specimens.

For each type of specimen, four laboratory specimens of JESS-steel construct were prepared using two cylindrical pieces. The diaphysis of tibial bone was simulated by using a cylindrical stainless steel rod 250 mm long (outer diameter 22 mm with two mm wall thickness) and to simulate the proximal region of tibia a 35 mm long hollow cylindrical rod (outer diameter of 60 mm and inner diameter of 52 mm) was used. The higher diameter cylinder was used for proximal region to ensure that the effective length of k-wires should remain intact.The JESS was attached to both cylinders in a way shown in Fig. 1.2. The larger diameter cylinder simulating upper tibia was attached to the upper half circular rings using three stainless steel k-wires and smaller diameter cylindrical rod was attached to the JESS with the help of three pins inserted into the z shaped link along the medio-lateral direction and one half pin of along the anterior-posterior plane. The laboratory specimen is shown in Fig. 1.2. A 15 mm gap was maintained between the rod ends to represent the metaphysealtibial fracture. The gap was maintained at 15 mm as we wanted to measure the stiffness of fixator construct rather than that of the steel tubes.

Fig. 1.2 JESS specimen under axial compression

Since the aim of the study was to characterise the mechanical properties of the device therefore other components of the bone-fixator assembly such as bone and interfragmentary gap were idealised and approximated so that experimental test can be conducted.

1.3 TESTING OF JESS IN AXIAL COMPRESSION

The testing of JESS specimens under axial compression and under 3-point bend were carried out as per the guidelines suggested by Solomin [18]. The JESS-steel specimens were propped up vertically between the compression plates of a 10 ton capacity Universal Testing Machine (INSTRON 3382 at Central Institute for Plastic Engineering & Technology, Lucknow) for axial compression loading (Fig. 1.3).

The construct was subjected to an axial compressive load at 1 mm/min and the loading was gradually increased. As soon as the axial deformation at bone fragment site reached to a value of about 1.0 mm the universal testing machine was stopped as the loading of the frame beyond fragment displacement of 1 mm is not recommended [18]. Load on the JESS specimen was released and the specimen was again loaded to same axial deformation. This way, each specimen was tested five times under axial compression load. Axial stiffness of the fixator can be characterized by the rigidity coefficients of distraction and compression K_A:

$$K_A = F_A / u$$

where u is the fragment displacement in the axial direction due to axial compressive force FA. The unit for measuring the axial stiffness is Newtons per millimetre (N/mm). Table 1.2 shows the average axial stiffness of each specimen.

1.4 TESTING OF JESS UNDER THREE POINT TRANSVERSE

A special fixture was developed to test the JESS fixator under 3-point bend Fig. 1.3. The JESS specimen was supported on the vertical arms of the fixture in horizontal position.A transverse load was applied at the midpoint of the specimen at the rate of 0.1 mm/min and the loading was gradually increased. When the lateral deformation at bone fragment site reached to a value of 1.0 mm the machine was stopped. Load on the JESS specimen was released and the specimen was again loaded to same axial deformation. Each specimen was tested five times under transverse load. Figure 1.4 shows the JESS specimen supported on the transverse fixture and under 3-poiont transverse load condition. Transverse stiffness of the fixator may be characterized by K_B:

$$K_B = F_B\ L/\theta$$

where L is the length of the arm on which the transverse forces FB create transverse moments (distance from the point of application of the force to the site of rigid fixation of the module and θ is the angle of fragment rotation under the influence of the transverse forces, during abduction, adduction, flexion or extension. The unit for measuring transverse stiffness of the fixator is Newton millimetres per degree (N mm/degree).

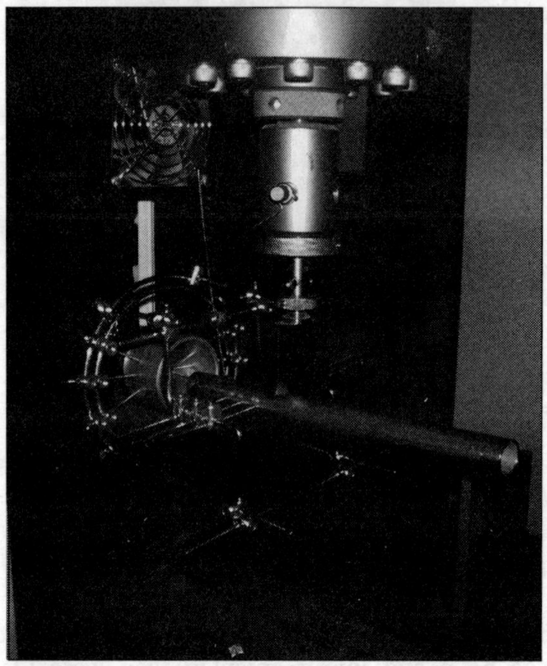

Fig. 1.3 JESS specimen under 3-point transverse loading

1.5 RESULTS

The stiffness was calculated from the slope of load-deformation curve obtained by laboratory testing. The stiffness of the JESS construct was 32.5093 ± 1.19 N/mm for the specimen having 2 mm size k-wire and 45.80 ± 2.26 N/mm for specimen with 3 mm size k-wires. The transverse stiffness of the specimen with 2 mm size k-wires was 554.49 ± 56.86 N-mm/degreeand 868.08 ± 62.77 N-mm/degree for specimen with k-wire of size 3 mm.

Table 1.1: Specifications of laboratory specimen of JESS

S. No.	*Part name*	*Quantity*	*Specifications*
1.	Inner ring	01	Mean diameter = 135 mm Rod diameter = 4 mm
2.	Outer ring	01	Mean Diameter = 155 mm Rod diameter = 4 mm

Contd.

3.	Support Rod	04 (two front and two rear)	Rod diameter = 4 mm Front Rod length = 130 mm Rear Rod length = 70 mm
4.	Z-link	02	Rod diameter = 4 mm
5.	k-wires	03	Specimen 1:Diameter 2 mm Specimen 2: Diameter 3 mm
6.	Half pin	01	Diameter 2 mm
7.	Pins	03	Diameter 2.5 mm Length 56 mm

Table 1.2: Average axial stiffness of JESS specimens

S. No.	*Specimen*	*Axial Stiffness N/mm*	
		k-wire = 2mm	*K-wire = 3mm*
1.	JESS-steel sample I	33.775 ± 1.618	48.3420 ± 7.155
2.	JESS-steel sample II	30.683 ± 0.410	45.6920 ± 2.901
3.	JESS-steel sample III	33.820 ± 0.487	42.8800 ± 3.229
4.	JESS-steel sample IV	31.759 ± 1.049	46.3180 ± 2.357

Table 1.3: Average transverse stiffness of JESS specimens

S. No.	*Specimen*	*Transverse Stiffness (Nmm/degree)*	
		k-wire = 2mm	*K-wire = 3mm*
1.	JESS specimenI	470.03 ± 27.82	800.27 ± 97.74
2.	JESS specimenII	696.91 ± 73.09	803.21 ± 64.71
3.	JESS specimen III	512.99 ± 55.47	880.41 ± 65.76
4.	JESS specimenIV	538.02 ± 71.06	988.43 ± 22.88

1.6 DISCUSSION

The axial and transversestiffness of the JESS construct was calculated in a Universal Testing Machine. Any axial deformation more than 1 mm at the fracture site is not good [18] therefore, the axial stiffness of the JESS frame was evaluated for an average fragment deformation of 1 mm. The other clinically relevant finding from this study is the quantification of axial and transversestiffness of JESS. The axial stiffness of the JESS was found to be 32.51 N/mm with 2 mm size k-wires and 45.80 N/mm with 3 mm size k-wires. The increase in axial stiffness by changing the size of k-wires from 2 mm to 3 mm was about 40%.The transverse stiffness of the JESS specimen under 3-point transverse was found to be 554.49N mm/degree for specimens with 2 mm k-wires and 868.08 Nmm/ degree for specimens with 3 mm k-wires. The effect of changing the size of k-wires from 2 mm to 3 mm on transverse stiffness was an increase of about 55%. However, this mechanical advantage has to be balanced as the 3 mm k-wires will do more soft tissue damage compared to a 2 mm k-wire (Fig. 1.4a & b).

The present study is perhaps the first to calculate the axial stiffness of the JESS device used in the treatment of tibial fractures and suggests due to its lower stiffness compared to other external fixation frames, the current design of JESS is not suitable for full load bearing activities of the patients. During the later stages of fracture healing, however, due to consolidation of callus tissues in the gap the loading will be shared by the bone as well as the JESS, therefore some weight bearing activities may be recommended. Further studies are required to demonstrate the mechanical performance of a JESS such as stiffness and interfragmentary movements during various stages of fracture healing.

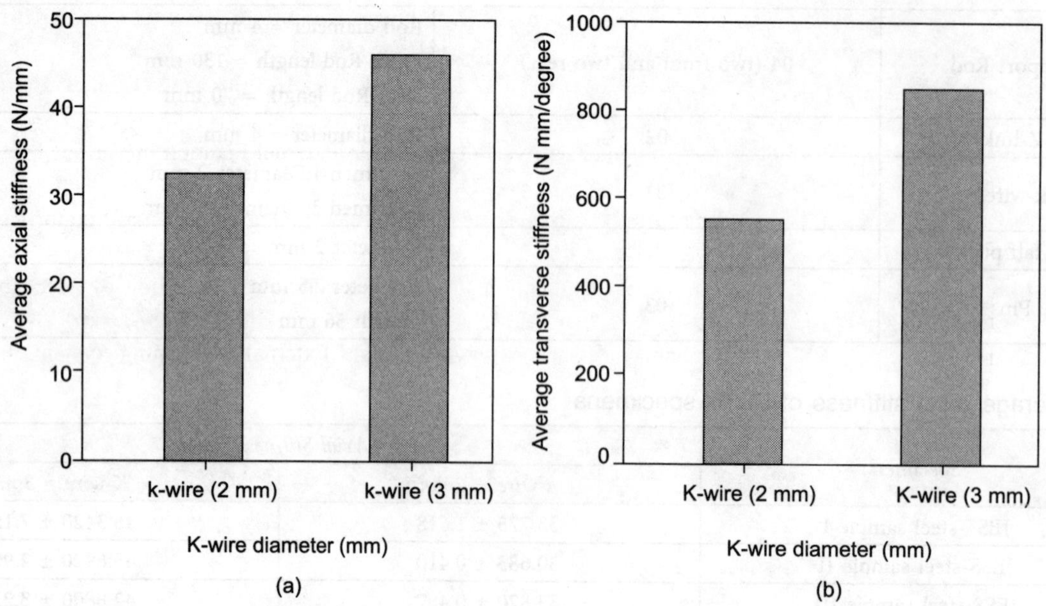

Fig. 1.4 (a) Average axial stiffness of JESS specimens (b) Average transverse stiffness of JESS specimens

1.7 LIMITATIONS OF THE STUDY

The results of the study are based on a four JESS construct used in stabilisation of tibial fractures. The JESS is available in slightly different modules by different manufacturers in India. Further, different configurations are used in clinics depending upon surgeon's own judgment and practice. Therefore, the above results may not be generalized for all types of the JESS configuration. Nevertheless, the present study quantifies the mechanical characteristics and stiffness of the JESS.

REFERENCES

1. Prabhu R.A. legend forever-Dr BrijBhusan Joshi. Indian Journal of Orthopaedics 2009; 43(3): 312.
2. Gulati S., Joshi B.B., Milner S.M. Use of Joshi External Stabilizing System in postburn contractures of the hand and wrist: a 20-year experience. Journal of Burn Care and Rehabilitation 2004; 25: 416-420.
3. Salafia A., Chauhan G. Joshi External Stabilising System (JESS) in proximal interphalangeal joint (PIP) contractures in leprosy. Indian Journal of Leprosy 1997; 69:331-9.
4. Thomas S., John C., Johnny T.P. Intra-articular distal radial fractures – external fixation or conventional closed reduction. J. orthopaedics 2007; 4(2) e39.
5. Suresh S., Ahmed A., Sharma V.K. Role of Joshi's external stabilisation system fixator in the management of idiopathic clubfoot. Journal of Orthopaedic Surgery 2003: 11(2): 194–201.
6. Joshi B.B. Joshi's External Stabilization System (JESS): a simple mini external fixator for the management of hand trauma and its sequels. Injury 1997; 28: 244-244.
7. Singh A., Srivastava R.N., Jah M., Kumar A. Ligamentotaxis for complex calcaneal fractures using Joshi's external stabilization system. Indian J Orthopaedics 2008; 42: 330-335.
8. Joshi B.B., Laud N.S., Warrier S., Kanaji B.G., Joshi A.P., Dabake H. Treatment of CTEV by Joshi's External Stabilization System (JESS). In: Kulkarni G.S., editor. Textbook of Orthopaedics and Trauma, 1st edition. New Delhi: Jaypee Brothers Medical Publishers Ltd; 1999.
9. Anwar M.H., Arun B. Short term results of Correction of CTEV with JESS Distracter J. Orthopaedics 2004; 1(1) e3.
10. E. Yilmaz, O. Bllhan, L. Karakurt, N. Arslan, E. Serin. Mechanical Performance of Hybrid Ilizarov External Fixator in Comparison with Ilizarov Circular External Fixator. Clinical Biomech., 2003, 18: 518-522.
11. E. Hasenboehler, W.R. Smith, L. Laudicina, G.C. Philips, P.F. Stahel, S.J. Morgan. Fatigue Behaviour of Ilizarov Frame versus Interlocking Nail in a ComminutedTibial Fracture Model: A Biomechanical Study. J Othrop. Surg. Res., 2006, 1:16-21.

12. D.G. Bronson, M.L. Samchukov, J.G. Birch, R.H. Browne, R.B. Ashman. Stability of External Circular Fixation: A Multi-Variable Biomechanical Analysis. Clinical Biomech., 1998, 13: 441-448.
13. H. Stein, R. Mosheiff, F baumgart, R. Frigg, S.M. Parren, J. Cordey. The Hybrid Ring Tubular Extrenal Fixator: A Biomechanical Study. Clinical Biomech., 1997, 12 (4): 259-266.
14. H.A. Schrøder, R.E. Weeth, T. Madsen. Experimental Analysis of Hoffmann External Fixation in Various Mountings. Clinical Biomech., 2001, 16: 151-159.
15. J. Doyle, P. Hayes, G. Fenlon. Experimental Analysis of Effects of Pin Pretensioning on External Fixator Rigidity. Archives OthropTraum. Surg., 1988, 107: 377-380.
16. T.N. Gardner, H. Simpson, J. Kenwright. Rapid Application Fracture Fixators: An Evaluation of Mechanical Properties. Clinical Biomech., 2001, 16: 151-159.
17. R. Kumar, Anupam Gupta, V.P. Sharma, Sanjay Mishra. Strength of Joshi External Stablishing System. J. Orthop. Surg., 2011, 19(1): 72-5.
18. Solomin L.N. The Basic Principles of External Fixation Using the Illizarov Device, Springer, 2005; 341-349.

2

Ergonomic Evaluation and Economics of Improved Harvesting Technology for Farm Women in Odisha

Bibhu Santosh Behera[1], Anama Charan Behera[2] and Rudra Ashish Behera[3]

[1]*Orissa University of Agriculture and Technology, Bhubaneswar, Odisha, India*
[2]*Department of Economics, D. B. College, Turumunga, Keonjhar, Odisha, India*
[3]*Vellore Institute of Technology, Balasore, Odisha, India*

Email: b.behera88@gmail.com

ABSTRACT

Many agricultural tasks demand high level of strenuous activity. Investment to improve agricultural worker's (Farmer and farm women) health would be justified on both humanitarian and economic ground. It is also quite true that maximum farm activities were carried out by farm women and these activities not only demands maximum physi-ological workload as well as energy expenditure also due to the highly drudgery involved in certain farm activities which ultimately leads to health problems and also ill effect on efficiency which reduce output of the activities. It causes considerable physical and mental fatigue and other health problem. The root cause of their suffering is ignorance about improved technologies, age-old method of doing the work. Hence, a study was conducted to assess ergonomically the efficiency of improved technology –Naveen Sickle for harvesting of wheat crop. Forty farm women were selected from adopted cluster area of Odisha (perticularly North Odisha) to assess and compare the impact of improved technology over conventional one. Drudgery index, RPE, degree of difficulty, muscular skeletal problems, efficiency and output, etc. were the main parameter of conducted study. Studies have pointed out that farm activities are time and labour intensive, monotonous, repetitive and more drudgery prone are generally performed by women. The result shows that improved method not only reduces drudgery by 28 per cent involved in the activity but also saves energy and time as well as minimizes muscular skeletal problems also. The work output was also found higher 138 m2 area with improved technology as compare to conventional method. In this way re-sults indicates that this method is useful to the farmers as well as to the particular labour also. Additional 286.7 m2 area covered per day and saves Rs. 506/ha. by the farmer. Farm women completed the work faster and saved nearly 3 hr. /day which can be utilised for another work. In terms of money calculated she could earn additional income Rs. 1450/month due to additional area covered per day.

Keywords: Drudgery, efficiency, muscular problem, farm women, technology, ergonomic.

2.1 INTRODUCTION

Women in India are the major work force in agriculture and perform almost all the agricultural activities. In India, out of 30 million women work force, 20 million live in rural areas. The rural women play a significant role in agriculture and other agro based activities. The daily work schedule of rural women is very demanding and arduous as per Suma Hasalkar et al. (2005). It is estimated that during peak period, women work every day for about 8-9 hours in agriculture and 4 hours in household activities and there are certain agricultural operations in which female agricultural workers are considered better than male workers as studied by Bhopl and Pattai (1998). Women carry-out many jobs as weeding, transplanting, harvesting, threshing and storing grains, tending animals and providing fuel and water etc. These tasks often have serious consequences for women due to the uncomfortable technologies of performance. A study conducted on farm women by Bibhu et al. (2005) stated that during wheat harvesting activity from morning till evening women usually adapts squatting posture and they continue to work in this posture for long duration without adapting any other posture due to which they reported severe pain in lower back and knees. Work-related musculoskeletal disorders develop slowly over months and years of repeated stresses. The risk factors them selves are ubiquitous, found in most jobs. Left unaddressed, musculoskeletal disorders can result in lifelong pain and permanent disability. It occurs due to the highly drudgery involved in the activities. Hence, Drudgery involved in the activity and ergonomic study of farm women is necessary. Drudgery is a term used to represent the dissatisfactory experiences that constrain work performance in any activity (Technical module/AI-CRP- FRM/ DRWA/ 2001) and Ergonomics is the new discipline in developing countries applied to the assessment of workload in various activities performed by women in home and farm (Suma Hasalkar et al (2005). Hence, keeping this fact in mind the present study was conducted by setting the objectives, viz. to personal and social background of the farm women, to assess the Drudgery Index of the farm activity-harvesting and to calculate the body muscular skeletal problems of the farm women while completing the harvesting activity.

2.2 METHODOLOGY

The study was conducted in three village, viz. Turumunga, Hatibari and Neuliposi which comes under the adopted cluster area of KVK-Keonjhar (Odisha). In Rabi season (Yr. 2009, 2010, 2011 and 2012) randomly 40 women were selected who were exclusively performed the farm activities. In the said villages. On Farm Trials (OFT) and Frontline Demonstrations (FLD) were conducted for reducing drudgery of farm women at the time of harvesting of wheat. Comparative study was carried out with improved technology (Naveen Sickle) and existing tool (local sickle) for assessing the drudgery involved in harvesting of crop (Fig. 2.1). Well prepared pro-forma was used for assessing the data.While collecting the data especial attention was given on the selected sample size as regards to their physical fitness and prevalence of any serious health hazard.

Fig. 2.1 Showing drudgery tools and women using drudgery at the time of harvesting of wheat

2.3 THEORITICAL FRAMEWORK

Tables 2.1 and 2.2 showing General profile and social background of the farm women (Housing etc.)

2.3.1 Blood Pressure

Blood pressure was measured with the help of sphygmomanometer (Digital) in mm. Hg. Systolic-120 + 10 (range 110-130) and Diastolic 80 + 10 (range 70-90)

2.3.2 Physical Fitness

Oral body temperature was measured with the help of thermometer in F.

2.3.4 Age Group

Criteria of age group were selected below 40 years purposively.

2.3.5 Degree of difficulty

Degree of difficulty was measured on five point scale ranging from very easy to very difficulty.

2.3.6 Time

Time were made in One hr. (60 minutes) observations the practices for assessing and for comparing the efficiency.

2.3.7 RPE

Perceived exertion was measured on five point scale ranging from very easy to very difficulty.

2.3.8 Drudgery Index

It was measured by using fallowing formula: $X + Y + Z * 100/3$

2.3.9 Improved Technology

Especially designed farm tool for harvesting of wheat – Naveen Sickle (source CIAE 1998) was used for the study. It was compared with the Naveen practice Sickle (local-sickle).

2.3.10 Specific Features of Tools

Light in weight, serrated blade, economical, easy to handle, covers more area, more suitable to wheat crop, vegetables and green fodder, etc., specially designed handle cuts the crop from bottom level, saves time, women friendly farm tool

Table 2.1: General profile and social background of the farm women

Parameters	*Respondent*	*%age*
26.0 (20-35) 35<	24	60
	12	30
1 – 45 (45-50) 50<	26	65
	04	10
Joint family	12	65
No. of Members		
4-5	8	45
7<	12	30

Table 2.2: Types of houses of women farmers

Types of Houses	*Families*	*%age*
Semipakka	8	45
Pakka	10	25

Data in Tables 2.1 and 2.2 pertaining to the assessment of total profile and social background of the farm women was depicted work being accomplished and the effect of earning per labour in Table 2.1. It was observed that 60 per cent farm women and saving to the farmer who employs the labours for harvest were belongs to 26-30 years age group and 30 per cent wereing, clearly indicates the effectiveness of improved sickle over in the 31-35 years age group where as only 10 per cent were the existing traditional sickle. On one hand it is evident that ain the 20-25 years age group. In case of weight also maximum labour covers an additional area of 91.9 m per day and thus the farm women's were dwelt in the 41-45 kg weight (65%), 25 quite less number of days to cover one hectare area but on the per cent were under the malnourished category and only 10 per contrary the same labour covers an additional area of 0.86 ha cent were in the normal weight, i.e. 10 per cent. In the table it per month which ultimately results in an additional income of was observed that nuclear family system was more common in Rs 1450. Similarly, for the farmer who employs the labour for the rural areas also (70%) as compare to joint family system harvesting is also at a benefit because his work gets done faster (30%). In the rural area type of house was also a status symbol and thus he too saves Rs 507 per hectare. Thus, the technology as result shows 25 per cent families were living in Pukka house is farmer as well as labour friendly in all aspects of work and whereas 45 per cent were living in semi pakka house and onlyeconomics. 30 per cent were living in kachha house.

Table 2.3: Feedback of the farm women regarding existing and improved technology

Existing Tool	*Improved Tool*
Heavy to handle	Light weighted, Easy to handle, Economical
Frequently sharpness of blade is required	Frequent sharpness of blade is not required and it can cuts the crop from bottom level
Maximum stress is required on hand muscles	Less stress on hand muscles etc.
Squatting posture is required for long period leads to Leg muscles problems.	Squatting is required for less period
Limitations & Constrains: Poor Output at the end of day Covers less area Produces Health Hazards Limitations and Constrains:	**Coves more area:** Saves time and money Practice is required Useless for left hand persons Not easily available in the market

Table 2.4: Showing results of income generated by traditional & improved sickle

<table>
<tr><th>Results</th><th>Area Covered/ day/m2</th><th>Required/ hectare</th><th>Area/ covered/ Daily</th><th>Additional area covered/month Month /Hectare</th><th>Total area covered</th><th>Daily wages/ Total area covered/ month</th><th>Additional Income/month</th><th>Total additional Income enerated</th></tr>
<tr><td>Traditional Sickle result</td><td>More</td><td>1 ha</td><td>91.9 meter/day</td><td>Nil</td><td rowspan="2">92.76</td><td rowspan="2">Rs. 680/hectare</td><td>Nil</td><td>Nil</td></tr>
<tr><td>Improved Sickle results</td><td>Less</td><td>1 ha 92.76</td><td>91.9 meter/day</td><td>0.86ha/month</td><td>Rs. 507/Ha</td><td>Rs. 1450/month</td></tr>
</table>

2.4 ECONOMICS OF IMPROVED TECHNOLOGY

Data in Table 2.4 pertaining to the assessment of total work being accomplished and the effect of earning per labour and saving to the farmer who employs the labours for harvesting, clearly indicates the effectiveness of improved sickle over the existing traditional sickle. On one hand it is evident that a labour covers an additional area of 91.9 m per day and thus requires less number of days to cover one hectare area but on the contrary the same labour covers an additional area of 0.86 ha per month which ultimately results in an additional income of Rs 1450. Similarly, for the farmer who employs the labour for harvesting is also at a benefit because his work gets done faster and thus he too saves Rs 507 per hectare. Thus the technology is farmer as well as labour friendly in all aspects of work and economics.

2.5 EQUATIONS

Drudgery Index:

It was measured by using fallowing formula: $X + Y + Z * 100/3$

2.6 RESULTS AND DISCUSSION

In Keonjhar reason wheat and gram is the main rabi crop. Harvesting of the crop is done manually and this is exclusively carried out by farm women. In the OFF and ON campus training programmes it emerged during the discussion that harvesting of the crop was more exerting, time consuming and due to the squatting posture maximum muscular pain in the body was improved practice (Naveen Sickle). It means near about 28 per cent drudgery could be reduced in improved practice. Reducing of drudgery it indicate that improved practice saves time, gives more output and it reduces body muscular skeletal problems also. It also indicate that women friendly farm tools minimizes maximum problems of the farm women but these equipment are not popular in rural area because the new techniques or technologies' are still not reached at bottom level particularly at rural area.After completing the study in general discussion was conducted with the farm women for receiving the feedback regarding the improved method (Naveen sickle) and existing practice (local sickle) is depicted in Table 2.3. Based on the experience and feedback of the farm women it was crystal clear that Naveen sickle is quite better than local sickle. Local sickle is very heavy to handle due to the heavy weighted handle and it covers less area and require more time and labour for harvesting of the crop. Squatting posture was adopted for long time creates more skeletal problem and it leads to body hazards to the particular ladies. On the contrary there were several positive points were elaborated by the farm women regarding improved method—Naveen sickle. The first and most important point was pointed out by the farm women that Naveen sickle was very easy to handle because of the light weight. The another important point was noted by them that it cuts the crop from bottom level and it helps to avoid the burning problem of residues at field level only. It cuts the crop very sharpen due to the serrated blade and covers more area and saves time and money. It minimizes the body muscular problems also. Only one strong point negatively pointed out regarding the Naveen Sickle that it was useless to the left handed person. Further they stressed out that it should be easily available to the faced by the farm women and it should be economic to all groups. Data on 40 farm women randomly selected for the study were Economics of Improved Technology: incorporated in the following tabulation format. The general data in Tables 2.1 and 2.2 pertaining to the assessment of total profile and social background of the farm women was depicted work being accomplished and the effect of earning per labour. It was observed that 60% farm women, who were employed as a labour for harvest were belongs to 20-25 years age group and 30 per cent were between 45-50 years of age group. Also, showed the effectiveness of improved sickle over the existing traditional sickle. On one hand it is evident that ain the 20-25 years age group. In case of weight also maximum labour covers an additional area of 91.9 m per day and thus the farm women's were dwelt in the 41-45 kg weight (65%), 25 quite less number of days to cover one hectare area but on the per cent were under the malnourished category and only 10 per contrary the same labour covers an additional area of 0.86 ha cent were in the normal weight i.e. 10 per cent. The improvement of per month income, ultimately results in an additional income of was observed that nuclear family system was more common in Rs 1450. Similarly, for the farmer who employs the labour for the rural areas also (70%) as compare to joint family system harvesting is also at a benefit because his work gets done faster (30%). In the rural area type of house was also a status symbol and thus he too saves Rs 507 per hectare. Thus, the technology as result shows 25 per cent families were living in pukka house is farmer as well as labour friendly in all aspects of work and whereas 45 per cent were living in semi pukka house and only economics. 30 per cent were living in kachha house.

2.7 CONCLUSION

The above findings showed that improved method of farm equipment is women friendly and very helpful in minimizing the drudgery of farm women while harvesting of crops. In this study, it was observed that Naveen Sickle is better than the local sickle because it covers more area, minimizes the drudgery (reduces by 28 per cent), perceived exertion was also low, saves time and money expent on labour, minimizes body muscular problems and very easy to handle. The only drawback observed was that this Naveen Sickle was not helpful to the left handers. The most important point which emerged from the end-users was that these advanced technologies could be adopted by the farm women but it will take time because these new techniques or women friendly farm tools have still not reached at the rural level. For reaching to unreached person some efforts are required to diffusion of new technologies.

REFERENCES

1. AICRP in Home Science, ICAR 2001. Published Progress Report (1996-2001). New Delhi: ICAR, pp. 108-109.
2. Bhople, R.R. and Pathai, A., 1998, Socio-economic dimensions of farm women labour. Rural India, 1: 192.
3. Jyotsana K R, Singh K, Mehta M 2005. Ergonomic Evaluation of Rural Women While Performing Wheat Harvesting Activity. Journal of Human Ecology, 18(4): 309-311.
4. Suma Hasalkar Shobha Huilgol Medha Hosakeri P. Kavita Chhayadevi A. Badiger (2005): 'Evaluation of Workload of Farm Women in Various Agricultural Activities' Karnataka J. Agric. Sci., 18 (3): (865-868) 2005
5. ICAR News Letter (2011-13), ICAR Pub.
6. Behera, B. S. (2013) Correlates of Adoption of Vegetables by Tribal Farmers of Keonjhar District of Odisha (Thesis, OUAT, Bhubaneswar).
7. Aum Sarma, 2004. Mechanisation for major oil seds and cereals in dry lands, winter school on mechanisation of dryland agriculture, CRIDA, Hyderabad. pp. 363-371.
8. Behera, B.K., S. Swain and S.K. Mohanty, 2007. Ergonomic evaluation of push-pull type weeders with women operators. Journal of Agricultural Engineering 44(3); 39-43.

9. DARE, 2006, Gender issues for technological empowerment of women in agriculture, ICAR, Annual report 2006-07, pp. 184-189.
10. DARE, 2007, Gender issues for technological empowerment of women in agriculture. ICAR, Annual report 2007-2008. pp. 116-118.
11. Neetu Sharma, 2002. Perception of farm women about feasibility of drudgery reducing farm implements. Annals of biology. 18(2); pp. 209-210.
12. Shirahatti, S. S., Ayyanagowdar, M. S. and Polisgowdar, B. S., 2007, Computer based water allocation to water users' associations- a model for water management officials, Karnataka Journal of Agricultural Sciences, 20(2) 335-337
13. Singh S. P. and L. P. Gite, 2007. Ergonomical evaluation of a hand operated paddy winnower by women workers. Journal of Agricultural Engineering, 44(4): 67-71.
14. Sirisha, D., R. Manian and K. Kathirvel, 2008. Development and evaluation of direct paddy seeder for assessing the suitability to rural women. A.M.A, Agricutlural Mechanisation in Asia, Africa, and Latin America, 39(4); 41-45.
15. Suman Singh, Puja Mathur and Madhu Rathore, 2007. Weeders for drudgery reduction of women farm workers in India. Journal of Agricultural Engineering 44(3); 33-38.

3

Development of Pulsatile Cardiovascular Flow Simulator and Assessment of Flow Through Stenosed Blood Vessel

Arindam Bit[1], Rajat Kabiraj[2] and Himadri Chattopadhyay[2]

[1]Assistant Professor, Dept. of Biomedical Engineering, National Institute of Technology, Raipur, India
[2]Director, School of Bioscience & Engineering, Associate Professor, Dept. of Mechanical Engineering Jadavpur University, Kolkata, India

Email: arinbit.bme@nitrr.ac.in

ABSTRACT

Blood flow through elastic vessels are considered to be a continuous process, while heart is treated as a pumping station as well as chamber for the temporal storage of blood for being filtered. Being colloidal solution, blood behaves like a non-Newtonian fluid at low shear rate, which often found at special geometry of the vessel and at the site of diseased blood vessel. Stenoses is a pathological disorder in blood vessels due to arthrosclerosis, and thus alter the lumen diameter of the vessel. This leads to significant changes in the flow dynamics through the effected blood vessel, and pulsatile nature of flow of blood adds to the complexity in the flow domain. Analysis of such a critical condition is of prime focus to most of the researchers. Also, designing of pulsatile pump for cardiovascular simulation is a challenging job, and thus focussed in this paper. Differential peristaltic pumps are used to in a parallel lumped manner in order to generate the pulsatile nature of flow similar to that generated as the outcome of cardiovascular flow. Flow domain has been created by plexiglass, considering the wall of the vessel to be in-elastic with respect to the radial deformation forces exerted by the flowing fluid on the wall. 40% solution of glycerine in distilled water at room temperature of 220 C is considered to be simulated as blood where the density of blood is comparable to the fluid density. However, viscosity of the above solution is realised as Newtonian model which shiftes to non-Newtonian behaviour at low shear rate. Flow domain at pre-stenoses, and post-stenoses region have been studied at progressive downstream method at different axial lengths by LDV. In order to enrich the study on effect of flow due to stenoses of different degrees, radial comparison of different flow parameters have been done at each distinct axial length. A detail study regarding Turbulent Kinetic Energy distribution, Pulse averaged statistics, time-averaged flow statistics, and fluid shear stress distribution with variation of degree of stenoses at different Reynolds number have been studied in the paper.

Keywords: Stenoses, peristaltic pump, LASER doppler velocimetry, turbulent kinetic energy, Reynolds number.

3.1 INTRODUCTION

Stenoses is a common vascular disorder causing blockage in arterial lumen, thereby narrowing lumen size. It occurs predominantly through plaque formation in the inner lining of the arterial wall. Haemodynamics plays a significant role in understanding the arterial wall thickening process as well as the effect of stenoses on the flow structure. The transport process in stenosed blood vessels involves different complex issues such as the pulsating nature of flow field and an appropriate rheological description of the dynamic fluid. Pulsatile flow of blood through artery had been considered by Siddiqui et al. [16] and Tu and Deville [20] by implementing a sine function, whereas in both cases non-Newtonian characteristics of blood had been emphasized. Neofytou and Drikakis [12] made tackled the problem by considering three different viscosity models of blood, namely Casson, Power Law and Quemada models for comparative study of the blood flow through stenoses. They had considered a stenotic geometry which is a function of upstream length of stenosis as well as the degree of stenoses. Foutrakis et al. [7] found that proteolytic degradation of aortic wall connective tissue, mechanical wall stress, molecular genetics, and inflammation and immune response are principal causes of aortic aneurysm. From flow mechanics standpoint, changes in geometry of aneurysm will induce changes in blood flow pattern, and thus hemodynamic stress distribution along the inner wall. Finite element analysis has been incorporated for addressing the smaller stresses derived from viscous and inertial components of flowing blood, assuming the flow to be laminar by Budwig et al. [2], and viscosity consideration to be Newtonian.

Although experimental approach of investigation of flow dynamics in blood vessels with aneurysm is limited, still LDA, PIV, and ultrasonic transducer have been used to address the problem considering simple constant flow or sinusoidal pulses at the inlet of the tube. Bruce et al. [1] had conducted an experiment to construct the phenomena of LASER Doppler Velocimetry in order to measure standard deviation of velocity for moving surfaces. Axial velocity in common carotid artery for different pulses was measured by Laser Doppler Annemometer by Gijsen et al. [8]. Spatial distribution of axial velocity in diseased and healthy vessels was monitored in the study with its transient analysis. In another study, Siouffi et al. [17] used Pulse Doppler Ultrasonic Velocimeter to measure velocity field of oscillatory flow rate at different Reynolds number at different degree of stenoses. Influence of Reynolds number on oscillatory flow rate at certain degree of stenoses was also observed in the study. Thus evolution of influence length was predicted in the same study. In a similar study done by Deplano et al. [6], temporal evolution of Wall Shear Stress was done at different Reynolds number. And thus viscous dissipation for different characteristic length at temporal variation was observed at different flow rates. LDV experiment was carried out in order to investigate the blood flow phenomena through long channel during purely pulsatile cycle based on a purely sinusoidal pressure gradient (Beratlis et al. [5]). Phase – averaged statistics of the stream-wise velocity at different time steps was monitored to for calculating mean and rms axial velocity. Banerjee et al. [4] used external pressure and flow sensors to monitor pressure drop coefficient, and pressure recovery factor. In another experiment conducted based on Particle Image Velocimerty (Brunette et al. [3]), three dimensional display (coronal and saggital cross-sections) representative of the global velocity was made. Three dimensional shear stresses were calculated within the phantom having transverse velocity vector at the neck of stenoses. Talukdar and Drikakis (2009) had used Pulsed Doppler Ultrasound system in order to monitor temporal variation of mean Doppler frequency for different degree of stenoses. Averaged Power Spectra for 60 milli seconds around peak flow for different degree of stenoses was monitored, and variation of relative disturbance with degree of stenoses was addressed.

In this paper, pulsating flow generator mimicking physiological flow profiles through large arteries have been illustrated, which is developed in the laboratory. Later, in order to validate the output of the flow generator pumping system, it is compared with the numerical results obtained by finite element analysis. LASER Doppler Velocimetry was used to generate LASER source in order to create the Field of View Focal Point for the measurement of transient responses of dynamic fluid in the optically sensitive tube. Measurement of the transient velocity profiles of the seeding particles in the fluid was being done by the processing unit. A comparative study of the outcomes from both experiment and numerical approach yield a benchmark model for the experimental setup. Details of the development of the flow generator along with the comparative study are represented in the paper.

3.2 METHODOLOGY

This section can be divided into two parts: numerical designing, and experimental benchmark development. Numerical part includes geometry development for finite element modelling, function development to represent the physiological pulsatile flow for inlet boundary condition, non-Newtonian viscosity modelling of blood. Experimental benchmark includes development of physiological pulsatile flow generator, measurement principle for flow visualisation.

3.2.1 Numerical Modelling

The axi-symmetric computational domain is given in Fig. 3.1, and the stenoses profile are defined by Neofytou and Drikakis [12] as follows:

$$y = 0,\ y = 1\ \forall x \in [0,7] \cup [11.6,40] \qquad \text{...(1)}$$

$$y = 0,\ y = 1 - \left(a - R + \sqrt{R^2 - \left[\frac{b}{2} - (x - L_u)\right]^2}\right) \qquad \text{...(2)}$$

where $R = \dfrac{\left[a^2 + \left(\frac{b}{2}\right)^2\right]}{2a}$ $\forall x \in (7,11.6)$ and $a = 0.56, b = 4.66.$

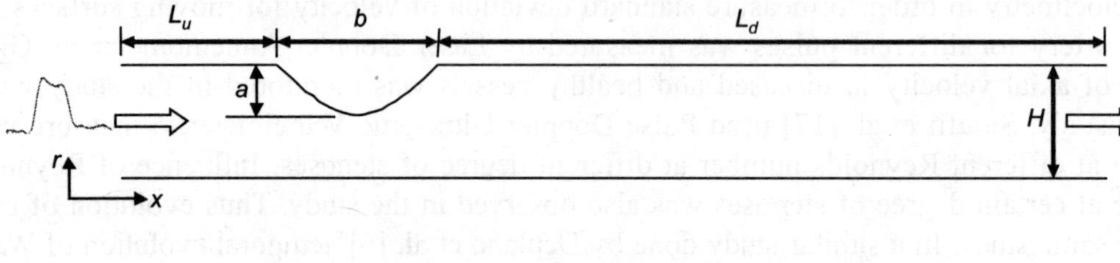

Fig. 3.1 Stenosed blood vessel geometry definition

The blockage of the vessel, i.e degree of stenoses is defined as the ratio of maximum height of stenosed part a to the vessel inner radius R.

Discretization of the models has been done by Quadrilateral geometrical mesh, with pave meshing. A finer mesh was developed at the region of aneurysm and stenosis with a gradient geometrical decrease in mesh size, while a linear constant mesh size was considered away from the geometry of aneurysm. Localised pressure deflections leading to compressibility of fluid due to fluid structure interactions is neglected and incompressibility property has been considered for the simplicity of the model. This reduces the mass conservation continuity equation which is given by equation 3 as:

$$\frac{1}{r}\frac{\partial}{\partial r}(rv) + \frac{\partial}{\partial x}(u) = 0 \qquad \text{...(3)}$$

Similarly, the conservation of momentum equation along axis of symmetry as well as along the radial direction consists of a transient term, convective term, a negative pressure term, and diffusion term, and can be represented by the equations 4 & 5 as:

Axial Momentum equation

$$\frac{\partial u}{\partial t} + \frac{1}{r}\frac{\partial}{\partial r}(ruv) + \frac{\partial}{\partial x}(u^2) = -\frac{1}{\rho}\frac{\partial p}{\partial x} + \frac{1}{r}\frac{\partial}{\partial r}\left[\frac{\mu_{app}}{\rho} r \left(\frac{\partial u}{\partial r} + \frac{\partial v}{\partial x}\right)\right] + \frac{\partial}{\partial x}\left(\frac{2\mu_{app}}{\rho}\frac{\partial u}{\partial x}\right) \qquad \text{...(4)}$$

r – Momentum equation

$$\frac{\partial v}{\partial t} + \frac{1}{r}\frac{\partial}{\partial r}(rv^2) + \frac{\partial}{\partial x}(uv) = -\frac{1}{\rho}\frac{\partial p}{\partial r} + \frac{1}{r}\frac{\partial}{\partial r}\left(\frac{2\mu_{app}}{\rho} r \frac{\partial v}{\partial r}\right) - \frac{2\mu_{app}}{\rho}\frac{v}{r^2} + \frac{\partial}{\partial x}\left[\frac{\mu_{app}}{\rho}\left(\frac{\partial u}{\partial r} + \frac{\partial v}{\partial x}\right)\right] \qquad \text{...(5)}$$

Wall of the blood vessels in both cases have been considered rigid wall, implementing a no slip boundary condition at the wall, and a zero gauge output pressure has been considered at the output of the vessel due to the presence of the whole system (stenosed blood vessel) in the submerged body fluid. 4 D MRI Phase Contrast image has been considered for the realistic mosaic pulsatile blood flow profile as the input mass flow rate of this problem. Transient profile obtained from the above data has been numerically modelled by generating a Gaussian function and is given by equation 6.

$$u(t) = -0.0214 \times \exp\left[\left(\frac{-(t-0.5458)}{0.008372}\right)^2\right] + 0.04842 \times \exp\left[\left(\frac{-(t-0.3374)}{0.1284}\right)^2\right]$$

$$+0.0407 \times \exp\left[\left(\frac{-(t-0.4749)}{0.1268}\right)^2\right] + 0.08353 \times \exp\left[\left(\frac{-(t-0.6215)}{0.234}\right)^2\right]$$

$$+0.09432 \times \exp\left[\left(\frac{-(t-0.0.002087)}{0.3171}\right)^2\right] + 0.08364 \times \exp\left[\left(\frac{-(t-1.006)}{0.314}\right)^2\right] \quad ...(6)$$

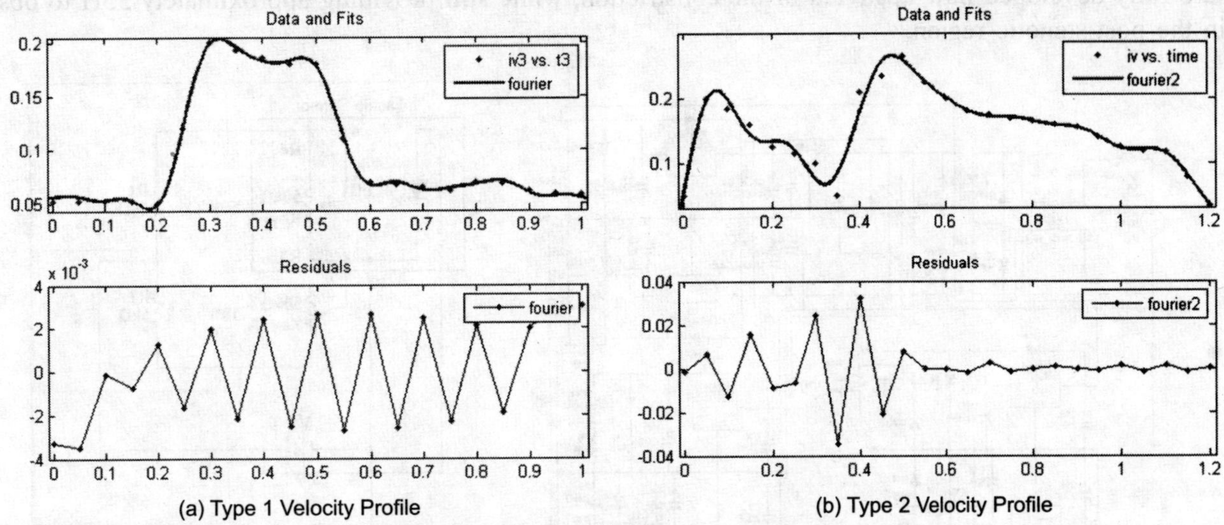

Fig. 3.2 Two different inlet velocity profiles

Considering the blood viscosity as non-Newtonian, Power Law model proposed by Walburn and Schneck [21], hematocrit and total protein minus albumin was taken into account. It is known as Power law model, also referred to as 'Best three variable model'. It is represented by the system of equations 7 and 8.

$$\overline{\tau} = k\left|\overline{\gamma}\right|^{n-1}\overline{\gamma} \quad ...(7)$$

and

$$\mu\left(\left|\overline{\gamma}\right|^*\right) = k\frac{U_\infty^{n-1}}{l^{n-1}}\left|\overline{\gamma}\right|^{*n-1} \quad ...(8)$$

For normal blood samples the parameters are $k = 14.67 \times 10^{-3}$ Pa.s, and $n = 0.7755$ as described by Ku D.N et al. [10]. If $n > 1$, the fluid is known as shear-thickening, while if $n < 1$ it is known as the shear thinning fluid. The characteristic parameter for a power-law-model-based flow is Re_{PL}.

Further considering the viscosity of concentrated disperse system, Quemada [15] proposed a model based on shear rate and haematocrit. System of equations of shear stress and effective viscosity in tensorial form and dimensionless form are described in equations 9 and 10 as:

$$\overline{\tau} = \mu_F\left(1 - \frac{1}{2}\frac{k_0 + k_\infty\sqrt{\left|\overline{\gamma}\right|/\gamma_c}}{1+\sqrt{\left|\overline{\gamma}\right|/\gamma_c}}\phi\right)^{-2}\overline{\gamma} \quad ...(9)$$

and

$$\mu\left(\left|\overline{\gamma}\right|^*\right) = \mu_F\left(1 - \frac{1}{2}\frac{k_0 + k_\infty\sqrt{\left|\overline{\gamma}\right|^*/\gamma_c^*}}{1+\sqrt{\left|\overline{\gamma}\right|^*/\gamma_c^*}}\phi\right)^{-2} \quad ...(10)$$

where $\gamma_c^* = \dfrac{\gamma_c}{U_\infty/l}$, and the parameters are given $\mu_F = 1.2 \times 10^{-3}$ Pa.s as is the viscosity of plasma (suspending medium) and for haematocrit $\phi = 0.45$ the values of the parameters are $\gamma_c = 1.88s^{-1}$, $k_\infty = 2.07$ and $k_0 = 4.33$. The characteristic parameters for a Quemada-model-based flow are Re_{QU} and γ^*_c.

3.2.2 Experimental Setup

The experiments were conducted in a closed-loop pulsatile flow facility designed in-house. Geometry of the test section was maintained to be of same parameters as considered for numerical study. The symmetric stenosis was machined from brass in the shape of two cylindrical segments (radius of curvature of 12.5 mm and a height of 3.13 mm), giving a blockage ratio of 12.5%, 25% and 50%. The stenosis was placed 110H from the channel entrance to ensure fully developed flow upstream of the constriction, while still providing approximately 35H to observe the flow in the post-stenotic region.

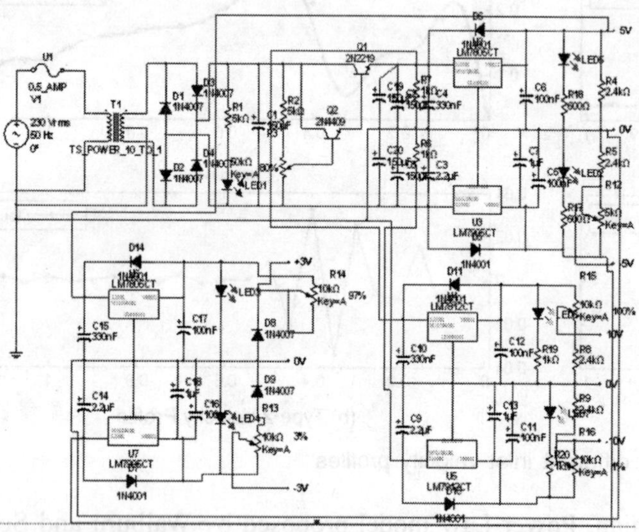

Power Supply of differential motor control board

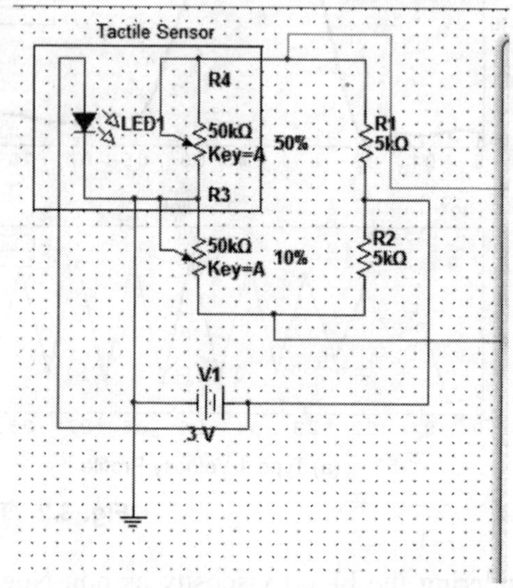

Sensor Unit of differential motor control board

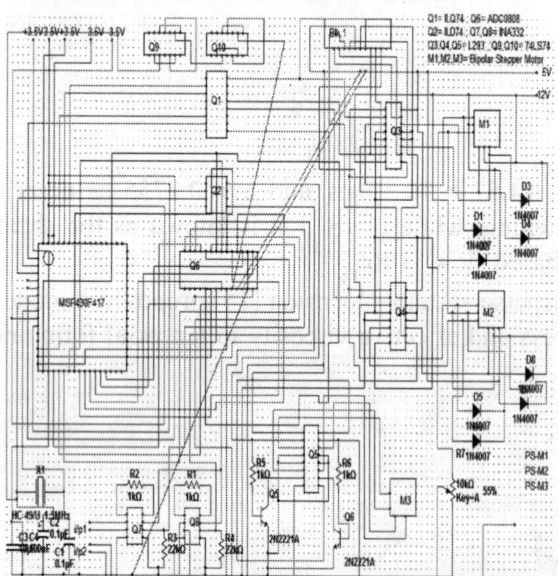

Processing unit of differential motor control board using MSP430mf417

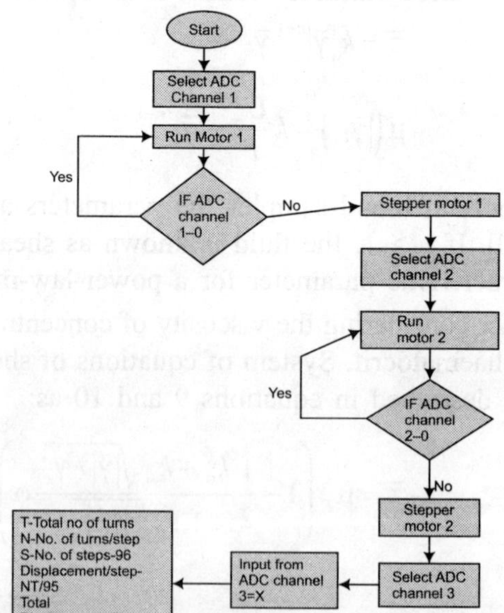

Flow Chart of differential motor control board

Contd...

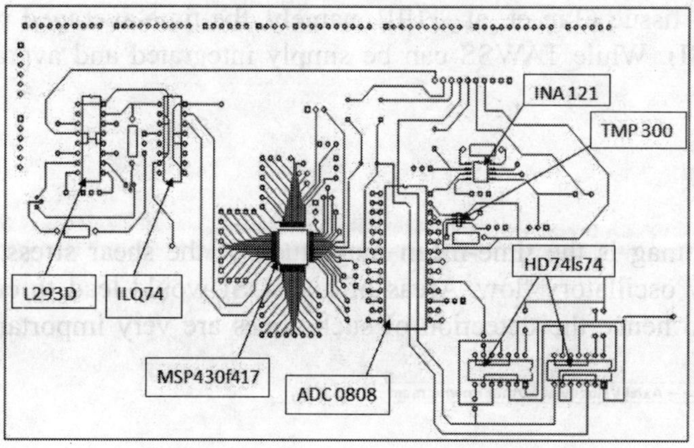

PCB Designing of differential motor control board

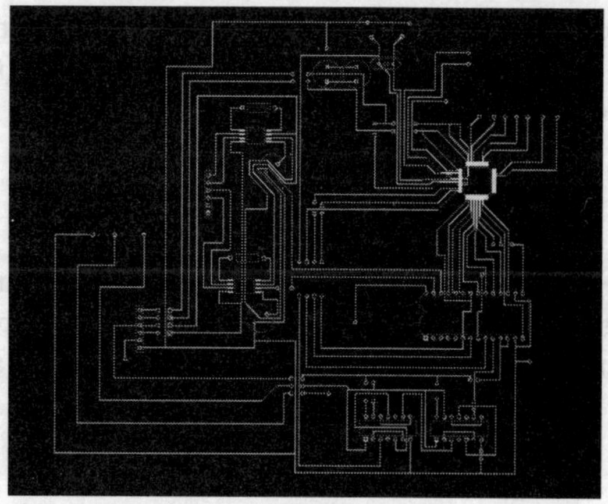

PCB design of microcontroller board

Fig. 3.3 Circuit components of differential physiological pulsatile flow generator

Detailed measurements of the flow characteristics were made using a dual-component LDV (514.5 nm for green, and 488.0 nm for blue). The measurement system consisted of a Dantec instrument (BSA P50 processor, Fibre Flow 60 x 63 transceiver with beam expander) and computer controlled traverse system. The flow was seeded with a solution of hollow borosilicate glass spheres (Sphericel 110P8, by Potters Industries) nominally 10 um in diameter with a specific gravity of 1.1. The particles were added in sufficient concentration of allow a mean sampling rate of 150 Hz. A solution of glycerine (40%) in water at 22 degree Celsius was considered as artificial blood. An automatic stirrer was used at a constant rpm for mixing seeding particles in the solution without any deposition in any of the chamber.

Flow generator was made to generate physiological pulses in order to achieve the same in flowing fluid. Peristaltic pumping system with differential grid arrangements was made to produce differential pulsating frequencies, which forms the fundamental harmonics for the production of physiological pulses. Differential Strength (temporal amplitude) of the pulse was achieved by two steps. First step was by regulating the volumetric flow rate, i.e., by using tubing system of differential diameters. Secondly, differential length of the tubing system was used in order to control the variations in temporal strength of the pulses.

Controlling unit of the generator was made of mixed signal processor assembly system. It consists of MSP430 series microcontroller, and interfacing modules to generate power, to drive the motor unit, to consider input commands respectively. Detail circuit diagram of the system components have been reported in Fig. 3.3.

3.3 RESULTS

The fluid mechanical equations were solved using the SIMPLE formulation (Patankar, 1980 [13]). In order to achieve higher accuracy, QUICK scheme is used to discretize the convective terms. Fixed time step size of 0.0005 sec to 0.00005 sec with second order accuracy in time domain was used requiring 2400-24000 time steps for one cycle. The momentum interpolation method due to Rhie and Chow (1983) was used in a collocated grid arrangement which can satisfactorily overcome the problem of pressure oscillation. The technique involves appropriate interpolation of velocity field at the interface of the cell faces. It has been demonstrated that such momentum interpolation is equivalent to solving a continuity equation with an added fourth derivative of pressure which does not change the formal second order accuracy of the basic discretization process. The procedure was also successful implemented in a previous work by Nandi and Chattopadhyay [11]. In order to check the faithful repetition of result after grid convergence test, a multiple cycle simulation had been carried out for reproducibility of the result, and it has been reported in Fig. 3.4.

From the calculated data, important fluid-mechanical parameters of interest such as shear stress, vortical structures, pressure profile and velocity distribution are be evaluated. Two important parameters need special mention in this connection having relevance to the health of arterial tissue (Tan et. al., [19]), namely, the time-averaged wall shear stress (TAWSS) and the oscillatory shear index (OSI). While TAWSS can be simply integrated and averaged over a cycle, OSI is defined as

$$OSI = \frac{1}{2}\left(1 - \frac{\tau_{mean}}{\tau_{mag}}\right) \qquad ...(11)$$

where τ_{mean} is the time-mean wall shear stress and tmag is the time-mean magnitude of the shear stress. OSI has a range between 0 and 0.5 where 0.5 defines purely oscillatory flow. Areas of high OSI would lead to endothelial dysfunction and artherogenesis (Tan et. al. [19]) and hence the detection of such zones are very important.

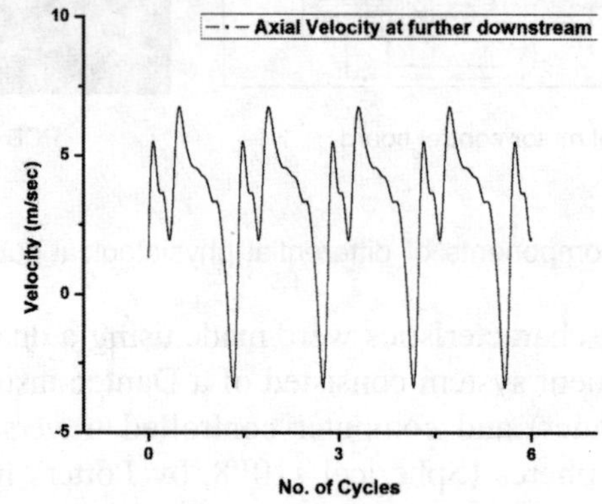

Fig. 3.4 Repetitive cycle of results reflecting axial velocity profile at near-outlet downstream

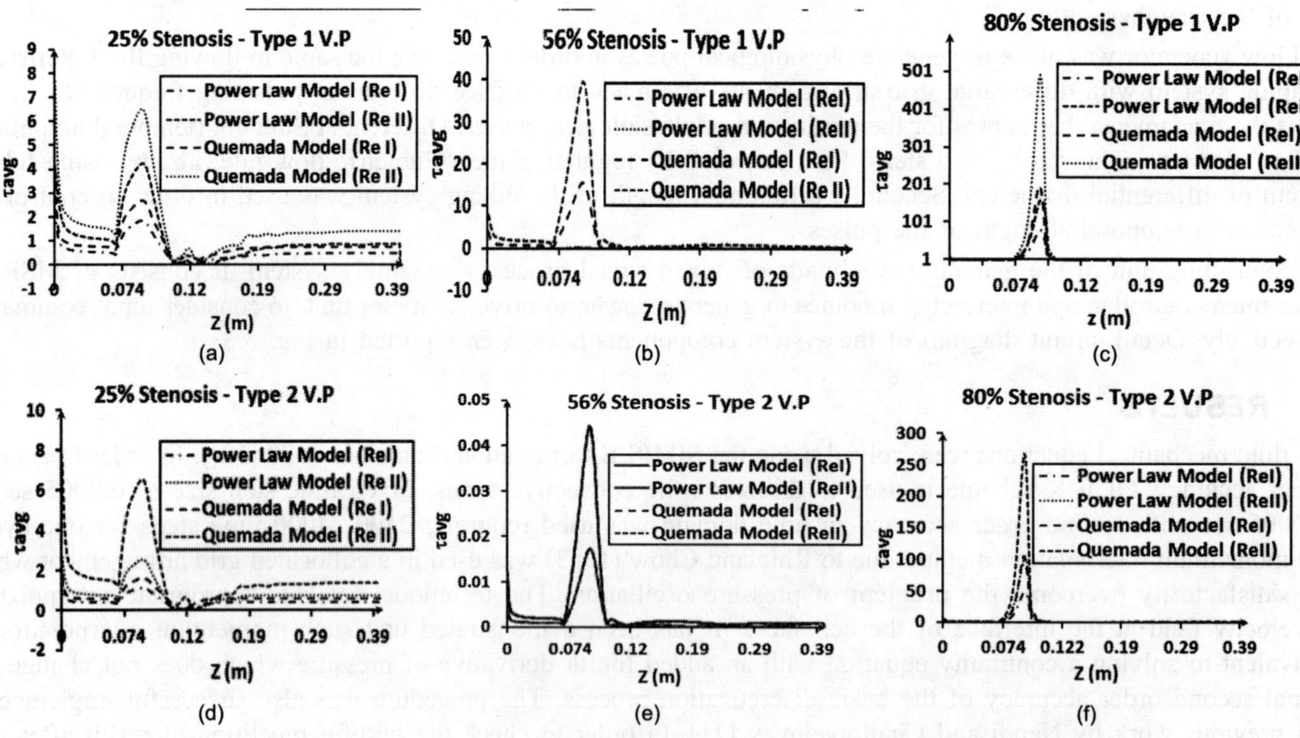

Fig. 3.5 Effect of Reynolds number on average WSS distribution using type 1 and type 2 inlet velocity at different degree of stenoses for different rheological models (Power Law Model and Quemada Model) (Fig. (a)–Fig. (f))

In Fig. 3.5, a comparative study of average shear stress on the wall of the vessel at different degree of stenoses for the same two non-Newtonian viscosity models at two different Reynolds number. It is observed that high shear stress results from flow at higher Reynolds number, and the magnitude of the peak of average shear stress increase sharply from 7 to 40 and 500 with the increase in degree of stenoses from 25% to 56% and 80% respectively. For both type of inlet velocity profiles Quemada viscosity model exhibit higher value for average wall shear stress than that of Power Law model.

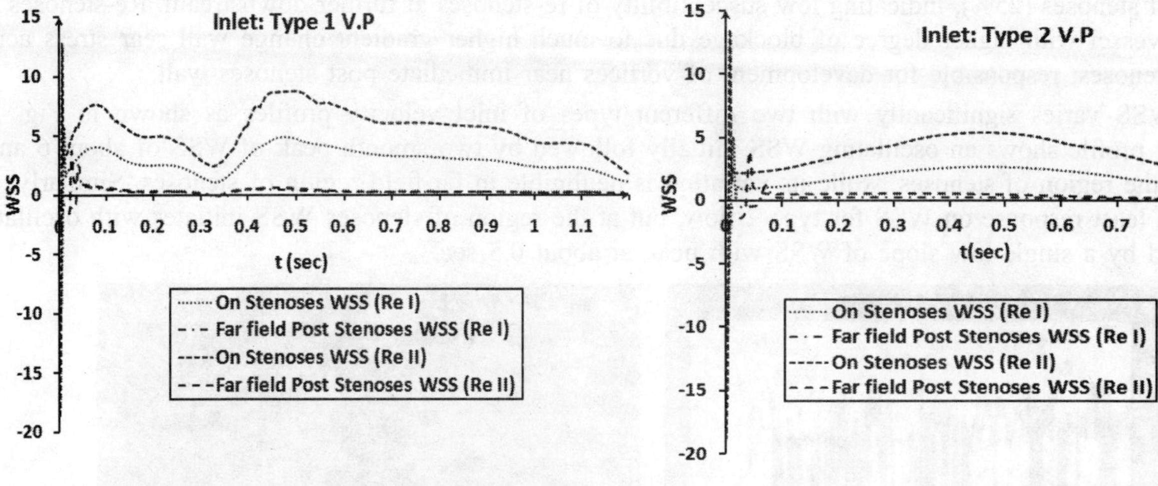

(a) (b)

Fig. 3.6 Temporal WSS for the first cycle (1.0 sec) in case of 80% stenoses for power law Model at different reynolds number (a) Inlet: Type 1 Velocity profile and (b) Inlet: Type 2 Velocity profile

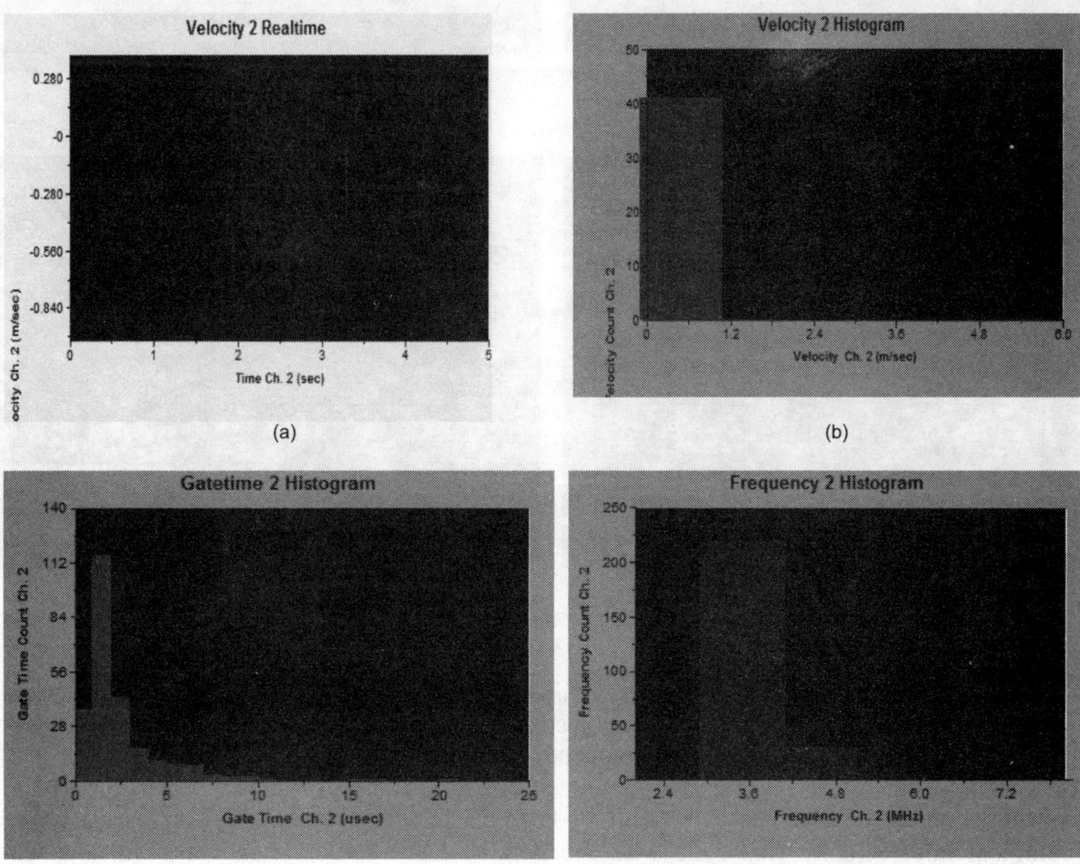

(a) (b)

(c) (d)

Fig. 3.7 Experiment result of LDV transreciever for flow through 12 .5% stenoses at very low resolution for approximately 5 seconds (a) Realtime axial velocity (b) Realtime velocity histogram (c) Gate time histogram of the axial velocity (d) Frequency histogram

It has been found that region just immediate after stenoses are prone to very low wall shear stress, irrespective of the flow rate as shown in Fig. 3.5. This is linked to the existence of recirculation pattern at such locations. At high degree of stenoses, post stenoses wall has low value of wall shear stress (about 0.5), development of prone region for re-stenoses at downstream of flow. Development of wall shear stress at post stenoses wall has been found at low degree of stenoses (25%), indicating low susceptibility of re-stenoses at further downstream. Re-stenoses can also occurs in vessel with higher degree of blockage due to much higher gradient change wall sear stress across the region of stenoses, responsible for development of vortices near immediate post stenoses wall.

Temporal WSS varies significantly with two different types of inlet velocity profiles as shown in Fig. 3.6. Type 1 velocity profile shows an oscillating WSS initially followed by two smooth peak of WSS of about 6 and 8 respectively at the region of stenoses, while its variation is negligible in far-field region of stenoses. Similarly, far-field region has least response on WSS for type 2 flow, but at the region of stenoses WSS initiates with oscillating nature, followed by a single low slope of WSS with peak at about 0.5 sec.

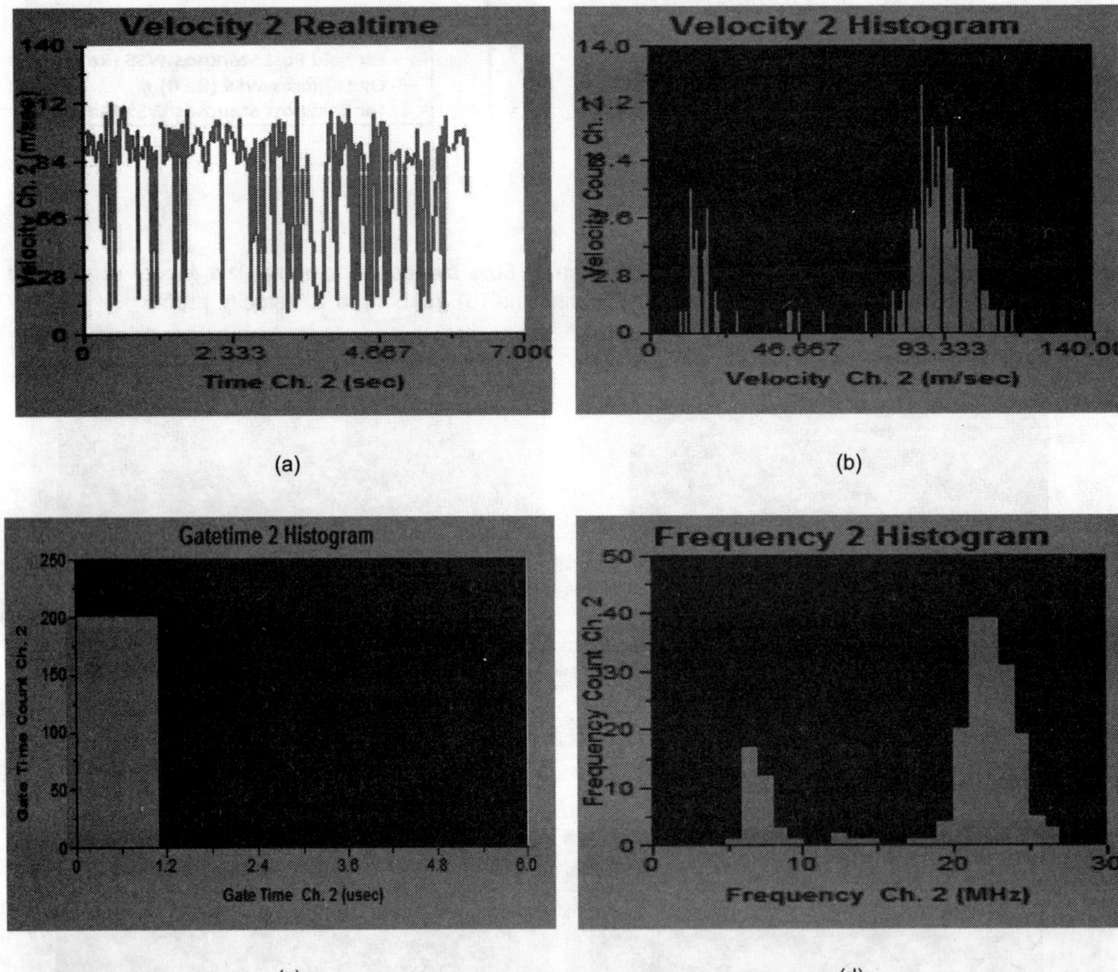

Fig. 3.8 Experiment Result of LDV Transreciever for flow through 12.5% stenoses at High Re for approximately 5 seconds (a) Realtime axial velocity (b) Realtime velocity histogram (c) Gate time histogram of the axial velocity (d) Frequency Histogram

An intensive study of the flow of glycerin solution through stenosed artery (of 12.5% and 25% stenoses) by applying physiological inlet velocity was done with the help of LDV setup. Transient Flow rate was maintained at different Mean Reynolds number (Low and High). Plot of Realtime axial velocity, velocity histogram, gatetime histogram, and frequency histogram was reported for different flow rates at different degree of stenoses (Fig. 3.7 – 3.9). It had been observed that for particular degree of stenoses (12.5%), the solution exhibits different particle behavior at different flow rates. From Fig. 3.7 and Fig. 3.8 it can be predicted that at lower flow rate distribution of frequency histogram concise to the mean flow of the particles, while at higher flow rates (as in Fig. 3.8) there is dynamic range of distribution of frequency histogram, indicating dynamic behavior of seeding particles, leading to turbulence nature of the flow. It also exhibits variation in velocity histogram, whereas gatetime histogram behaves in opposite manner.

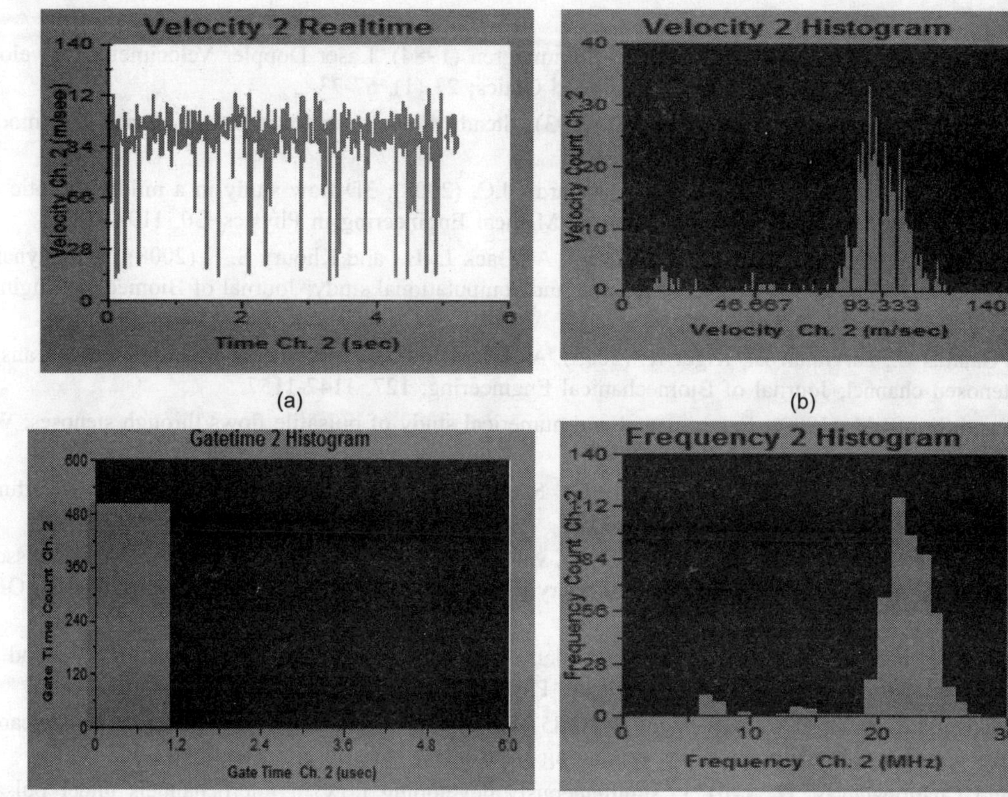

(a) (b) (c) (d)

Fig. 3.9 Experiment result of LDV transreciever for flow through 25% stenoses at high resolution for approximately 5 seconds (a) Realtime axial velocity (b) Realtime velocity histogram (c) Gate time histogram of the axial velocity (d) Frequency histogram

Flow behaviour of the solution maintaining a constant flow rate was also observed at different degree of stenoses (12.5% and 25%) as shown in Figs. 3.8 and 3.9. Flow at higher degree of stenoses exhibits higher values for gate time histogram, which indicates greater disturbances due to higher degree of stenoses. As all the measurements were carried out in post stenotic regions, longer characteristics length of the disturbance can also be concluded from higher value of gate time histogram. It is also observed that frequency histogram distribution increased by almost three folds at higher degree of stenoses (two fold increase in degree of stenoses). Same phenomena were observed for axial velocity histogram.

In the entire cases realtime velocity plot reflects the physiological pulses, indicating faithful construction of physiological pulses by differential flow generator developed.

3.4 CONCLUSION

The present work combines pulsating flow and non-Newtonian rheology for investigating flow structure in stenosed artery both numerically as well as experimentally. From numerical study, the location and movement of recirculation bubbles are found to be dependent on the degree of stenoses. As the geometry of blockage induces narrow duct to the flow of the fluid, it leads to the development of laminar to transition flow. Considering the viscosity property of the fluid to be non-Newtonian, especially considering hematocrit component of the fluid (blood) in case of Power Law model and Quemada model, contribution to the steep increase in WSS at downstream of post stenoses region is of significance. Similarly, with the development of vortices, oscillatory nature of WSS decreases, leading to low OSI. And temporal low value of OSI leads to spatial aggregation of macromolecules at the wall, leads to development of re-stenoses at further downstream. However, main emphasis was given to the development of the physiological pulsating flow generator, experimentally it was found that the characteristics length of disturbance sustain for a prolonged temporal range as degree of stenoses elevated.

REFERENCES

1. Bruce E. Truax, Frank C. Demarest, and Gary E. Sommargren (1984); Laser Doppler Velocimeter for velocity and length measurements for moving surfaces; Journal of Applied Optics; 23 (1), 67-73.
2. Budwig R., Elger D., Hooper H. and Slippy J., (1993), Steady flow in abdominal Aortic aneurysm models, Journal of Biomechanical Engineering., 115, 419–423.
3. Brunette J., Mongrain R., Laurier J., Galaz R., and Tardif J.C. (2008); 3D flow study in a mildly stenotic coronary artery phantom using a whole volume PIV method; Journal Medical Engineering in Physics; 30, 1193-1200.
4. Banerjee R. K., Ashtekar K. D., Helmy T. A., Effat M. A., Back L. H., and Khoury S. F. (2008); Hemodynamic diagnostics of epicardial coronary stenoses: in-vitro experimental and computational study; Journal of Biomedical Engineering (online); 7: 24, 1-24.
5. Beratlis N., Balaras E., Parvinian B., Kiger K. (2005); A numerical and experimental investigation of transitional pulsatile flow in a stenosed channel; Journal of Biomechanical Engineering; 127, 1147-1157.
6. Deplano V., and Siouffi M. (1999); Experimental and numerical study of pulsatile flows through stenoses: Wall shear stress analysis; Journal of Biomechanics; 32, 1081-1090.
7. Foutrakis G.N., Yonas H. and Sclabassi R.J., (1999), Saccular aneurysm formation in curved and Bifurcating arteries, American Journal of Neuroradiology, 20, 1309-1317.
8. Gijsen F.J.H., Palmen D. E. M., van der Beek M.H.E., van de Vosse F.N., van Dongen M. E. H. and Janssen J. D. (1996); Analysis of the axial flow field in stenosed carotid artery bifurcation models – LDA experiments; Journal Of Biomechanics; 29 (11), 1483-1489.
9. Huang C., Chai Z., and Shi B., (2013), Non-Newtonian effect of Hemodynamic characteristics of Blood flow in stented cerebral aneurysm, Communication in Computational. Physics, 13, 916-928.
10. Ku D. N., Gidden D.P., Zarins C.K., and Glagov S. (1985), Pulsatile flow and atherosclerosis in the human carotid bifurcation: Positive correlation between plaque and low oscillating shear stress, Arteriosclerosis, 5, 293-302.
11. Nandi, T. and Chattopadhyay, H., (2012), Simultaneously developing flow in microchannels under pulsating inlet flow condition, International Journal of Transport Phenomena, 13, 110-120.
12. Neofytou, P. and Drikakis, D, (2003), Effects of blood models on flow through a stenosis, International Journal of Numerical methods in fluid, 43, 597-635.
13. Patankar S.V., (1980). Numerical Heat Transfer and Fluid Flow. McGraw-Hill, New York.
14. Park S.M., Min Y.U., Kang M.J., Kim K.C., and Ji H.S. (2010); In Vitro hemodynamic study on the stenotic right coronary artery using experimental and numerical analysis; Journal of Mechanics in Medicine and Biology; 10 (4), 695-712.
15. Quemada D., (1977), Rheology of concentrated disperse systems III. General features of the proposed non-newtonian model. Comparison with experimental data, Rheologica Acta, 17, 643-653.
16. Rhie C.M., Chow W. L., (1983), Numerical Study of the Turbulent Flow Past an Airfoil with Trailing Edge Separation, AIAA Journal. 21, 1525-1532.
17. Siddiqui, S. U. Verma, M. K. Mishra, S. and Gupta, S, (2009), Mathematical modeling of pulsatile flow of Casson's fluid in arterial stenosis, Journal of Applied Mathematics & Computation, 210, 1-10.
18. Siouffi M., Deplano V., Pelissier R. (1998); Experimental analysis of unsteady flow through a stenosis; Journal of Biomechanics; 31, 11-19.
19. Talukder N., and Giddens D.P. (2009); Model studies on Arterial flow disturbance related to severity of arterial stenosis; International Journal of Cardiovascular Medicine and Science; 6 (1), 33-37.
20. Tan F.P.P., Soloperto G., Bashford S., Wood N. B., Thom S., Hughes A., and Xu X. Y., (2008), Analysis of flow disturbance in a stenosed carotid artery bifurcation using two-equation transitional and turbulence models, Journal of Biomechanical Engineering, 130, 061008-1-12.
21. Tu, C. and Deville, M, (1996), Pulsatile flow of non-Newtonian fluids through arterial stenoses, Journal of Biomechanics, 29, 899-908.
22. Walburn F.J., and Schneck D.J. (1976), A constitutive equation for whole human blood," Biorheology, 13, 201-219.

4

Mechanical Analysis of Cause Production of Deformities in Case of Perthes' Disease of Hip

J. Sahoo, R. Patra, H. C. Das, A. K. Dash and B. P. Patro

Department of Orthopaedics & Department of Mechanical Engineering
SOA University, Bhubaneswar–751003, Odisha, India

Email: drjsahoo@gmail.com

ABSTRACT

Perthes' diseases is a self limiting avascular necrosis of capital femoral epiphysis of idiopathic in nature. In this condition the head of the femur, the neck and the trochantor undergo different changes in configuration leading to deformities and ultimately leads to disability in function. Many a speculation has been advocated about the changes but nothing has stood test of time so far. A biomechanical preposition is framed observing the clinical, radiological and functional disabilities into consideration which seems to depict a laudable explanation for formation of deformities and can influence the management.

Keywords: Drudgery, efficiency, muscular problem, farm women, technology, ergonomics.

4.1 INTRODUCTION

Perthesis disease is an avascular necrosis of capital epiphysis of femur. So far the aetiology of avascular change has not been confirmed and as such it remains as idiopathic. However, different deformities are observed around upper end of femur in course of disease process in a particular period after which a degerative state appears in hip joint. The deformities seen radiologically are

Horizontal metaphysical plate (Fig. 4.1)

Subluxation of head of femur (Capital epiphysis) from acetabular socket (Fig. 4.2).

Irregular shape of the capital epiphysis (Fig. 4.3)

Prominent trochanter with beaking (Fig. 4.4)

Coxavara (Decrease neck shaft angle of femur) (Fig. 4.5)

Affected limb is short

In Addition to the functional capability of the affected lower limb is deranged leading to limping abduction and external rotation restricted and unable to bear weight on the affected side during stance phase with healthy side raised, off the ground (Trendelenburg Test positive).

Retrospective Preliminary Work: Clinically diagnosed patients having perthes disease were subjected for radiological examination. From the X-Ray picture following measurements were taken.

Measurement of angle of (metaphysical line) to the horizontal.

Measurement of neck shaft angle.

Level of the tip of the trochantor in relation to the centre of head of femur.

Position of capital epiphysies in Perkin: square is located. (Fig. 4.6)

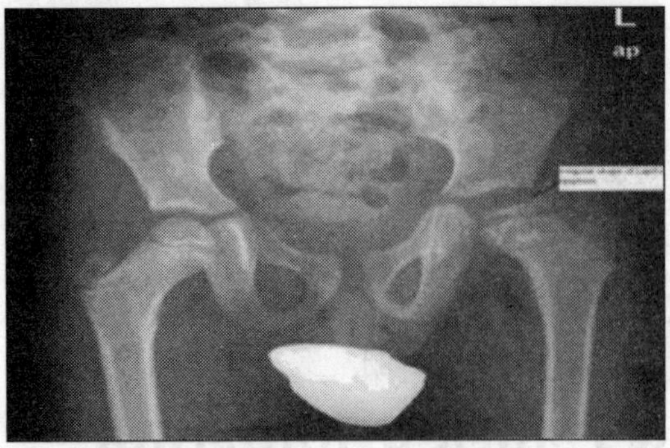

Fig. 4.1 Horizontal epiphysis

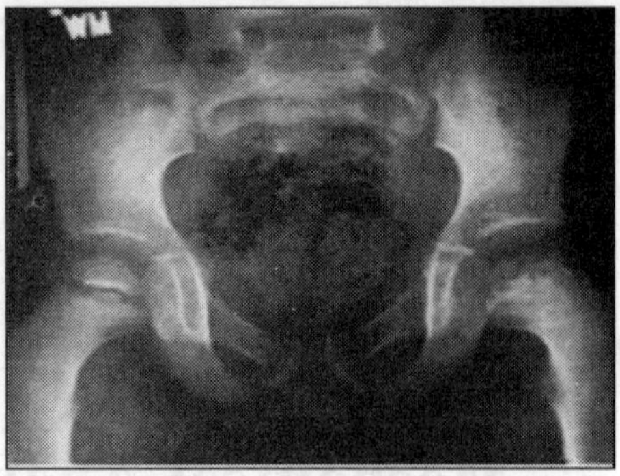

Fig. 4.2 Irregular avascular

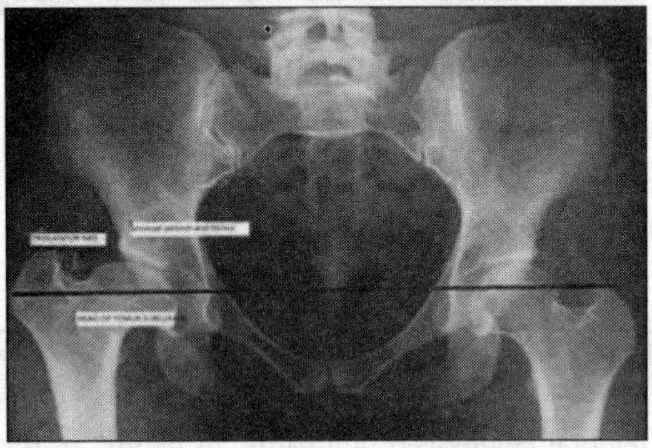

Fig. 4.3 Breaking of trochanter

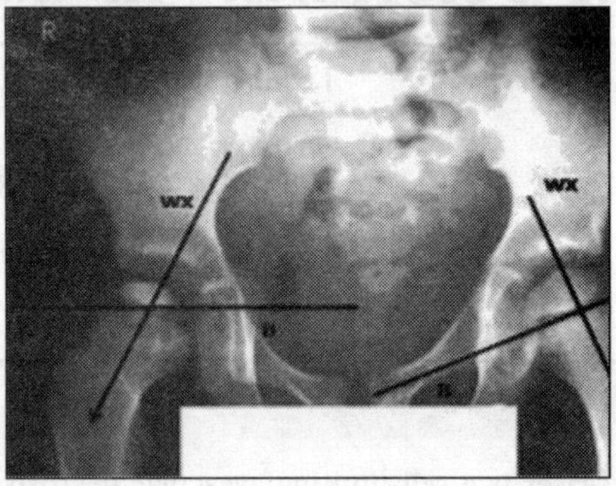

Fig. 4.4 Weight transmission axis through femoral epiphysis normal (Left) and Perthes' disease (Right)

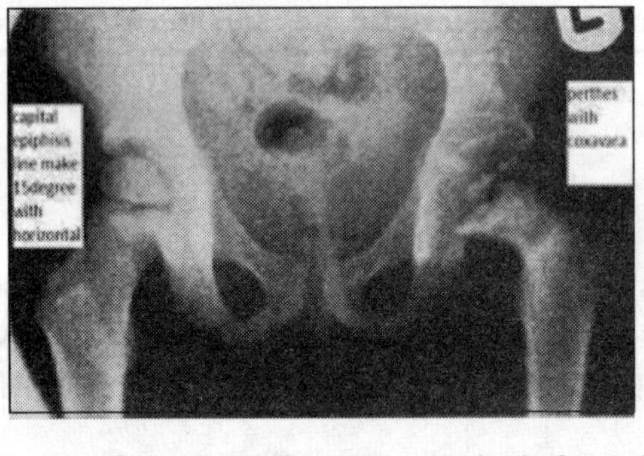

Fig. 4.5 Coxavara. (Decrease neck shaft angle of femur) with irregular capital femoral epiphysis

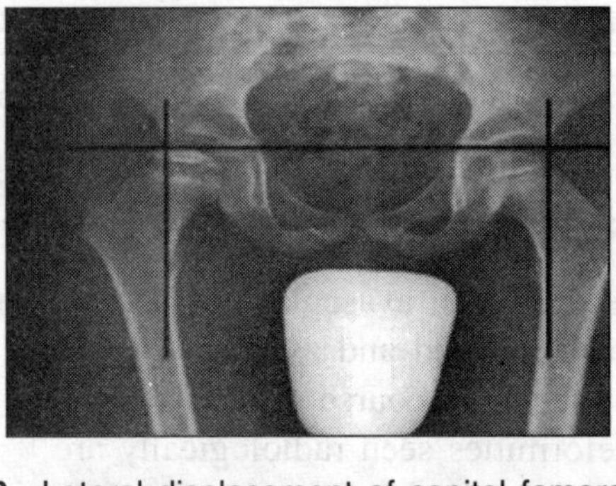

Fig. 4.6 Lateral displacement of capital femora epiphysis in perkins square in case perthers' disease

4.2 OBSERVATION

In every case the metaphyseal line sustains lesser degree of angulations with horizontal to the horizontal, i.e less than 100. Thus the metaphyseal line appears to be in horizontal position in comparison to other side as shown in Fig. 4.4. Neck shaft angle is reduced being compared with other side. The tip of the trochanter is observed to be at a higher level and since the position of tip of trochanter with the centre of head of femur isn't horizontal, angulated upwards and laterally. The capital epiphyses in the X-Ray was found to be sifted to outer aspect in Perkin: square.

4.3 BIOMECHANICAL ANALYSIS

Upper end of the femur is under control of muscular contractions around, in coronal and sagital plane for keeping it in equilibrium. The multiplication of force and force moment in opposite direction to be equal leading to the capital epiphysis, i.e head of the femur to remain inside the acetabular cavity. In addition to, the transmission of body weight runs perpendicular to the oblique cross section of head of femur, i.e metaphyseal plate and along the neck axis as shown in Fig. 4.2. Among the muscles the adductors and adductors play in coronal plane. Force moment of both the muscles equate each other to keep the head of the femur inside the acetabulum, well contained as observed in stance phase in coronal plane. The same mechanical position also keeps the metaphysical plate of capital epiphysis with about 150 angulations' to the horizontal, as a result of which weight transmission takes place in a perpendicular line to the centre of epiphysial plate (Fig. 4.4). Assuming adductor force becoming more to the abductor force, the capital epiphysis will shift with rotation laterally thus leading to weight transmission to be concentrated on the inner side of the metaphyseal plate since angle subtend to horizontal plane by metaphysical plate decreased and it tends to be horizontal. Translation of capital epiphysis towards outside leads to inner aspect of capital epiphysis to remain along the line of body weight transmission in the acetabular cavity. According to Wolf's law form follows the pressure; thus leads rounded shape of inner aspect of capital epiphysis and fattening of lateral aspect which sublaxate and capital epiphysis takes a irregular shape.

This in total decrease the growth potentiality of capital epiphysis which fails to grow in peripasu to trochantor; which will have normal growth. The growth potentiality being represented as forces acting upwords at two places of the neck, i.e. head of femur and trochanter normally maintaining a relationship. The disparity in this situation having less force on head side with normal force on trochanmter leading to leaver arm going upwords in lateral side(trochatoric side). Thus, prominence of trochantor with beaking and neck shaft angle is decreased perpetuating coxavara. Prominent trochantor with beaking leads to decrese of effective length of abductor which remains in a passive contractile state and cann't exhibit active contraction to accomplish movement (abduction) so abduction force becomes weak thus it facilitates the adductor force to increase and with increased adduction, the capital epiphysis subluxates out further. The limb remaining in adducted position looks short, in addition to fallacious growth at upper end of the femur also contributes to the shortening of length of femur in particular and limb in general to same extent as well.

Casecade of Biomechanical Analysis

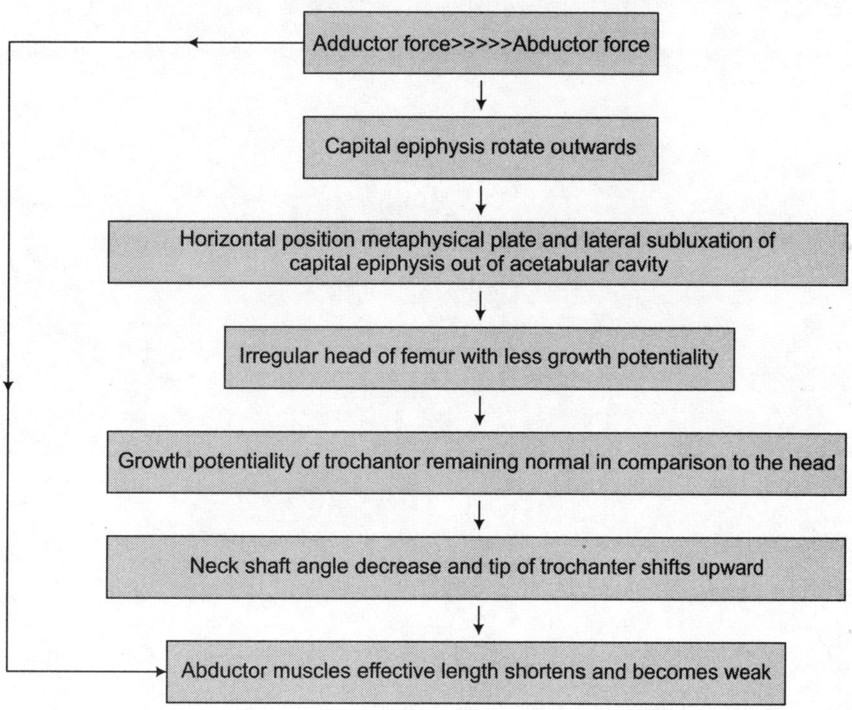

4.4 CONCLUSION

The biomechanical analysis could throw light on cause production of deformities in upper end of femur in perthes' disease of hip in children having adduction force to be more than abductor force. Aetiologys of avascular change being still obscure, the correction of disparity between adduction and abduction force could bring a solution for prevention of deformities provided the problem is nipped in bud.

REFERENCES

1. Kim, H.K. Pathophysiology and new strategies for the treatment of Legg-Calve-Perthes disease. J Bone Joint Surg (A) 2012, 94: 659-69.
2. Joseph, B. (2011). Prognostic factors and outcome measures in Perthes disease". Orthopedic Clinics of North America 42: 303–15.
3. Perry, D.C., Bruce, C.E.; Pope, D. et al. (2012). Legg-Calvé-Perthes disease in the UK: Geographic and temporal trends in incidence reflecting differences in degree of deprivation in childhood. Arthritis & Rheumatism 64: 1673–1679.
4. Perry, D.C.; Machin, D.M.G.; Pope, D. (2012). Racial and Geographic Factors in the Incidence of Legg-Calve-Perthes' Disease: A Systematic Review. Am J. Epidemiol 175: 159–166.
5. Daniel A.B., Shah H., Kamath A., Guddettu V., Joseph B. (2012) Environmental tobacco and wood smoke increase the risk of Legg-Calvé-Perthes disease. Clin Orthop Relat Res. 2369-2375.
6. Demko J., McLaughlin R. (2005). Developmental orthopedic disease. Vet Clin North Am Small Anim Pract 35; 1111–1135.
7. Sankar W.N., Horn B.D., Wells L., Dormans J.P. Legg-Calve-Perthes disease. In: Kliegman R.M., Stanton B.F., St. Geme J.W. III, et al., eds. Nelson Textbook of Pediatrics. 19th ed. Philadelphia, Pa: Elsevier Saunders; 2011: chap 670. 3.

5

Ergonomic Evaluation of Different Length Sweeping Brooms in Indian Women

C. S. Ram[1], Indu Kashyap[1] and R. K. Saxena[2]

[1]*Department of Physiotherapy I.T.S. College of Physiotherapy, Ghaziabad.*
[2]*Department of Biological Engineering, Shobhit University, Meerut*

Email: csram@its.edu.in

ABSTRACT

Manual sweeping brooms are used daily in Indian households. Study was conducted to study and compare the effects of the use of sweeping broom of different lengths on heart rate (HR), Borg's scale of rate of perceived exertion (RPE), pain (VAS), body part discomfort scale (BDPS) and attitude assessment in healthy Indian women. The results found in the study showed that the introduction of the broom with adjustable length significantly reduces RPE, pain and body part discomfort in lower back. The result also suggested that the females rated adjustable broom much better over the traditional broom in terms of the attitude assessment. Results suggests that Long Handle Sweeping Broom (LHSB) can provide comfortable and better cleaning tool for reducing workload and awkward postures during floor cleaning.

Keywords: Ergonomics, cleaning tools, body parameters, Indian women.

5.1 INTRODUCTION

Cleaning is an important activity in day-to-day domestic as well as professional lives[1]. Cleaning is considered a high physically demanding job, resulting in high cardiovascular load, high frequency of awkward postures and, as such, it is rated as a highly strenuous job. In several studies the relationships between a poorly designed workplace, poor working posture and diseases of the musculoskeletal system have been demonstrated. Cleaning in India is one of most important task for an Indian women as a part of her household work. There are no such data available for cleaning activities and cleaning related injuries in India. To clean the floor, Indian women use a different kind of sweeping broom which is made of some tree branches or leaves and its average length is between 70 cm and 80 cm. Floor cleaning requires strenuous activities in order to clean. As a result cleaners have to adapt awkward postures while cleaning floor[2]. In this study an attempt is made to identify the problems and develop solutions related to the cleaning processes with special emphasis on cleaning tools

A study conducted by Burdorf et al., concluded that the main contributing factor for musculoskeletal disorders is poor working posture[3], which according to Aaras et al can result in minor back problems to severe handicap.[4] Saari et al concluded that awkward postures may lead to high demands of energy expenditure and develop high work stress on cleaners which can lead to musculoskeletal disorders, such as upper limb disorders, lower limb disorders and less job satisfaction which can affect the productivity and quality of the work[5] The bending of body while cleaning floor requires more demands of oxygen consumption and heart rate[6] Karen Messing mentioned that the cardiovascular load can be quite heavy.[7] Johansson and Ljunggren recorded the heart rate of female cleaners during office or toilet cleaning at 123 beats/minute[8] Hagner and Hagberg found a high level of oxygen consumption (up to 40% of the VO_2 max) among cleaners doing floor mopping under experimental conditions[9] Sogaard found that relative cardiovascular strain of female school cleaners measured in the workplace was 53% of the VO_2 max.[10]

5.2 METHODOLOGY

5.2.1 Sample Size

Thirty Indian women aged between 25-45 years meeting inclusion/exclusion criteria from rural areas of UP and Delhi. Inclusion Criteria (1) Know how to use sweeping broom (traditional and adjustable brooms) (2) Ability to understand the questionnaire and give answers to them. (3) No history of pre-task exertion, discomfort and pain.

Exclusion Criteria: Any history of self-reported musculoskeletal problem, respiratory problem, cardiac problem, illness, psychosocial/cognitive illnesses. Study sampling method used was convenient sampling. Subjects were given random allocation of tasks by chit method.

Fig. 5.1 Showing traditional broom

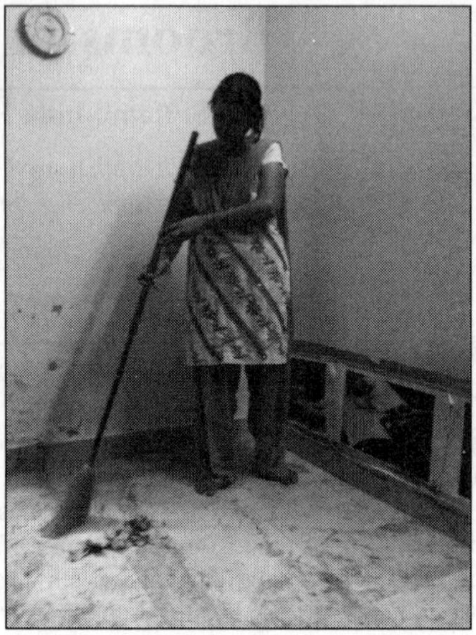

Fig. 5.2 Showing adjustable handle broom

5.2.2 Instrumentation

Two brooms, one traditional broom of approximately 70-80 cm in length and another with adjustable length., Stopwatch, Worksheets, Pulse oxymeter (nidex), Dry sand and papers, Chair (for rest purpose).

5.2.3 Variables

Pain (VAS), Heart rate (bpm), Rate of perceived exertion (RPE) on Borg scale (0-10 scale), Body part discomfort scale (BPDS) rating (0-5 scale) and Attitude assessment (i.e., comfortable to use) (1-5 scale), Length of broom sticks.

5.2.4 Procedure

All the selected subjects were requested sign informed consent form prior to study commencement. General clinical physiotherapy assessment was done for each subject. Subjects meeting inclusion and exclusion criteria were selected for the study and assigned randomly (Group 1 & Group 2) into either of the two tasks first and then assigned to the second task after rest duration of 30 minutes. Subjects were rested for 15 minutes before the commencement of the first task and pre task measurements of two dependent variables, i.e. HR and RPE were taken. Task A was sweeping the floor with a traditional short length broom (non-adjustable) and task B was sweeping the floor with adjustable broom (relatively long length broom). Thus Group 1 comprised of traditional short length broom group and Group 2 comprised of adjustable broom (relatively long length broom) group. Sweeping task (A or B) was done for 7 minutes of each cleaning task. An area of 80 m^2 was cleaned by the subjects during the 7 minutes of either of the cleaning test. Dry sand and paper were used as cleaning materials to clean. After commencement of the task (A or B), the post task measurements of dependent variables which were to rate their perceived exertion on Borg's scale (0-10) just 30 seconds before the end of each test, HR rating, pain on VAS were taken and their body part discomfort to be marked on the given body chart. After finishing both tests, cleaners rated their attitude for subjective assessment of the aspect: "comfortable to use" on the scale of 1 to 5 (1 is very bad and 5 is very good). The data collected was used for statistical analysis. A paired t-test was used to determine the differences between HR. The Wilcoxon singed rank test was used to test the correlation between the Pain on VAS, RPE, BPDS and Attitude scale.

5.3 RESULTS

Data of Table 5.1 showing anthropometric, physiological and psychological measurements of subjects.

Table 5.1: Mean values of body parameters of subjects after sweeping

Variable Handle	*With Adjustable Handle Broom (Group 1)*		*With Traditional Handle Broom (Group 2)*		*P Value*
Post Task–between group analysis	*Mean*	*SD*	*Mean*	*SD*	*Mean*
HR	91.23	±11.36	93.1	±13.74	0.16
Borg's RPE	0.47	±0.74	1.02	±1.14	0.001
BDPS upper back	0.03	±0.18	0.00	±0.00	0.31
BDPS lower back	0.50	±0.86	1.00	±1.23	0.002
Pain –VAS upper back	0.30	±0.83	0.37	±0.96	0.31
Pain–VAS lower back	0.47	±0.86	1.10	±1.40	0.009
Attitude Scale	4.33	±0.71	3.43	±1.33	0.001

Wilcoxon statistical test was conducted to evaluate whether subjects had greater rating of Perceived Exertion with adjustable broom or traditional broom. It indicated a significant difference ($p = <0.001$). Thus, adjustable broom produced less exertion. (Fig. 5.3).

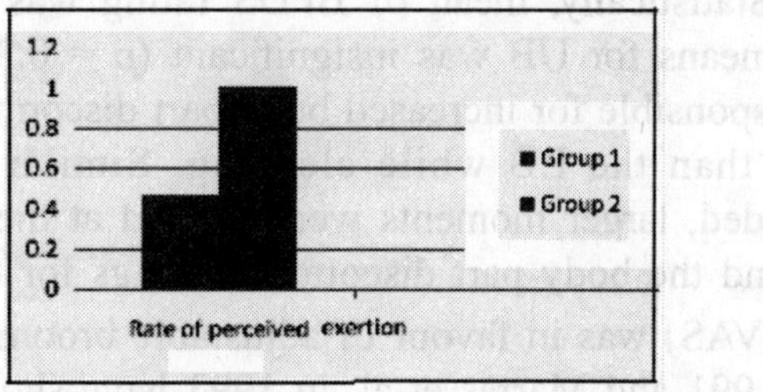

Fig 5.3 Comparisons of post task Borg's rating of perceived exertion adjustable broom (group 1) and traditional broom (group 2)

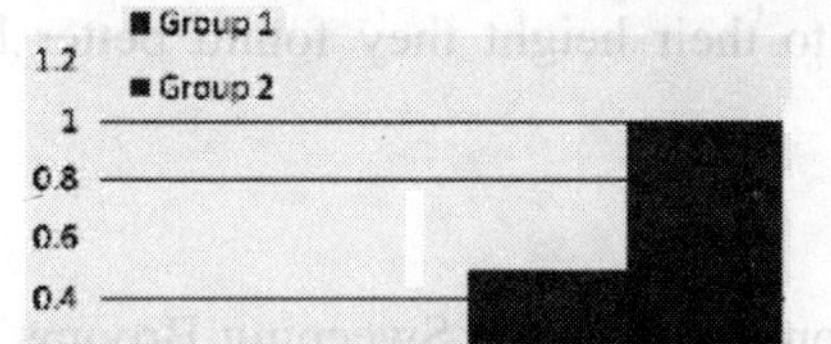

Fig 5.4 Comparisons of post task pain (VAS) scores of upper back (UB) and lower back (LB) absolute value adjustable broom (Group 1) and traditional broom (Group 2)

There was also a significant difference (p = 0.002) between the groups post tasks for Body Part Discomfort Scale in Lower back indicating adjustable handle produced less lower body discomfort while brooming. (Fig. 5.4)

There was also significant difference between Post Task Pain Scores (VAS) in Adjustable Handle and Traditional Broom (Fig. 5.3). Adjustable handle broom produced less lower body pain while brooming. There was also significant difference between Post Task Attitude Scale in Adjustable Handle and Traditional Broom. (Fig. 5.3). Subjects were comfortable, i.e higher Attitude Scale Score with adjustable broom than with traditional broom.

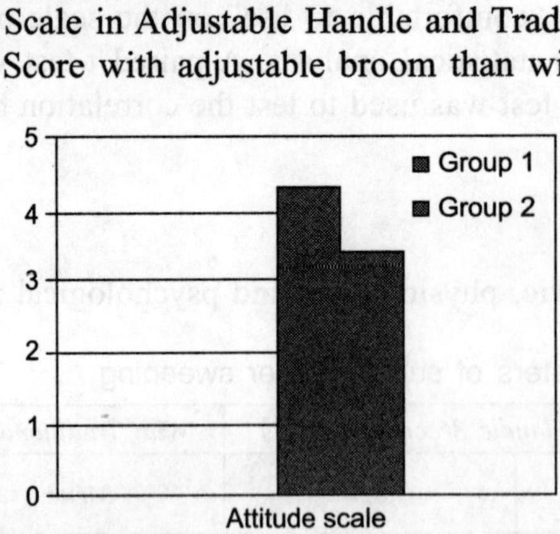

Fig. 5.5 Comparisons of post task attitude scores with adjustable broom (Group 1) and traditional broom (Group 2)

5.4 DISCUSSION

5.4.1 Rate of Perceived Exertion

In the current study, the subjects rated traditional broom more strenuous on the Borg scale than the adjustable broom while cleaning. Most likely, very frequent and excessive bending of the upper part of the body of the females created greater biomechanical loads on the back compared to the adjustable broom, and muscles had to work with higher forces against the center of gravity while bending. This could be the possible reason behind the perceived higher exertion compared to the adjustable broom.

Body Part Discomfort Scale: Statistically, mean of BPDS rating was in favour of adjustable broom for LB (p = 0.002) but the difference in means for UB was insignificant (p = 0.31). The bending position at lower back causing larger moment might be responsible for increased body part discomfort at the LB as compared to UB which comparatively was straightened than the LB while cleaning. Similarly in the research done by S.S. Ulin et al. in 1992 with the arms extended, larger moments were created at the elbow and shoulder, and consequently, the ratings of perceived exertion and the body part discomfort ratings for the arms increased.[20]

Pain Scores-Statistically, Pain (VAS) was in favour of adjustable broom for LB (p = 0.009). Many studies such as work done by Punnet et al. in 1991 and Marras et al. in 1993 have shown that periods of static or cyclic trunk flexion have an increased risk of low back pain 22, 23. As per study done by Rupesh Kumar in 2001, the trunk flexion postures are associated with increased risk of low back pain (LBP).6

Attitude scale-Between groups analysis was done and cleaners found the adjustable broom significantly better than the traditional broom (p = 0.001). It is likely that females' experience postural load often and when they used the adjustable broom setting according to their height they found better back posture which reduces the load on them.

5.5 CONCLUSION

Study has been done to assess the ergonomics of Indian Sweeping Brooms to understand difficulties faces by Indian women, and to provide better sweeping broom for cleaning the floor. Two types of sweeping brooms (1) Traditional broom (Length: 70-80 cm) and (2) Adjustable broom (Length: 100-200 cm) were used in this investigation. Time dependent study was performed and measurements of physiological parameters such as Pulse rate, Oxymetery index, Pain on VAS, RPE, BPDS and psychological Attitude scale were done. Results of study showed that adjustable length broom significantly reduces rate of perceived exertion, pain and body part discomfort in lower back in studied subjects. Result also suggests that the women rated adjustable broom much better over the traditional broom in attitude assessment study. Hence, long handle broom can provide better and comfortable cleaning tool for reducing workload and awkward postures during cleaning floor compared to traditional broom.

REFERENCES

1. Rupesh kumar (2002); Ergonomic assessment of two different cleaning tools; International Cyberspace Conference on Ergonomics, Sept: 2002.
2. Rupesh Kumar (2006); Ergonomic Evaluation and Design of Tools in Cleaning Occupation; Doctoral Thesis, Doctoral Thesis, Lulea University of Technology, 2006: No. 16, 1402-1544.
3. Burdorf, A., Govaert, G., Elders, L. (1991). Postural load and back pain of workers in the manufacturing of prefabricated concrete elements, Ergonomics 34, 909-918
4. Åaras, A., Westgaard, R.H., Strandew, E. (1988). Postural angles as an indicator of Postural load and muscular injury in occupational work situations, Ergonomics: 31; 915-933.
5. Saari, J., (1987). Management of housekeeping by feedback. Ergonomics 30, 313 – 317.
6. Kumar, R., (2001). Development of an Ergonomic Cleaning Tool and Its Evaluation. M.Sc. Thesis, Luleå University of Technology, Luleå, Sweden. 2001: 143 CIV.
7. Karen Messing, Jeanne Mager Stellman; Encyclopaedia of Occupational Health and Safety: Industries and occupations, Volume 3(Chapter Indoor cleaning services; pp. 100.3-100.2.
8. Johansson S.E., Ljunggren G. (1989); Perceived exertion during a self-imposed pace of work for a group of cleaners; Appl Ergon. 1989 Dec; 20(4): 307-12.
9. Hagner, I-M and M Hagberg. (1989). Evaluation of two floor-mopping work methods by measurement of load. Ergonomics 32 (4): 401-408.
10. Sogaard, K. 1994. Biomechanics and Motor Control during Repetitive Work: A Biomechanical and Electromyographical Study of Floor Cleaning. Ph. D. thesis. Copenhagen, Denmark: Department of Physiology, National Institute of Occupational Health.
11. Sogaard, K., N Fallentin, and J Nielsen. (1996). Workload during floor cleaning. The effect of cleaning methods and work technique. Eur J App Physiol. 1996; 73(1-2): 73-81
12. NIOSH (1999). Survey of Occupational Injuries and Illnesses, National Institute of Occupational Safety and Health. Washington, DC. Publication Number 99-101.
13. Saari (1987); Management of housekeeping by feedback; Ergonomics Vol. 30, Issue 2, 19-27.
14. McGill S.M. and Brown S. (1992) Creep response of the lumbar spine to prolonged full flexion. Clin Biomech. Vol. 7: 43-6.
15. Solomonow M., Baratta RV, Zhou BH, Burger E., Zieske A., and Gedalia A. Muscular dysfunction elicited by creep of lumbar viscoelastic tissue. J Electr. Kines. 2003; 13: 381-96.
16. Tatiana de Oliveira Sato, Helenice Jane Cote Gil Coury; Evaluation of musculoskeletal health outcomes in the context of job rotation and multifunctional jobs; Applied ergonomics, Vol. 40, Issue 4, July 2009, pp. 707–712
17. Gunnar Borg (1985); Perceived exertion, ASCM
18. Corlett & Bishops (1997); Book on Human ergonomics
19. Sally L. Collins et al; August 1997; The visual analogue pain intensity scale: what is moderate pain in millimetres?; Pain; Vol. 72, issues 1-2, pp. 95-97
20. Sheryl S. Ulin, Examination of the effect of tool mass and work postures on perceived exertion for a screw driving task; International Journal of industrial ergonomics, 12(1993)105-115; Elsevier
21. Punnett L. and Wegman D.H. Work-related musculoskeletal disorders: the epidemiologic evidence and the debate. J Electr. Kines. 2004; 14: 13-23.
22. Marras W.S., Lavender S.A., Leurgans S.E., and et al. The role of dynamic three-dimensional trunk motion in occupationally-related low back disorders: The effects of workplace factors, trunk position and trunk motion characteristics on risk of injury. Spine 1993; 18: 617-28.

6

Role of CAD/CAM for Development of Tailor-made Knee Implant

Ronak Shah and H. J. Nagarsheth

Department of Mechanical Engineering
Sardar Vallabhbhai National Institute of Technology
Dumas Road, Ichchhanath, Surat–395007, Gujarat, India

Email: ronak1009@gmail.com

ABSTRACT

Knee implant surgeries are most preferred for the lower extremities replacement due to its endurance and acceptability. Every knee joint patient have different configuration and structural anatomy of lower extremity, hence it is necessary to develop individual specific knee implants, so it may provide an individual physiological compatibility for standard implant for the natural functioning of the knee joint. Thus, developing the parametric CAD model of the knee joint is the solution. The parametric CAD model can be directly used for manufacturing using integrated CAD/CAM technology like rapid prototyping, advance CNC machining, non-conventional machining like EDM. The parametric CAD model can also help to analyze the performance of the implant by FEA and make necessary design modification at the upstream of development process itself. This saves the lead time and cost of developing the knee prosthesis.

Keywords: Biomechanics, CAD/CAM, finite elemental analysis, implant, prosthesis, rapid prototyping, titanium, tribology.

6.1 INTRODUCTION

Biomechanics is the study of structure and functions of various biological systems such as plants, animals, organs, etc. based on the mechanical concepts and principles. Among the various fields of the biomechanics, orthopedic and musculoskeletal is the most challenging and critical field; especially development of implants. Knee joint is the largest synovial and complicated joint in the body and subjected to the high stresses. The damage to knee bones and tissues occur because of arthritis, accident, impact or other factors which necessitates knee arthroplasty. The role of knee implants is to restore or maintain the natural functioning of the knee joint and reduce patients' trauma. The commonly used materials for the knee implant manufacturing are titanium alloys, alloy of Co-Cr-Mo or Ni-Cr alloys with vanadium, aluminum and other solutes in low proportion for femoral and tibial components while the tibial insert and patella component are made of ultra high molecular weight polyethylene (UHMWPE) largely. It is common practice today that various implants manufacturing companies produce the implants as per standard sizes and configuration so as to have mass production and deliver the product quicker and at competitive prices. As the anatomical structure of each patient is different and the standard implants tend to change the natural functioning (or kinematics) as reported by various surgeons and patients. Thus, need for tailor made patient specific implant.

The present paper will discuss the method of developing customized implants. It concentrates on use of advance CAD/CAM technologies to develop a parametric CAD model and later discuss advance manufacturing methods like CNC machining and rapid prototyping and also how CAD model can help in surgical planning and assistance.

6.2 PRESENT PROBLEMS IN KNEE IMPLANTS

There are various problems associated with the present knee implant designs. The various problems identified are discussed in this section. The key problem in major implant failures is aseptic loosening. It is caused by accumulation of wear debris resulting from initial wear causing rapid wear of UHMWPE insert (Carr and Goswami, 2009). The joint loosening leads to separation of femoral and tibial component which increase pain and sometimes seizure of knee joint motion. Figure 6.1 shows the wear of implant components retrieved from a patient due to joint loosing issue. Factors mainly contributing to the aseptic loosing of joint are improper design and material selection. Improper design includes incongruity and shape simplification while material selection includes large variation in modulus of elasticity, hardness, other properties (Bahraminasab et al., 2013).

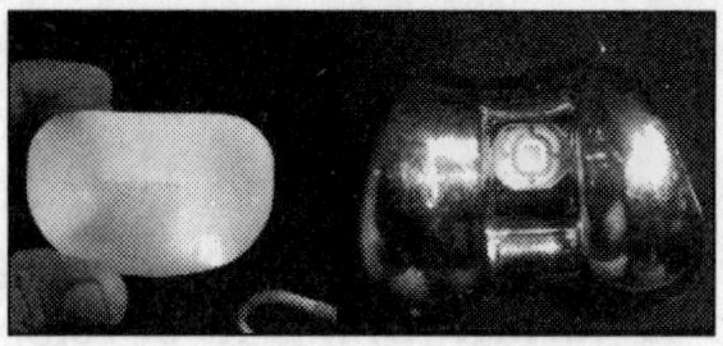

Fig. 6.1 Implants retrieved showing various wear symptoms

Biocompatibility of the implant material is one of the prime concerns which restrict the selection of implant material. The titanium and its alloys are found to be the ideal material for femoral component because of its biocompatibility properties and high strength but its density is in range of 4.43-4.48 g/cc (Welsch, Boyer and Collings, 1993) which is greater than the bone density. This causes problem with normal gait of the person as the leg having implant will be heavier than the other non-implanted leg.

Change in knee kinematics occurs due to simplification in the implant designs and use of generic shapes. The symptoms of change in kinematics are reduction in range of motion, changes in gait, increase in wear of implant components and instability of joint. Heever et al. (2011) have reported that there is a direct correlation between ligament function and the articulating surface of the normal knee and changes to any of these structures can affect the other.

As mentioned standard implant parts are also creating long-term problems as they do not match the specific requirement of the patient. Problems such as oversized or undersized joints are frequently noticed causing premature failure. Oversized joints do not provide stability as the ligaments are over tensed while undersized joints will increase contact pressure and reduces tension in ligaments. Also the selection of standard implants depends on the surgical procedure adopted by the surgeon.

Finally the problems associated with manufacturing of implant includes high degree of accuracy, precision, surface finish, requirement of sophisticated 5-6 axis CNC machines and need of sophisticated controlled environment which ultimately increases the cost of the implant.

6.3 DESIGN OF KNEE IMPLANTS

The development of the knee implant CAD model is challenging task of implant development process. With the recent advancements in the computer technology and CAD packages, the complicated shapes and difficult geometries can be considered in the design of the knee implant which helps in achieving the natural kinematics of the joint. The problem associated with the knee implant CAD model is non-availability of generic shapes and standard design procedure. The knee implant consists of femoral, tibial and patella components. Two contact joints associated with knee are patellofemoral (joint between patella and femur) and tibiofemoral (joint between tibia and femur) which are to be considered while designing as its plays important role in knee kinematics.

6.3.1 Reverse Engineering

The most promising method of developing the CAD model of implant components is "Reverse Engineering". RE will help to develop the CAD model rapidly when the geometrical information of the object is not known. The geometrical information of the bones can be extracted by X-rays, CT scan or MRI data. The X-rays gives only 2D information whereas CT scan and MRI provide 3D information. It helps in retrieving the geometrical information and further modifying the existing implant design by developing its CAD model from the physical model available.

Lin et al. (2005) suggests a systematical approach to do reverse engineering of knee implant to accurately obtain the CAD model which helps in reducing the lead time for developing. The RE process is shown in the Fig. 6.2.

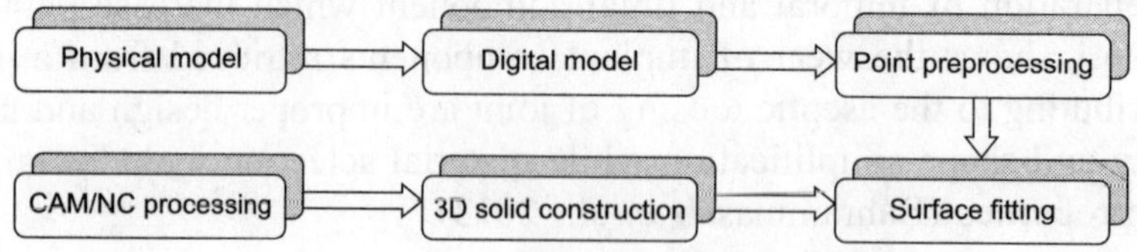

Fig. 6.2 Reverse engineering process (Lin et al., 2005)

The RE process starts with digitalization where point cloud data is obtained using the contact/non-contact type scanners. The point cloud data obtained is added with some noise and geometry of fixtures if used which is to be removed in pre-processing step. Next step is surface fitting to develop surface model of implant by using some mathematical model to curve fit the points and generate small surfaces which will be blended together. Finally, surface model is converted to solid model. Various commercial software packages such as Pro/E, Rapidform, Catia are capable for reverse engineering.

6.3.2 Mathematical Modelling of Implant Shape

The femoral shape needs to be regenerated as accurately as possible to obtain near to natural shape of the implant to maintain the natural kinematics. Various mathematical models are developed to represent the tibiofemoral and patellofemoral joint. The different patients' X-ray study revealed that the lateral femoral is larger than medial, i.e. the condyles are highly asymmetrical; therefore simpler methods cannot portray the condyles sufficiently. So, condyles are required to be handled individually. Lee & Guo (2010) performed comparative analysis of three mathematical models representing the tibiofemoral joint. Model 1 considers two sequential circles rolling a flat plane while other two models consider mathematically differentiable ellipses based models with and without sliding at the contact. Without sliding elliptical model considers ellipse rolling on flat surface while other ellipse model considers ellipse rolling on ellipse which create small sliding action. The latter model is more accurate and practically verified and the joint gait analysis and joint retrieved showed both rolling and sliding motions.

Heever et al. (2011) proposed other method of mathematically reconstructing the shape of femoral component at sagittal plane and transverse plane. The curve fitting by using 4 different curves: single radius model, dual-radius model, polynomial model, and B-spline model were analysed. The Fig. 6.3 represents the comparative accuracy of various mathematical curves. It is observed that in B-spline model fitted the original data most accurately in both sagittal and transverse planes.

6.3.3 Parametric Modelling

There are various parameters that influence the kinematics of the joint which are to be considered while developing the basic parametric CAD model. The basic design variables such as radii of curvature of the femoral component in the frontal and sagittal planes, the spacing of the condyles, and the frontal and sagittal conformity with the UHMWPE control the design of prosthesis majorly. Willing and Kim (2011) explained the limitation of few design variables for CAD model and optimization of femoral and tibial insert components and thus identified 7 different variables namely femoral distal sagittal radii of curvature of condyles (FDSR), femoral posterior sagittal radii of curvature of condyles (FPSR), femoral frontal radii of curvature of condyles (FFR), distance from most distal point on condyles to mid-line in frontal plane (W), UHMWPE insert sagittal radii of curvature of condyles (ISR), UHMWPE insert frontal radii of curvature of condyles (IFR) and AP length of UHMWPE insert condyles (L).

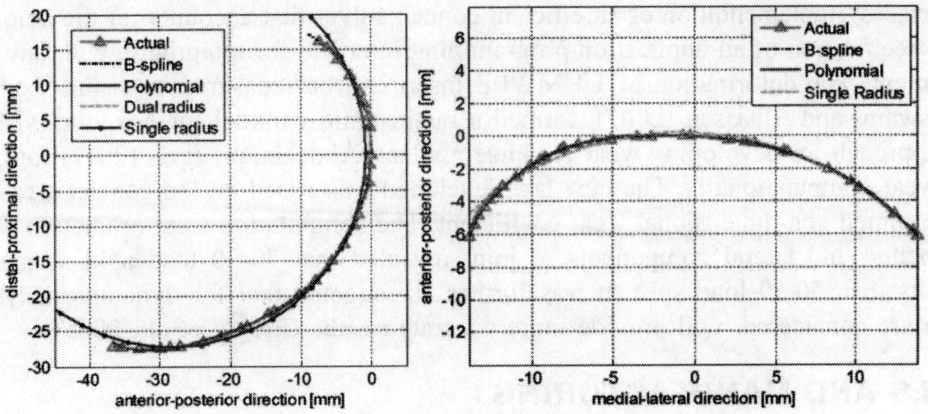

Fig. 6.3 Comparison of mathematicalo model to approximate femoral condyle shape in sagittal and transverse plane respectively (Heever et al., 2011)

With increase in the variables, the more control over the design is done and thus the most ideal and accurate tailor made designs of implant can be achieved. The parameters listed above are considered as the most basic and important parameters governing the CAD model.

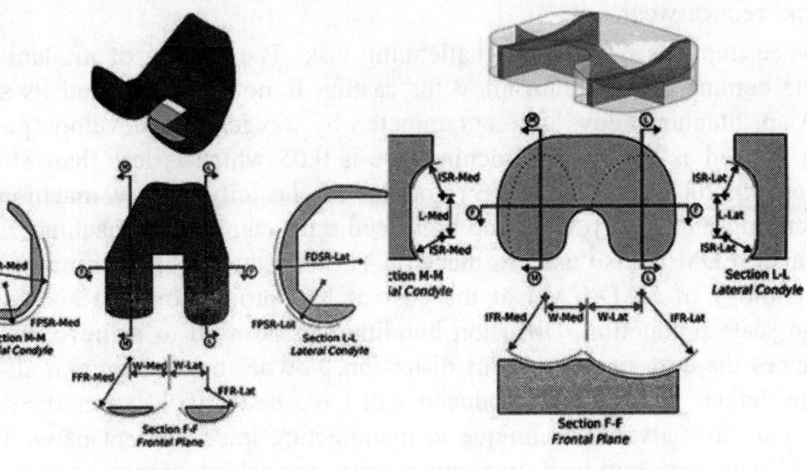

Fig. 6.4 Parameters of knee implant CAD model

6.4 FINITE ELEMENTAL ANALYSIS

The finite elemental analysis of the CAD model developed will help to check the performance of the implant at the early stage of development. The FEA can be static or dynamic. The static stress-strain analysis study by Kubíček and Florian (2010) have shown that medial condyle is loaded heavily compared to the lateral condyle and the displacement of meniscus took place in both lateral and medial regions. Further it was observed that maximum value of contact pressure between femoral cartilage and tibial cartilage and meniscus was of magnitude 2.25 MPa. Lesso-Arroyo, et al. (2004) also conducted static analysis of 3D CAD model in stages to study the individual performance of each element. The analysis was divided into stages of increasing complexity in models: individual model, simplified model, reduced model and final model so that more accurate approximate results can be obtained. Compressive load on joint increases with increase in flexion angle and the contact nature changes from area to line to quasi line with increase in flexion angle (Chu, 1999).

The dynamic analysis includes mainly includes wear predictions by multi-body contact analysis. Bei and Fregly (2004) conducted the multi-body simulation of knee contact mechanics using quasi-static with deformable contact or dynamic with rigid contact. They suggested a systematic approach of performing dynamic analysis of knee implant which requires preparation of the articular surface geometry, development of efficient methods to calculate distances between contact surfaces, implementation of an efficient contact solver that accounts for the unique characteristics of human joints, and specification of an application programming interface for integration with any multibody dynamic simulation environment. The deformation of UHMWPE insert can be measured by using 'bed of spring' or 'half elastic theory'. Goswami and Alhassan (2007), formed a mathematical model for hip joint wear measurement and suggested similar approach for developing wear for knee implant. Abdelgaied, et al. (2011) formed a new wear law and predicted the wear computationally. The new law developed was based on the contact area and an independent experimentally determined non-dimensional wear coefficient. The computation wear prediction with differential load consideration on medial and lateral components of joint revealed that 70-30 load split at medial-lateral predicts more accurate results than 50-50 load split. It was further investigated that for dynamic analysis if 85% gait and 15% stair activities are considered, will provide more accurate results (Fregly et al., 2005).

6.5 MATERIALS AND MANUFACTURING

Titanium alloys are the most common and ideal material used for implant manufacturing because of various advantages offered by it such as high strength/weight ratio, high corrosion resistance and biocompatible (inert to reactions with tissues). Previous studies on the biocompatibility of Ti show that human bone forms intimate contact with the bone in a process called osseointegration (Oshida Yoshiki, 2007) which gives structural support to implant after fixation. The implant surface which is in contact with the tissues is modification instead of whole implant material to reduce the manufacturing cost and maintain the strength of implant. Surface treatments include coatings and heat treatment. Coating of implant has additional benefits such as reduce discharge of foreign particles, increase adhesion with bones and reduce wear.

Manufacturing of knee implants is the next challenging task. The casting of implant with titanium alloy is not wise option because the commercial titanium alloy for casting is not available and its strength reduces compared to wrought titanium. Also, titanium alloys get contaminated by oxygen and develops porous object. Machining of Ti is the most suitable method as the work hardening rate is 0.05 which is less than SS (0.45), copper (0.54) and brass (0.49) (Oshida Yoshiki, 2007) . As the MOE (modulus of elasticity) is low, machining by taking care like low cutting speed, surplus cutting fluid, sharp tools and high feed rates can help to machine Ti without any problem and having good surface finish. EDM is also used to machine Ti blank due to high accuracy and precision requirement with the advanced technology of CAD/CAM at the cost of low production rate and high production cost which is not suitable for large scale production. Diffusion bonding can be used to achieve the near net shape (NNR) of the implant, which reduces the cost and avoid part distortion. Powder metallurgy can also be used for the NNR as it eliminates the casting defects and implant produced will have better yield strength. Rapid metal prototyping of implant is the modern and most advance technique to manufacture knee implant parts. This will produce implants at comparative price and with very low lead time and maintaining the required accuracy. Heat treatment is used to reduce the residual stresses and obtain the correct micro structural composition of alloys.

6.6 CONCLUSION

The present study and investigation revealed that there is large scope of improvement in the knee implant. The parametric CAD model will eliminate in selection of joint from the standard sizes available and will help to stabilize joint more confirmatively. The method of reverse engineering is non-invasive and reliable method for developing the CAD model for extracting the geometrical parameters of knee bones. The CAD model developed will also help in investigation of various issues of implant such as stem dimensions, cruciate retaining/sacrificing joint decisions and surgical planning and assistance. The advance manufacturing systems like rapid prototyping, 4-5 axis CNC machines, EDM are more accurate and precise and integrating them with CAD systems will help to reduce the lean time drastically.

The parametric CAD model for bone and implant developed for the patient can be used for planning the surgery and use of robotic surgical system. This will provide minimal invasive knee arthroplasty with high degree of accuracy and proper positioning and alignment of implants.

ACKNOWLEDGEMENTS

The authors wish to acknowledge the valuable inputs from Dr. Bharat Mody, Director & Chief Surgeon, Welcare Hospital, Vadodara.

REFERENCES

1. Abdelgaied, A., Liu, F., Brockett, C., Jennings, L., Fisher, J., & Jin, Z. (2011). Computational wear prediction of artificial knee joints based on a new wear law and formulation. Journal of Biomechanics, 44, 1108–1116.
2. Bahraminasab, M., Sahari, B.B., Edwards, K.L., Farahmand, F., & Arumugam, M. (2012). Aseptic loosening of femoral components–materials engineering and design considerations. Materials & Design.
3. Bei, Y., & Fregly, B.J. (2004). Multibody dynamic simulation of knee contact mechanics. Medical 2
4. Carr, B.C., & Goswami, T. (2009). Knee implants – Review of models and biomechanics. Material and Design, 30, 398-413.
5. Chu, T. (1999). An Investigation on Contact Stresses of New Jersey Low Contact Stress (NJLCS), Knee Using Finite Element Method. Journal of Systems Integration, 9, 187-199.
6. Fregly, B.J., Sawyer, W.G., Harman, M.K., & Banks, S.A. (2005). Computational wear prediction of a total knee replacement from in vivo kinematics. Journal of Biomechanics, 38, 305–314.
7. Gerhard Welsch, Rodney Boyer, E.W. Collings (1993), Materials Properties Handbook: Titanium Alloys, ASM International, Materials Park, OH.
8. Goswami, T., & Alhassan, S. (2007). Wear rate model for UHMWPE in total hip and knee arthroplasty. Materials and Design, 29.
9. Heever, D.J., Scheffer, C., Erasmus, P., & Dillon, E. (2011). Mathematical Reconstruction of Human Femoral Condyles. Journal of Biomechanical Engineering, 133.
10. Kubićek, M., & Florian, Z. (2010). Stress Strain Analysis of Knee Joint. Engineering Mechanics, 16, 315-322.
11. Lee, K.M., & Guo, J. (2010). Kinematic and dynamic analysis of an anatomically based knee joint. Journal of Biomechanics, 43(7), 1231-1236.
12. Lesso-Arroyo, R., Sánchez Jiménez, J., Castro, R.R., & Lopez, F.B. (2004). Biomechanical Behavior of the Knee Joint, Using ANSYS. International ANSYS Conference Procedings.
13. Lin, Y.P., Wang, C.-T., & Dai, K.R. (2005). Reverse engineering in CAD model reconstruction of customized artificial joint. Medical Engineering & Physics, 27, 189–193.
14. Oshida Yoshiki (2007), Bioscience and Bioengineering of Titanium Materials, Elsevier, Oxford, UK.
15. R. Boyer, G. Welsch, and E.W. Collings, (1994) Materials Properties Handbook: Titanium Alloys, ASM International, Materials Park, OH.
16. Willing, R., & Kim, I. (2011). Design optimization of a total knee replacement for improved constraint and flexion kinematics. Journal of Biomechanics, 44, 1014–1020.

7

Study of the Biomechanical Performance of Implants used for Fixation of Fractured Trochanter

Amiya Kumar Dash, Shakti Prakash Jena and Debyendu Mondal

Deparment of Mechanical Engineering
Siksha O Anusandhan University, Bhubaneswar, India

Email: amiyadash@soauniversity.ac.in

ABSTRACT

It is essential to determine the optimal treatment of inter-trochanteric fractures due to their high incidence and related public health burden. Debate remains as to whether dynamic hip screws (DHS) or proximal femoral nails (PFNs) are best practice, and this computational analysis seeks to collect information relevant to this query.In inter-trochanteric fractures DHS implant is more prevalent of paediatric femur fractures in recent years than PFN implant. Steel may be expected to be the material of implant to have superior fracture stability due to its higher elastic modulus. The purpose of the study was to verify these observations computationally and thus explain why DHS implant may be better suited for surgical fixation of inter-trochanteric fractures. The below work has been focused to decide on the basis of experimentation and research, which of the two PFN or DHS is more suitable for the fixation.

Keywords: Proximal femur nail (PFN), Dynamic hip screw (DHS), trochanter, total deformation, equivalent stress.

7.1 INTRODUCTION

Trochanter: One of the bony prominences toward the near end of the thigh bone (the femur).The trochanters are points at which hip and thigh muscles attach. The greater trochanter gives attachment to a number of muscles and the lesser trochanter receives the insertion of several muscles. Proximal Femur Nail (PFN) is an intramedullary implant for the treatment of unstable pertrochanteric and intertrochanteric femoral fractures. Dynamic Hip Screw (DHS) is a type of orthopaedic implant designed for fixation of certain types of hip fractures which allows controlled dynamic sliding of the femoral head component along the construct. The idea behind the dynamic compression is that the femoral head component is allowed to move along one plane; since bone responds to dynamic stresses, the native femur may undergo remodeling and proper fracture healing.

Haynes et.al [1] have studied the biomechanical cadaveric test and carried out to establish the influence of direct static loading situations on the modes of failure of the gamma nail, compared with the dynamic hip screw. In this study each failure mode has been isolated, to establish the loads of failure under various fracture configurations. It was suggested that the intramedullary gamma locking nail can be recommended over DHS in case of subtrochanteric fracture. Harrison [2] has performed an experiment on forces acting on the leg with particular attention to the forces at top of the femur. He observed that the force exerted on head of the femur by pelvis was very large that was 2.4 times the weight of the person and the force exerted on the trochanter by the abductor muscle was also large. He then concluded that these two forces were directed nearly opposite direction. According to him upper femur experiences large forces of compression and shear. Sparks et al. [3] have performed an experiment and found that contact pressures were measured in human hip joints using pressure sensitive film. They examined the effects of femoral angle on contact mechanics in greater trochanter loading and they found that the changes in the femoral abduction angle had a greater effect on the location. Perez et al. [4] experimentally demonstrated the improved biomechanical stability of titanium alloy. The aim of the study was to verify the observation and to explain that how titanium alloy would be better suited for surgical fixation of fractured femurs. They observed that titanium alloy nails distributed stress more evenly along the nail axis, resulting in lower peak magnitude. They also found that increased deformation of titanium alloy nail increases the contact area with the intramedullary canal wall, thus increasing the stability.Tzurbakis et al. [5] have conducted an experiment on posterior hip dislocation after a serious injury resulting from an accident. They have found that at surgery a deep wound infection can be artificially developed. They have concluded that hip dislocation after a trochanteric fracture internal fixation is rarely complicated. Infection irradiation and a second stage surgical repair of a hip joint can save life.Baumgaertner et al. [6] have conducted an experiment and found that unstable inter trochanteric hip fractures account for approximately one quarter of all hip fractures in the elderly and are increasing in frequency.

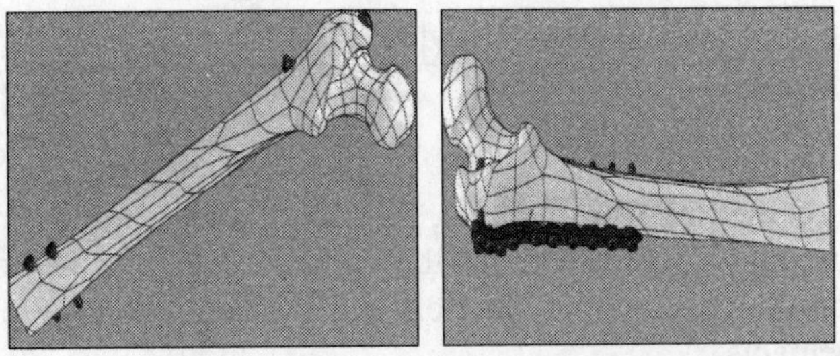

(a) (b)

Fig. 7.1 (a) Proximal femur nail in intramedullary femoral fractures (b) Dynamic hip screw (DHS) in hip fractures

Treatment goals include immediate mobilization while limiting complications. Unstable fracture patterns, problems with compression hip screw fixation, such as excessive fracture collapse and implant cut out, increases. For these fractures, adding a trochanteric stabilizing plate or using an axial compression hip screw or intra medullar hip screw is warranted. Filardi et al. [7] have studied to evaluate experimentally the behavior after fracture consolidation of two intramedullary gamma nail implants having one (G1) or two (G2) distal screws they had implanted nails into standardized synthetic femora, instrumented with strain gauge with the support of FEM. They found that G2 implant, although ensuring higher flexion and torsional stiffness could lead to localized contacts that occur between the tip of the nail and the femoral endosteum. They concluded that this might be one of the reason of the complication associated with the pain in the mid-portion of the thigh after implantation which had been reported in several clinical studies when gamma nail with the two distal screws are used. Wang et al. [8] have investigated that the stresses in the gamma nail which was used for different types of fractured femur. They had found that gamma nail used for subtrochanteric fractures will cause higher stresses at the lag screw and upper distal screw insertion in the hole when used for femur neck fractures.Asgari et al. [9] have conducted an experiment on the finite element modeling and comparison of the stress and strain analyses carried out for three different structures that are intact bone, stem less implant and stemmed one. The generic stem less implant reconstruction was numerically compared to the conventional stemmed implant and also to the intact bone as control solution. Two loading conditions were applied to the most proximal part of the models, while the most distal part was fixed for all degrees of freedom. The models were divided into two regions and studied along two paths of medial and lateral aspect. According to them the stem less implant had less deviation from the control solution of the bone in all regions and in both loading conditions, comparing to the large deviation of the stemmed implant from the intact bone.Schipper et al. [10] had investigated that during the treatment of unstable trochanteric fracture Gamma Nail was the most successful rather than PFN used during the operation. They have found out that PFN which is made of titanium metal of length 240 mm with (10-11 mm) dia which is directly inserted into femur bone without reaming but the GN which is made up of steel with length of 200 mm and dia 11 mm requires reaming for insertion.

7.2 METHODOLOGY

To fully understand the reasons behind the greater biomechanical stability afforded by properties of steel an element study was conducted. The following steps where followed over a period of 6 months and the results were recorded.

7.2.1 The bone model

The computer aided design (CAD) model of a three dimensional femur was downloaded as IGES surface definitions of the cortex from the mesh repository. This physical IGES model has been used in a number of biomechanical studies involving intramedullary nails. This "stabilized" model was then made from steel by applying identical boundary and load conditions. The bone model was 510 mm long a, a simulated transverse fracture with a gap of 5 mm, a canal diameter of 0.5 mm was made physically.

Table 7.1: Bone properties

Property	Value
Density	1750 kg/m³
Young's Modulus	16000 MPa
Poisson's Ratio	0.3
Bulk Modulus	13333 MPa
Shear Modulus	6153.8 MPa

7.2.2 The Mesh

After the bone and nail surface geometries were defined, the mesh was designed in ANSYS software after which an assembly model using nails and plates was used over the bone model. Prior to exporting the fully meshed femur model, the mesh underwent rigorous diagnostic evaluation and was remeshed as needed to ensure there were no crossing, inverted or distorted elements.

Nodes	13927
Elements	7050

7.2.3 Material Properties

For implant purpose same kind of material was chosen for the nails, plates and the screws.

Material	Structural Steel
Density	7850 kg/m3
Young's Modulus	200 GPa
Poisson's Ratio	0.3

7.3 RESULTS AND DISCUSSION

This section represents discussions made from the results obtained from the computational analysis.

7.3.1 For Dynamic Hip Screw (DHS)

7.3.1.1 Compressive Force

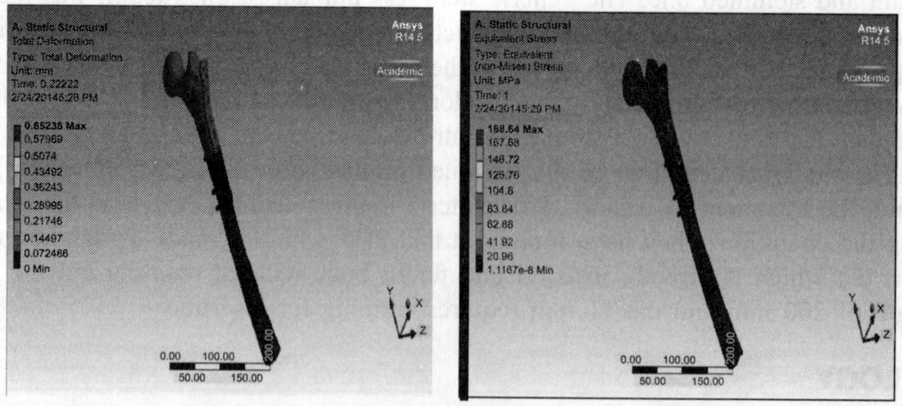

Fig. 7.2

7.3.1.2 Torsional Force

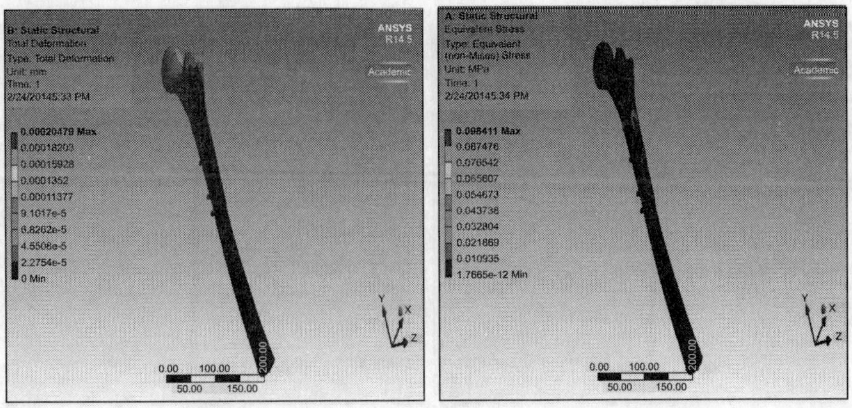

(a) (b)

Fig. 7.3

7.3.1.3 Combined Force

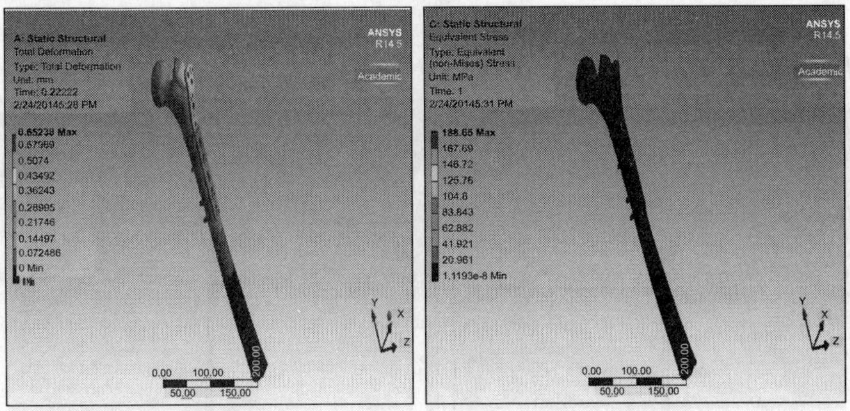

(a) (b)

Fig. 7.4

Fig. 7.2 (a) and (b) depicts the maximum total deformation and maximum equivalent stress at 300 N compressive force to be 0.65238 mm and 188.64 MPa respectively. However, Fig. 7.3 (a) and (b) depicts the maximum total deformation and maximum equivalent stress at 10 N.mm moment to be 0.00020479 mm and 0.098411 MPa respectively. Fiure 7.4 (a) and (b) depicts the maximum total deformation and maximum equivalent stress at 300 N compressive force and 10 N.mm moment combined force to be 0.65239 mm and 188.65 MPa respectively.

7.3.2 For Proximal Femur Nail (PFN)

7.3.2.1 Compressive Force

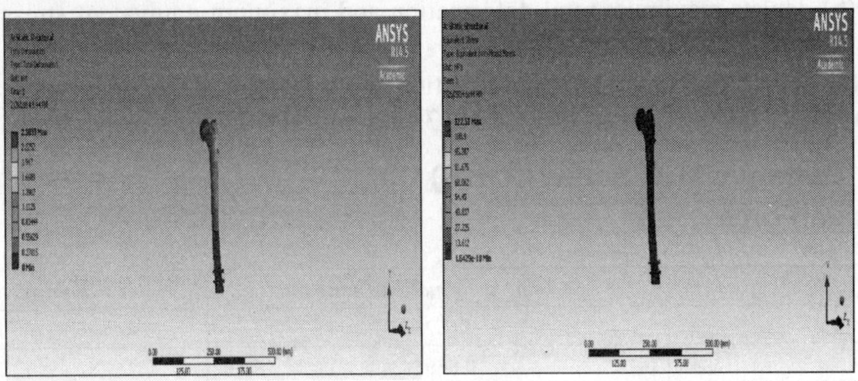

(a) (b)

Fig. 7.5

7.3.2.2 Torsional Force

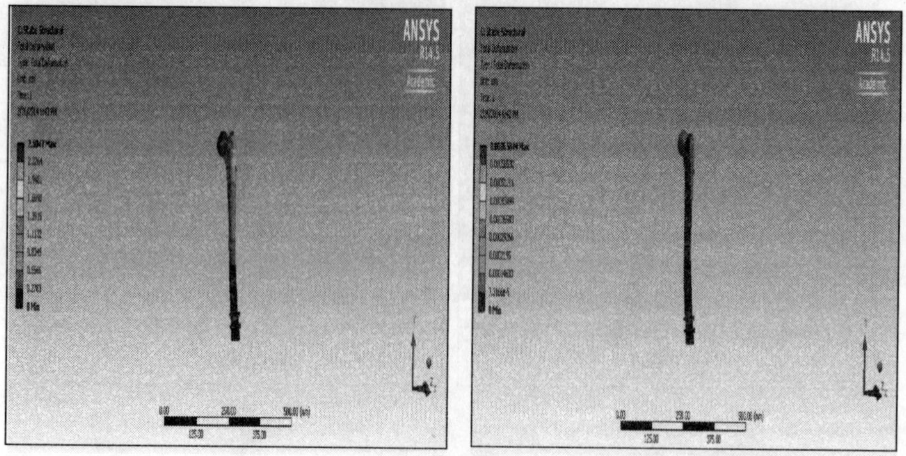

(a) (b)

Fig. 7.6

7.3.2.3 Combined Force

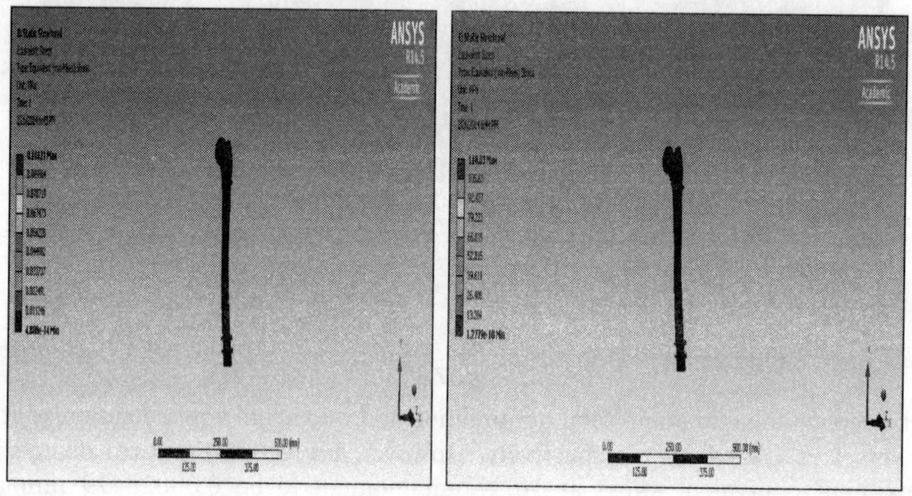

(a) (b)

Fig. 7.7

Figure 7.5 (a) and (b) depicts maximum total deformation and maximum equivalent stress at 500 N compressive force to be 1.502 mm and 73.507 MPa respectively.

Figure 7.6 (a) and (b) depicts maximum total deformation and maximum equivalent stress at 20 N.mm moment to be 0.00065849 mm and 0.10121 MPa respectively.

Figure 7.7 (a) and (b) depicts maximum total deformation and maximum equivalent stress at 500 N compressive force and 20 N.mm moment combined force to be 2.5047 mm and 118.83 MPa respectively.

7.3.3 Results for Dynamic Hip Screw (DHS)

7.3.3.1 Compressive Force

Force (N)	*Total Deformation (mm)*	*Equivalent Stress (MPa)*
300	0.65238	188.64
400	0.86984	251.52
500	1.0873	314.4

7.3.3.2 Torsional Force

Moment(N.mm)	Total Deformation (mm)	Equivalent Stress (MPa)
10	0.00020479	0.098411
15	0.00030718	0.14762
20	0.00040957	0.19682

7.3.3.3 Combined Force

Compressive Force (N)	Torsional Moment (N.mm)	Total Deformation(mm)	Equivalent Stress (MPa)
300	10	0.65239	188.65
400	15	0.86985	251.53
500	20	1.0873	314.41

7.3.4 Results for Proximal Femur Nail (PFN)

7.3.4.1 Compressive Force

Force (N)	Total Deformation (mm)	Equivalent Stress (MPa)
300	1.502	73.507
400	2.0027	98.01
500	2.5033	122.51

7.3.4.2 Torsional Force

Moment (N.mm)	Total Deformation (mm)	Equivalent Stress (MPa)
10	0.00029266	0.050605
15	0.00043899	0.075907
20	0.00065849	0.10121

7. 3.4.3 Combined Force

Compressive Force (N)	Torsional Moment (N.mm)	Total Deformation (mm)	Equivalent Stress (MPa)
300	10	1.5028	71.299
400	15	2.0038	95.067
500	20	2.5047	118.83

7.4 CONCLUSION

The results from the analysis revealed that

- For the DHS for the same compressive force, the maximum total deformation is less as compared to the PFN whereas the maximum equivalent stress induced is more with respect to the PFN.
- For the DHS for the same torsional force, the maximum total deformation is less as compared to the PFN whereas the maximum equivalent stress induced is more with respect to the PFN.
- For the DHS for the same combined force, the maximum total deformation is less as compared to the PFN whereas the maximum equivalent stress induced is more with respect to the PFN.
- For DHS and PFN for the same torsional force head of the trochanter is subjected to maximum total deformation whereas the maximum equivalent stress is at the screw and the bone interface.
- For DHS and PFN for the same compressive and combined force the top portion of the femur head and the upper trochanter is subjected to maximum total deformation whereas the maximum equivalent stress is where the screw and bone interface.

- Looking at the maximum total deformation the deformation magnitude is more for PFN as compared to DHS.
- Thus, in inter-trochanteric fractures DHS implant is more suitable of paediatric femur fractures in recent years than PFN implant.

REFERENCES

1. Haynes R.C.,. Poll, R.G., Miles A.W., Weston R.B. (1997) An experimental study of the failure modes of the Gamma Locking Nail and A.O Dynamic Hip Screw under static loading: a cadaveric study. 1997, pp. 446–453.
2. R. Hobbie, M. Harrison, "Forces Acting on the Leg", Willey, 1978, pp. 8-13.
3. D. Sparks, D. Beason, A. Eberhardt, F. Bandak," Contact Pressure Measurements in Hip joint: Effects of Femoral Angle in Greater Trochanter Loading", IEEE, 2002, Vol. 3, pp. 23-26.
4. Angel Perez, Andrew Mahar, Charles Negus, Peter Newton, Tom Impelluso, "A computational evaluation of the effect of intramedullary nail material properties on the stabilization of simulated femoral shaft fractures", medical engineering physics, 2008, pp. 750-760.
5. M. Tzurbakis, E. Morakis, G. Mouzopoulos, N. Lasaniunos, L. Georgilas, "Deep Wound Infection After A Trochanteric Fracture Internal Fixation Presenting with Hip Dislocation", Tzurbakis etal.cases journal, 2010, Vol. 3, pp. 19-22.
6. Baumgaertner, An Orthop, Koval, "Intertrochanteric Hip Ractures", 2004, Vol. 12, pp. 179-190.
7. Vincenzo Filardi, Roberto Montanini, "Measurement of local strains induced into the femur by trochanteric Gamma nail implants with one or two distal screws", 2007, pp. 38-47.
8. C.J. Wang, A.L. Yettram, M.S. Yao, P. Procter," Finite element analysis of a Gamma nail within a fractured femur", 1998, pp. 677-683.
9. S. Asgari, A. Hamouda, S. Mansor, H. Singh, R. Wiza, "Finite element modeling of a generic stem less hip implant design in comparison with conventional hip implants", Elsevier, 2004, Vol. 8, pp. 949-961.
10. I.B. Schipper, E.W. Steyerberg, R.M. Castelein, F.H.W.M. van der Heijden, P.T. den Hoed, A.J.H. Kerver, A.B. van Vugt," Treatment of unstable trochanteric fractures", 2004, Vol. 86-B, pp. 86-94.

8

Prediction of Peak Expiratory Flow Associated with Human Respiratory Mechanics Using Spirometric Investigations

A. Mythili[1], C. M. Sujatha[2], S. Srinivasan[1] and S. Ramakrishnan[3]

[1]*Dept. of Instrumentation Engineering, MIT Campus, Anna University, Chennai, India*
[2]*Dept. of Electronics and Communication Engineering, CEG Campus, Anna University, India*
[3]*Biomedical Engineering Group, Department of Applied Mechanics, IIT Madras, Chennai, India*

Email: mythiliasaithambi@gmail.com

ABSTRACT

Spirometry is the most commonly performed test to assess the dynamics of respiratory functions in human subjects. It measures airflow from fully inflated lungs through forced expiratory maneuver and generates large data set. Measurement of respiratory mechanics and lung function are crucial as they provide direct measure of chronic disease related to the lungs. However, spirometric investigations often result in incomplete data sets due to the inability of the children and patients to perform this test. Hence, there is a requirement for prediction of significant parameters from the available incomplete data set. In this work, Peak Expiratory Flow (PEF) is predicted using Extreme Learning Machine (ELM) regression. The measured spirometric parameters are given as inputs to the ELM which utilize single hidden layer feed forward network for prediction of PEF. ELM with sine and sigmoidal activation functions is used to train and test the prediction process. The performance of the model is evaluated by computing their prediction error and root mean square. Results show that ELM with Sine function obtained higher prediction accuracy for 30 hidden neurons. Further, it is observed that PEF is predicted with high accuracy and has better correlation with measured values for normal subjects. It appears that this method is useful in enhancing diagnostic relevance of spirometric investigations in case of patients who are not able to complete the test..

Keywords: Pulmonary function test, spirometry, peak expiratory flow, extreme learning machines.

8.1 INTRODUCTION

Pulmonary Function Test (PFT) is often used in assessment of lung function by measuring forced expiratory volumes and flow rates of respiratory system [1]. The spirometric pulmonary function test measures the rate of changing lung volumes during forced breathing maneuvers. Spirometry has been the preferred test of lung function because of its reproducibility and utility in diagnosing lung disease such as airway obstruction, emphysema and asthma. Spirometry is the most appropriate way to diagnose the respiratory disorders such as airway obstruction, emphysema and asthma. Spirometry is an essential investigation method of choice for a fast and reliable screening of Chronic Obstructive Pulmonary Disease (COPD). This test is used to diagnose, manage, and monitor patients with a variety of respiratory diseases and is increasingly advocated in primary care practice. The parameters generated by spirometry are influenced by height, weight, age, sex, environmental factors, ethnicity, prematurity, patient cooperation and effort, and technical factors [2]. These parameters are vital for the detection of declining lung function. Some of the significant parameters recorded from the spirometer are Forced Expiratory Volume of air in one second (FEV1), Forced Vital Capacity (FVC), Peak Expiratory Flow (PEF) and Forced Expiratory Flow at 25% of FVC (FEF25%). These spirometric parameters are clinically useful in the identification of the cause of airway obstruction and detection of early changes in lung function [3].

Peak expiratory flow is the useful diagnostic index to characterize pulmonary abnormalities [4]. PEF rate monitoring is often used alone in evaluating COPD and the response to a bronchodilator in the assessment of asthmatic subjects. Objective assessment of airway caliber based on FEV1 and PEF is a relevant and important aspect in the assessment of asthmatic subjects [5, 6].

The clinical value of spirometric measurements depends on the accuracy of the spirometer, performance of the breathing maneuver and use of relevant predicted normal values [7, 8]. Spirometric investigations being effort-dependent, requires understanding and motivation on the subject's part. It is often difficult to get an accurate result because physical and mental impairment may limit patient's ability to perform [9, 10 and 3]. Due to these reasons, spirometric measurements often result in incomplete data sets and would also have missing values [11, 12]. Hence, applying automated prediction methods which could estimate such missing parameters would be clinically important for diagnosing respiratory diseases [13, 14].

Artificial Neural Networks (ANN) has already been used in the prediction of respiratory data [15]. Radial basis function neural networks, neuro fuzzy inference system and support vector regression have already been employed for the prediction of FEV1 and FEV6 [16, 17]. However, conventional neural network methods have demonstrated difficulties in finding a good generalization performance. A new learning algorithm for Single Hidden Layer Feed-Forward Networks (SLFNs), called Extreme Learning Machine (ELM), has been proposed recently by Huang [18, 19], which can be used in regression and classification problems [20-22].

In this study, an ELM model with sine and sigmoid activation function has been employed for the prediction of peak expiratory flow using spirometric pulmonary function data.

8.2 METHODS

8.2.1 Data Acquisition

The spirometric pulmonary function test is assessed using flow-type spirometer. The patients made three forced expirations through the spirometer. The highest values are considered as acceptable value as per ATS/ERS standardization. The tests are performed in a sitting position. Room temperature is kept between 22 and 27 °C. Tests are performed for N = 100 subjects according to ATS/ERS guidelines for spirometry. Various spirometric parameters such as FEV1, FVC, PEF and FEF25% are generated. The spirometric parameters are subjected to regression using ELM.

8.2.2 Regression

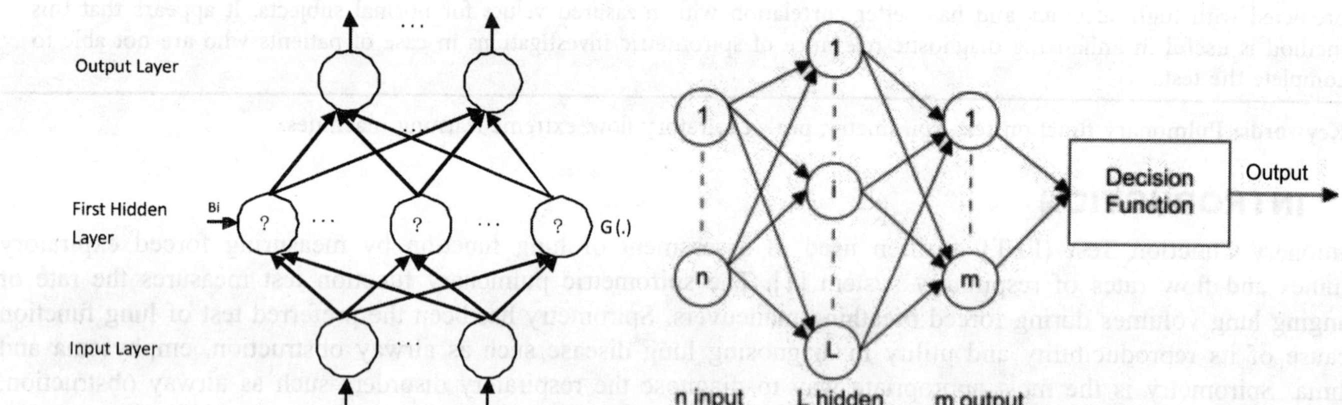

Fig. 8.1 Architecture of extreme learning machines

The structure of ELM network is shown in Fig. 8.1. ELM contains an input layer, hidden layer and an output layer. Single hidden Layer Feed forward Networks (SLFN) with K hidden neurons are trained with activation function $g(x)$ to learn N distinct samples (x_i, t_i), where $x_i = [x_{i1}, x_{i2}, \ldots, x_{in}]^T$ δ R^m represent 14 spirometric data of

100 subjects and $t_i = [t_{i1}, t_{i2}\ldots\ldots, t_{in}]^T$ δ R^m is the measured PEF value used as target during training. In ELM, the input weights and hidden biases are initially randomly generated and tuned in the training. The main goal of the training process is to determine the network weights $w_j = [w_{j1}, w_{j2},\ldots, w_{jn}]^T$, hidden layer biases $\beta = [\beta_1, \beta_2, \ldots. \beta_k]^T$,

This is to minimize the error function $\sum_{j=1}^{N}\left\|o_j - t_j\right\|$ where $o_j = [o_{j1}, o_{j2},\ldots,o_{jn}]^T$ represents the network output and $t_j = [t_{j1}, t_{j2},\ldots,t_{jn}]^T$ are its corresponding test target. The network model of ELM is mathematically modelled as

$$\sum_{i=1}^{M}\beta_i g(w_i.x_i + b_i) = o_j,\ j = 1, 2, \ldots N \qquad \ldots(1)$$

where is the activation function. More compactly the equation can be written as $H\beta = T$ form which the output weights are estimated as βδ = H†T where H† is the pseudo-inverse of H [23].

Different basis which include sigmoid and sine functions of ELM are considered. The training performance of the ELM model can be evaluated by the Root Mean Square Error (RMSE):

$$RMSE = \sqrt{\frac{1}{M}\sum_{Z=1}^{M}(S_z - Y_z)^2} \qquad \ldots(2)$$

where M is the total number of patterns, S_z is the target value, and Y_z is the ELM output value. The percentage of error between the predicted values and experimental values is given by:

$$\text{Error (\%)} = \frac{\text{Predicted value-experimental value}}{\text{experimental value}} \times 100 \qquad \ldots(3)$$

8.3 RESULTS AND DISCUSSION

The measured spirometric parameters are subjected to ELM regression. Typical response of a spirometer that depicts variation of airflow with lung volume for a normal subject is shown in Fig. 8.2(a). It is seen that this normal flow-volume curve has rapidly rising flow at the beginning of expiration and then it declines steadily in a linear fashion. The highest flow rate, obtained during the first part of expiration, is effort dependent and is approximately equal to one third of vital capacity. Figure 8.2(b) shows a typical flow-volume curve of subject with obstruction abnormality. In obstruction, the curve descends more quickly than normal and takes on a concave shape.

Table 8.1: Statistical analysis of spirometric pulmonary function data

Parameters	*Normal (50)*	*Abnormal (50)*
	Mean ±SD	Mean ±SD
Age	33.1±20.4	38.5±20.4
Height	153.7±11.5	150.1±9.84
Weight	55.6±15.4	56.6±16.8
PEF	5.19±1.64	3.46±1.52
FEV1	2.14±0.59	1.44±0.62
FVC	2.625±0.73	2.06±0.79

The statistical analysis on the spirometric pulmonary function parameters such as mean and standard deviation are presented in Table 8.1. The expected values for various lung volumes and flow rates are derived from healthy subjects and are adjusted for sex, age, height and race. The mean values of normal subjects are distinctly higher than that of the abnormal cases and the difference observed is always greater than 20%. Most of the normal subjects found to have less standard deviation, which explains consistency among the values. However, same degree of consistency is not observed in abnormal subjects, which indicates high inhomogeneity in the normal parameters.

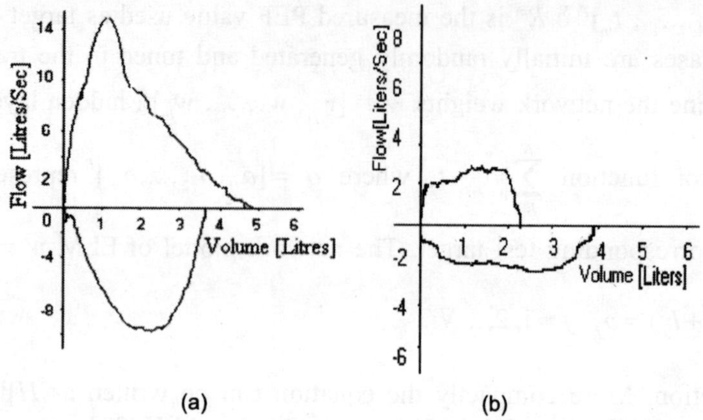

Fig. 8.2 (a) and (b) Typical flow-volume curves of normal and abnormal subjects

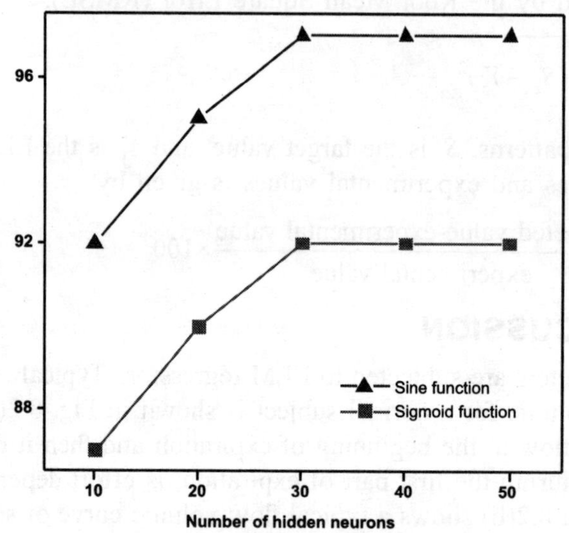

Fig. 8.3 Variation in prediction accuracy with number of hidden neurons

Spirometric measured parameters are given as input to the ELM regression model. ELM with sine and sigmoid activation functions are employed with the number of hidden neurons is varied from 10 to 50. In ELM prediction process both sine and sigmoid activation function could achieve higher accuracy for 30 number of hidden neurons. Statistical analysis of predicted and measure PEF values is presented in Table 8.2. It is observed that the measured and predicted PEF for normal and abnormal subjects are highly similar.

Table 8.2: Statistical analysis of pulmonary function data

	Mean ± SD	
	Measured PEF	Predicted PEF
Normal	0.545±0.192	0.552±0.166
Abnormal	0.278±0.157	0.342±0.115

Figure 8.4(a) and (b) show the deviations in the predicted PEF values from the measured values for normal and abnormal subjects respectively. It is observed that the predicted PEF values follow measured values with minimum variation in both normal and abnormal subjects. Further, it is found that the prediction of PEF is found to be accurate for normal subjects as the value of correlation coefficient is 0.93.

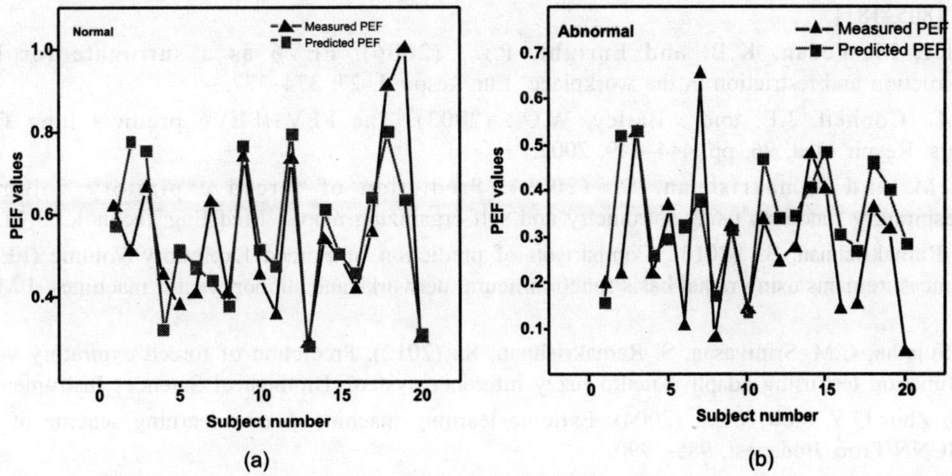

Fig. 8.4 Deviation of the predicted values from the measured PEF values for (a) normal and (b) abnormal subjects

8.4 CONCLUSION

Spirometry is the essential screening test for assessment of general respiratory health that provides important objective data to clinicians. The parameters obtained from spirometer are extremely diagnostic in the evaluation of respiratory functions. In this work, prediction of PEF is attempted using ELM model. The trained ELM model is employed for prediction using sine and sigmoid activation functions with varying number of hidden neurons. Results show that ELM model is capable of predicting PEF in both normal and abnormal subjects. ELM model with sine activation function predicted PEF with high prediction accuracy using 30 hidden neurons. It appears that this method of prediction of significant parameter could be clinically relevant in the diagnosis of pulmonary abnormalities with incomplete data set.

REFERENCES

1. Weir, N.A. Gulat, I.M. and Redlich, C.A.: (2005). Diseases of the lung and pleura. In: Rosenstock L, ed. Textbook of Clinical Occupational and Environmental Medicine. 2nd ed. Elsevier Saunders 285-9.
2. Kana Ram Jat.: (2013). Spirometry in children. Prim Care Respir J, 22(2), 221-229.
3. Gillette, C. Loughlin, E.C. Sleath, B.L,. Williams, D.M. and Davis, S.D. (2011). Quality of Pulmonary Function Testing in 3 Large Primary Care Pediatric Clinics in Rural North Carolina. N.C. Med J., 72(2), 105-110.
4. Shakeri, J. Paknejad, O. Moghadam, K.G. and Taherzadeh, M.: (2012). Logistic Regression Model for Prediction of Airway Reversibility Using Peak Expiratory Flow. Tanaffos, 11(1), 49-54.
5. Cross, D. Nelson, H.: (1991). The role of the peak flow meter in the diagnosis and management of asthma. J Allergy Clin Immunol, 87, 120-28.
6. Lebowitz, M.D.: (1991). The use of peak expiratory flow rate measurements in respiratory disease. Pediatr Pulmonol, 11,166-74.
7. Pierce, R.: (2004). Spirometer: an essential clinical measurement. Aus. Fam Phys, 34, 535–539.
8. David, P. Daryl, F. Jen, C. Alan, K. and Frank, C.: (2011). Earlier diagnosis and earlier treatment of COPD in primary care. Prim Care Respir J, 20(1), 15-22.
9. Lay Ekuakille, A. Vendramin, G. and Trotta, A.: (2010). Spirometric Measurement Post processing: Expiration Data Recovery. IEEE Sen J, 10(1), 25-33.
10. Roberts, N.J. Smith S.F. and Partridge, M.R.: (2011). Why is spirometry underused in the diagnosis of the breathless patient: a qualitative study. BMC Pul Med, 11-37.
11. Vandevoorde, J. Verbanc, S. Schuermans, D. Kartounian, J. and Vincken, W.: (2005). FEV_1/FEV_6 and FEV_6 as an alternative for FEV1/FVC the spirometric detection of airway obstruction and restriction. Chest, 127, 1560–1564.

12. Hankinson, J.L. Robert, O. Crapo, M.D. and Robert, L.J.: (2003). Spirometric reference values for the 6-s FVC maneuver. Chest, 124, 1805–1811.
13. Akpinar-Elci, M. Fedan, K.B. and Enright, P.L.: (2006). FEV6 as a surrogate for FVC in detecting airways obstruction and restriction in the workplace. Eur Respir J, 27, 374–377.
14. Enright, R.L. Connett, J.E. and . Bailey, W.C.: (2002). The FEV1/FEV6 predicts lung function decline in adult smokers. Respir Med, 96, pp. 444–449, 2002.
15. Sujatha, C.M. and Ramakrishnan, S.: (2009). Prediction of forced expiratory volume in normal and restrictive respiratory functions using spirometry and self-organizing map. J Med Eng Technol, 33(7), 538–543.
16. Kavitha, A. Ramakrishnan, S.: (2010). Comparison of prediction of Forced Expiratory Volume (FEV6) in flow-volume spirometric measurements using radial basis function neural networks and support vector machines. J Mech Med Bio, 10(4), 1-11.
17. Mythili, A. Sujatha, C.M. Srinivasan, S. Ramakrishnan, S.: (2012). Prediction of forced expiratory volume in spirometric pulmonary function test using adaptive neuro fuzzy inference system. Biomedical Sciences Instrumentation, 48, 508-515.
18. Huang, G.B. Zhu, Q.Y. Siew, C.K.: (2004). Extreme learning machine: a new learning scheme of feed forward neural networks. IJCNN Proc. Budapest, 985– 990.
19. Huang, G.B. Zhu, Q.Y. Siew, C.K.: (2006). Extreme learning machine: theory and applications. Neurocomputing, 70, 489–501.
20. Feng, G. Huang, G.B. Lin, Q. Gay, R.: (2009). Error minimized extreme learning machine with growth of hidden nodes and incremental learning. IEEE Trans. Neural Netw, 20, 1352–1357.
21. Huang, G.B. Zhu, Q.Y. Siew, C.K.: (2006). Universal approximation using incremental constructive feedforward networks with random hidden nodes. IEEE Trans. Neural Netw 17, 879–892.
22. Huang, G.B. Zhou, H. Ding, X. Zhang, R.: (2012). Extreme learning machine for regression and multiclass classification. IEEE Trans. Syst. Man Cyber, 42 (2), 513–529.
23. Huynh, H.T, Won, Y.: (2008) Evolutionary algorithm for training compact single hidden layer feedforward neural networks. In: IEEE International Joint Conference on Neural Networks, Hong Kong, 3028–3033.

9

Biomechanical Significance of Ankle and Foot in Human Life

R. K. Saxena, D. V. Rai and J. Manjhi
Department of Bioengineering
Shobhit University, Modipuram, Meerut – 250110, India
Email: rksaxena369@gmail.com

ABSTRACT

Ankle and foot (feet) in human being plays a meaningful and balancing role in smooth and frictionless free movement of weight bearing joints It provides locomotion over the ground with one foot in contact (walking), no feet in contact (running), or both feet in contact (standing) creates a major challenge to our balance control system. Also, about 70% of human body mass is located two-thirds of body height above the ground and it produces an unstable system, unless a control system is continuously acting. Aim of our presentation is to provide glimpse of general anatomy, mechanical basics and neuromuscular control of ankle and foot, which control and equilibrate in physiological gait and stability during several tasks. Ankle and foot (A&F) location, orientation and movement plays significant role in optimizing the sports performances and several other activities. Tennis, badminton, boxing and athletics and other sports foot movement and pressures plays pivotal role for winning the titles and money. In Golf, which is an million dollar industry foot properties has a crucial role for putting the ball in the hole. It is an excellent foot holding and motion plays crucial role in golf and other sports. Ankle and foot contribute significantly to the function of the whole lower limb. Foot supports the body in both standing and locomotion. Foot function can be altered significantly with any variation in alignment in the lower extremity or as a result of abnormal motion in the lower extremity linkage. Ankle and foot motion can affect the amount of pronation or supination. Early heel rise and toe walking are symptoms of disorders. Hence, foot pressure and motion are essential for optimization of the locomotion and important in several activities (In dancing, sports, running, walking etc.). The disorientation, variation and degeneration of the balance control of A&F system can cause immense loss and damage to people of sports and other professions. Also, in several pathologies of ankle and foot structures has desired researchers and clinicians to understand more about how the ankle and foot system works and how to quantify its status at any point in time. Also, this presentation may stimulate to study the ankle and foot of Indians populations, because of most of our population walks on barefoot.

Keywords: Biomechanics of gait, measurement techniques, functional anatomy of ankle and foot, rotation of motion, gait cycle.

9.1 INTRODUCTION

Ankle and foot (pleural: feet) joints are extremely complex locomotor system in its structural configuration and in functionals. In normal life, it serves as main unit for shock absorber and act as a driving force for propulsion of body movement. The foot can sustain enormous pressure (several tons in one-mile run) and provides flexibility and resiliency. Also, it helps in adjusting the body balance and stability in several activities when the rest of the body reacts during abnormal postural activities and weight/load bearing activity. If the ankle and feet are not stable, the rest of the body suffers. The human foot, the only part of the body that acts on an external surface in upright, unsupported positions, supports and balances the body during gait. With typical bony configuration, muscle and ligaments coordination, the foot can be compliant to cope with uneven ground surfaces to achieve a smooth motion and maintain dynamic stability. The human foot combines the amalgam of mechanical complexity and structural strength.

It is in fifties, it was realized that the ankle joint during axis differs during dorsiflexion and plantar flexion during motion and rotation. Barnett and Napier studied human toes and found that the medial and lateral curvatures of the talus are different, indicating that the axis of rotation of the ankles changes its position during the arc of motion (3). However, it was also suggested that dorsiflexion axis and plantar flexion axis used in movement can not occur in motion and rotation simultaneously (4). The foot and ankle are among the most common sites for both acute and chronic injuries in athletes and other physically active individuals (5). Ankle injuries, foot pain and muscle dysfunctions are common causes which are originate from the large impact forces and rotational moments during weight-bearing activities (6). Also, it has been observed that weak or damaged muscles of lower extremities may lead to many disorders of ankle and foot joint. Generally, motion of the foot is caused by the stronger muscles in the lower leg whose tendons connect in the foot. Contraction of the muscles and ligaments of the leg are the main reason which provide movement and motion to feet to stand, walk, run, and jump, etc. Human ankle and foot joint due to its construction; understanding of its strengths and limitations are important component for efficient, stable human movement of lower limb manoeuvrability and for performances.

Later on Donatelli (1985) studied the biomechanical properties of the foot and ankle joint and found that these components are important to the normal function of the lower extremity. He has also suggested that foot prevents the effect of external resistance and. normal arthro-kinematic movement and can influences the ability of the lower limb to attenuate the forces of weight bearing. It is important for the lower extremity to distribute and dissipate compressive, tensile, shearing, and rotatory forces during the stance phase of gait. Inadequate distribution of these forces could lead to abnormal stress and the eventual breakdown of connective tissue and muscle. The combined effect of muscle, bone, ligaments, and normal foot biomechanics will result in the most efficient force attenuation in the lower limb, (7).

Nuber (1989) studied running activity in competition or fitness of human subjects. It is found that breakdown injuries occur due to training errors, shoe wear or changing the intensity frequently (8). In order to understand and treat the pathologic situation, a fundamental understanding of the biomechanics of walking and running is essential. The treating practitioner must appreciate the distinct differences between the walking and running gait. These differences transcend a simple increase in speed of gait and include distinct changes in joint range of motion and electromyography activity. Armed with this knowledge, the practitioner treating a breakdown injury can work to a solution based on scientific understanding rather than anecdotal information.

In nineties Lehmann et al. (1992), Ostrosky et al. (1994) and Schneck (1998) studied normal gait cycle properties in peoples of different walk of life extensively, in order to understand physiology and biomechanics of normal gait and described that normal gait cycle has a time interval or sequence of motion between heel strike to heel strike of the same foot (9, 10, 11). The gait cycle has been broadly divided into two phases: stance phase and swing phase (Fig. 9.1). The phases gait cycle were further subdivided and discussed in terms of percentage of each with in the gait cycle). The stance phase is 60 percent of the gait cycle and can be subdivided into double-leg and single-leg stance. In double-leg stance, both feet are in contact with the ground. At an average walking speed, it represents 10 per cent of the entire gait cycle, but decreases with increased walking speed and ultimately disappears as one begins to run. At slower walking velocities the double-leg support times are greater (12).

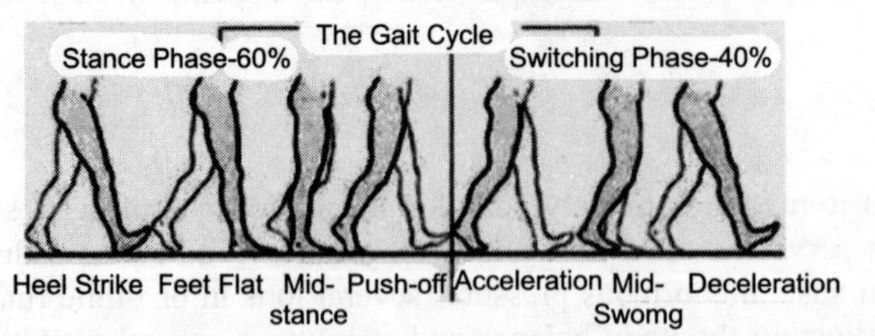

Fig. 9.1 Depiction of Gait cycle

Single-leg stance comprises up to 40 per cent of the normal gait cycle. The muscles that are active during the stance phase act to prevent buckling of the support limb. These include the tibialis anterior, the quadriceps, the hamstrings, the hip abductors, the gluteus maximus, and the erector spine. The swing phase is described when the limb is not weight bearing and represents 40 percent of a single gait cycle. It is subdivided into three phases: initial swing (acceleration), midswing, and terminal swing (deceleration). Acceleration occurs as the foot is lifted from the floor and, during this time, the swing leg is rapidly accelerated forward by hip and knee flexion along with ankle dorsiflexion. Midswing occurs when the accelerating limb is aligned with the stance limb. Terminal swing then occurs as the decelerating leg prepares for contact with the floor and is controlled by the hamstring muscles.

9.2 MEASUREMENT TECHNIQUES USED IN GAIT ANALYSES

In recent years, due to substantial technological and scientific advancement in the measuring techniques several new approaches for the analyses of the gait and motion of the ankle and foot joint have been explored for the understanding for the mechanics of walking, running and abnormal/pathological situations (Fig. 9.2.). Also, advance tools have provided methodologies for identifying the differences between normal and abnormal/pathological gaits. After basic goniometry for the assessment of motion to measure initial and subsequent range of motion in for the calculation of improvement/relief in cases of disability, at the visits to clinics for the respite from occupational injuries and to assessor to determine the level of permanent disability. Later on in eighties scientists and clinicians focused studies of gait analyses primarily on the measurement of joint kinematics and kinetics by electromyography (EMG), oxygen consumption and foot plantar pressures. The development of pneumatic force plates in the early twentieth century helped enormously for the understanding of kinetics of gait in more precise way. Force flow through the foot can be measured by a variety of means. The Electrodynogram was the first system available for office use and consisted of 7 sensors applied in standard positions on the foot can visualize forces beneath the foot. It also measures the timing of the phases and sub phases of gait. Computers programming played significant role in the development of experimental and theoretical planning and performance improvement especially in sports biomechanics and pathological situations. Now, computer technology extends its programs from planning to data capturing and data processing through to mathematical and statistical modelling and simulation and optimization (Lapham and Barlett, 1995). It has provided simple way to assess and monitor the orthopedic foot and ankle conditions to surgeons/scientists for the development of better and user-friendly prosthesis devices in order to provide maximum relief to patient with better understanding the methods and meaning by gait analysis research. In early 1900s, it has become possible to capture image of sequences of movement to reveal normal and abnormal details of human and animal locomotion that were not possible by watching the movement with the naked eye. 3D gait cinematography methodology using an 8 or more camera by fast cinematographic techniques captures complex motions in precise manner and analyze the signals in system immediately and can provide accurate and critical assessment of gait and motion abnormalities, for understanding of gait pattern analysis in normal and abnormal/pathological situations Also, retro-reflective markers (infrared, light emitting elements, mostly LEDs) wired or wireless can be placed on bony landmarks or specific anatomical positions in the kinematics models. These devices emit radiation when the tracking cameras are sensitive, and can be synchronized with the cameras and can generate pattern of movement in desired or specific task requirements. Also, surface electromyographic (EMG) data can be recorded in several dynamic muscles point bilaterally according to standardized electrode placement techniques and can be monitored simultaneously along with real-time gait cinematography in computer. The purpose of this article is to provide over view regarding progress in gait and motion technology, so in the field of ankle and foot joints nay generate pragmatic novelties for the development of user friendly prefect prostheses for amputees and help to patients of several pathological/abnormal disabilities of ankle and foot joint, .with a particular attention on the underlying methods and science involved for the to enhancement of human movement and health. Physicians, allied health professionals, engineers, biomechanists and scientists can be benefited working together to advance scientific knowledge, technical capabilities, and clinical practice in the field of human movement to improve functional outcomes and quality of life for individuals with any movement disorder at any age.

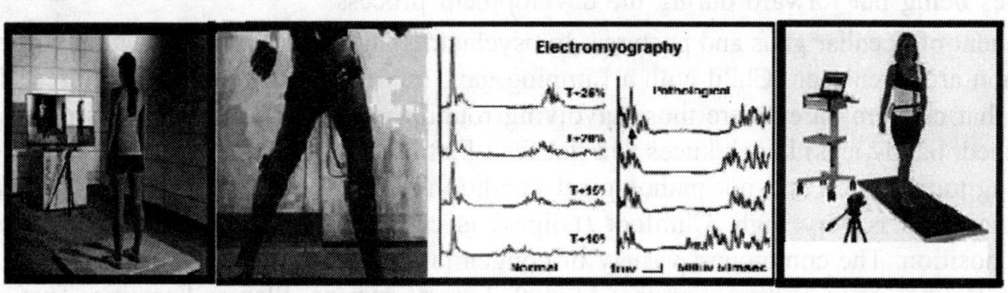

(a) (b) (c) (d)

Fig. 9.2 Force plate, electromyography and 3D real time gait analysis and measuring systems

9.3 APPLICATIONS OF GAIT AND MOTION ANALYSIS IN NORMAL AND ABNORMAL CONDITIONS

1. Application of gait and motion analyses in human being having pathological conditions such as cerebral palsy, Parkinson's disease and multiple sclerosis, spina bifida, and other early onset neuromuscular disorders. The study of coordination of the trunk can be measured in relation to carriage, gait, and station, and coordination in the lower extremities is evaluated by the heel-to-knee, heel-to-toe, and foot-to-buttock tests. We apply the latest technology and professional knowledge to measure, evaluate, and understand human movement, provide a forum for professional interaction and exchange of ideas, promote education, stimulate research, and disseminate current knowledge and best practices.
2. Balance impairment: Gait and balance disorders are common in older adults and are a major cause of falls in this population. Common causes include arthritis and orthostatic hypotension; however, most gait and balance disorders involve multiple contributing factors. Obesity: causes balance disorientation and peculiar gait pattern and shoes alignment problems.
3. Cerebral Stroke (Hemiplegia: where one part body get paralysed, causes difficulty in walking, poor balance and little coordination), paraplegia (cannot walk and if walks with considerable degree movement, many are dependent on wheelchairs or other supportive measures. Lower part paralysis, etc.) Spinal cord injury and traumatic brain injury
4. Diabetes, peripheral arterial disease, and other cardiovascular conditions affecting movement: Diabetes has also been associated with increased disability, poorer physical functioning, and an increased risk of being unable to do mobility-related tasks. Peripheral Artery Disease, Diabetes, and Reduced Lower Extremity Functioning. Vascular disease in other parts of the body. Acute myocardial infraction or stroke. Presence of occlusive arterial disease, etc.
5. Osteoarthritis, foot deformity, and other orthopaedic and rheumatologic conditions affecting movement; Arthritis of the midtarsal and tarsometatarsal joints, Impairments in mobility of the hip joint and the strength of the surrounding muscles, especially the hip abductor muscles. Foot orthoses due to knee Osteoarthritis, relationship between feet function and knee osteoarthritis foot function during walking in people with and without medial compartment knee osteoarthritis.
6. Occupational injury prevention, diagnosis, treatment and rehabilitation. Ankle to malleolar fractures is one of most common lower limb fractures, and the frequency has been increasing over the past few decades, especially in elderly women. Muscle co-contraction activation of agonist (prime mover) and antagonist (primer stabilizer) muscles across the same joint Idiopathic toe-walking Toe walking is defined as lack of heel floor contact at the onset of stance during gait. Idiopathic toe walking (ITW) is a term used to describe the condition in which children walk on their toes in the absence of any known cause.
7. Sports injury prevention, diagnosis, treatment and rehabilitation: Common sports injuries are fractures, ankle sprains, and plantar fasciitis, etc.
8. Limb deficiency and amputation, Prosthetics, orthotics, and assistive devices: Compensating a limb loss with prosthesis Designer of lower limb prostheses wants to know what specification of the device could provide the best approximation to the normal locomotion. Deep understanding of the latter is essential, and gait analysis may be a valuable tool for this. Once prosthesis is built, gait analysis provide help in comparing the wearer's performance with the new device and with the prior art, and in verification of the hypotheses being put forward during the development process.
9. Development of peculiar gaits and postures: In psychiatry, gait disturbances reflecting cortical and subcortical dysfunction are often seen. Child with a Limping gait, Intermittent Limping in children. The most common postures that concern parents are those involving rotation of the lower extremities particularly in-toed gait (pigeon-toed, bandy legs) knock knees and flat feet. Flatfoot" (pes planus) develops due to simple depression of the longitudinal arch and not pathological condition. The converse is pes cavus condition in which the longitudinal arch is very high. Clubfoot (talipes) is condition of foot where they appear twisted out of shape or position. The commonest variety of congenital clubfoot comprises plantar flexion, supination, and adduction (talipes equinovarus). Sprains: Medial, Lateral sprains, Plantar Fasciitis, Turf Toe (Sprain of the

connective tissues of the 1st metatarsophalangeal joint) Varus-refers to a body part turned inward or towards the midline, Forefoot/Rear foot Varus, Valgus-Body part turned outward or away from the midline: Forefoot Valgus.

9.4 FUNCTIONAL ANATOMY OF THE ANKLE AND FOOT JOINT

The foot and ankle is a hinge joint between the tibia and fibula a complex anatomical structure consisting of 26 irregularly shaped bones, 30 synovial joints, and more than 100 ligaments, tendons, and muscles acting on the segments (Fig. 9.3). The foot is considered to have four subdivisions: the hindfoot, midfoot, forefoot, and phalanges. Other than the talocrural joint (ankle), most of the motion in walking occurs at three of the synovial joints: the subtalar, midtarsal, and metatarsophalangeal joints. The joint capsule is thickened on each side by a strong ligament. The medial and lateral ligaments prevent anterior and posterior slipping of the talus. In injuries of ankle, lateral ligaments significantly weaken and liable for sprain. The shape of the surfaces at the ankle is such that ligaments provides grip to talus tightly in dorsiflexion. The tension that develops during the support of body weight is taken up by strong ligaments on the plantar aspect of the tarsus. Several small muscles of ankle and foot joints provide smooth and dynamic actions. Damage to some of these muscles can cause problems. Ligaments stretch during motion and return. When ligaments are stretched beyond its capacity, they get torn causes pain and injury.

The foot has two important functions: weight bearing and propulsion. These functions require a high degree of stability. In addition, the foot must be flexible, so it can adapt to uneven surfaces. The multiple bones and joints of the foot give it flexibility, but these multiple bones must form an arch to support any weight. The muscles of the foot are classified as either intrinsic or extrinsic. The intrinsic muscles are located within the foot and cause movement of the toes. These muscles are flexors (plantar flexors), extensors (dorsiflexors), abductors, and adductors of the toes. Several intrinsic muscles also help support the arches of the foot. The extrinsic muscles are located outside the foot, in the lower leg. The powerful gastrocnemius muscle (calf) is among them. They have long tendons that cross the ankle, to attach on the bones of the foot and assist in movement. The talus, however, has no tendon attachments. The foot and ankle contain over 100 muscles, mostly small in length and cross-sectional area, but many very important. The main functions of these muscles are to dorsiflex, plantar flex, evert, invert, extend the toes and flex the toes as well as pronate and supinate the foot. The foot also has also, 33 joints, and hundreds of tendons and ligaments. Not to mention huge amounts of proprioception receptions in these joints, tissues, and skin to give feedback to the body.

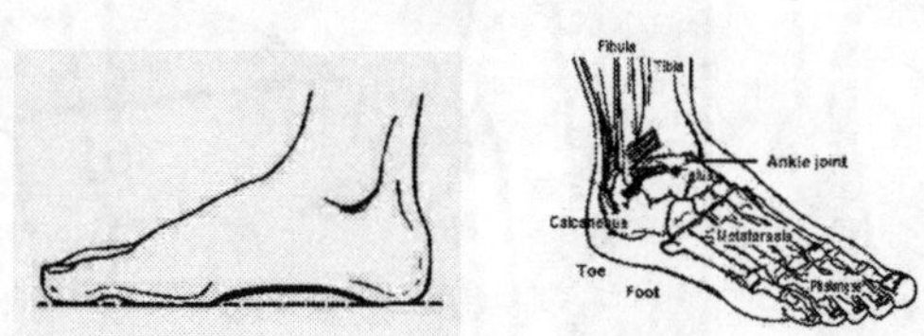

(a) (b)

Fig. 9.3 A & B. Bones, ligaments, tendons and muscles in ankle joint and foot

9.5 OTHER ANATOMICAL FUNCTIONS OF ANKLE JOINTS

Articulations of ankle allow gliding motion. The medial cuneiform and first metatarsal have an independent joint cavity. The foot joints allow flexion, extension, abduction, and adduction. The interphalangeal joints permit flexion and extension. The sesamoids of the big toe take the weight of the body, especially during the latter part of the stance phase of walking. The sesamoid mechanism is deranged in bunions and in hallux valgus. Arches of the foot develop on the requirement of the mechanical arrangement of the bones and are supported by ligaments. The arches are masked by a fat pad.

9.6 BIOMECHANICS OF GAIT OF THE FOOT AND ANKLE JOINT

In order to understand properly the role of any part of the human musculoskeletal system, it is important to fully understand its biomechanics. Biomechanics is the study of normal mechanics (kinetics and kinematics) in the musculoskeletal system by analysing forces and their effects on anatomical structures. The biomechanics of the foot and ankle are important for the normal function of the lower extremity. The foot is the terminal joint in the lower kinetic chain that opposes external resistance. Proper arthro-kinematic movement within the foot and ankle influences the ability of the lower limb to attenuate the forces of weight bearing. It is important for the lower extremity to distribute and dissipate compressive, tensile, shearing, and rotatory forces during the stance phase of gait. Inadequate distribution of these forces could lead to abnormal stress and the eventual breakdown of connective tissue and muscle. The combined effect of muscle, bone, ligaments, and normal foot biomechanics will result in the most efficient force attenuation in the lower limb. The normal mechanics of the foot and ankle result from the combined effects of muscle, tendon, ligament, and bone. The coordinated and unified effect of these tissues within the foot, ankle and lower extremity results in the most efficient force attenuation. To comprehend biomechanics of the foot during standing and walking, it is important to understand gait.

9.7 HUMAN GAIT AND POSTURE IN STANDING AND IN MOTION CONDITIONS

Humans are bipeds and locomote over the ground with one foot in contact (walking), no feet in contact (running), or both feet in contact (standing) creates a major challenge to our balance control system. Because two-thirds of our body mass is located two-thirds of body height above the ground we are an inherently unstable system unless a control system is continuously acting. There are three cardinal anatomical planes that pass through the body i.e. sagittal (anteroposterior), frontal (coronal), and transverse (horizontal) (Fig. 9.4). The saggital plane divides the body into the left and right side. The frontal plane divides the body front (anterior) and back (posterior) portions. The transverse plane divides the body into the upper (superior) and lowers (inferior) portions. Each plane is perpendicular to the other.

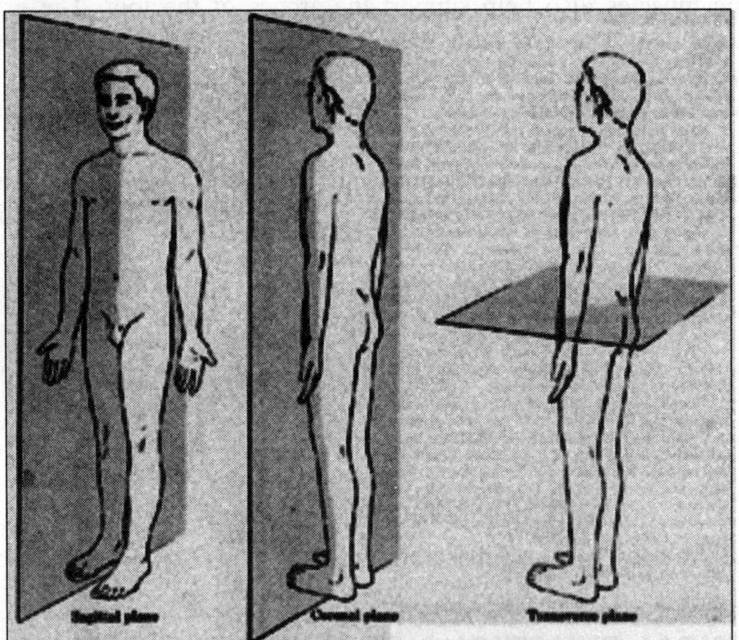

Fig. 9.4 Three cardinal anatomical planes in human

9.8 MOTION AND MOVEMENTS OF THE ANKLE AND FOOT DURING NORMAL AND ABNORMAL GAIT CYCLE

During gait period ankle joint has an oblique axis and has an angle 82° in frontal plane, causes it direction towards laterally. In transverse plane, axis is directly laterally & posteriorly about 20-30; Due to this oblique ankle axis is that foot externally rotates when the ankle is in maximal dorsiflexion; When the foot plantar flexes, it internally rotates with foot fixed on ground as in stance phase of gait; and there is forward tibial progression (ankle dorsiflexion) results in tibial internal rotation and when tibia is behind foot (ankle plantar flexion), it is externally rotated;

Foot during gait: External rotation of leg produces supination of the foot; and during weight bearing, external rotation of limb & subsequent forefoot supination are followed by pronation twist of forefoot to remain plantigrade. This tends to lock the foot, making it a rigid lever. Internal rotation of the leg produces a pronated foot and internal rotation of leg and its subsequent pronation are followed by supinatory twist of the forefoot to remain planti grade. This action unlocks mid tarsal joint and produces mobile mid foot noted at time of heel contact & first 15% of the gait cycle; also, inversion of the heel in the normal foot promptly occurs as wt is transferred from heel to forefoot when a person rises on the toes. Such inversion of the heel causes the mid foot to covert from a mobile structure to a rigid lever. When the heel is elevated during standing or at the time of push off, weight of the body is shared by all the metatarsal heads, ,in order to achieve this fair division of the body weight among the metatarsals, foot must supinate slightly and deviate laterally; Ankle joint and subtalar joint work together: Ankle combines dorsiflexion with abduction and plantar flexion adduction, subtalar joint combines dorsiflexion, abduction and eversion in one direction and plantar flexion, adduction, and inversion in the other direction; These combined subtalar motions are referred to as pronation & supination (Fig. 9.5); When the subtalar joint is fused, rotation is increased in the ankle and may cause arthritic change and when the ankle is fused, greater stresses are placed on the sub-talar and midtarsal joints. Dorsiflexion of the toes tightens the plantar aponeurosis and assists in inversion of the heel. Supinatory twist activates the locking mechanism in foot, thus converting a flexible foot into a rigid lever, an action that is necessary at push off condition.

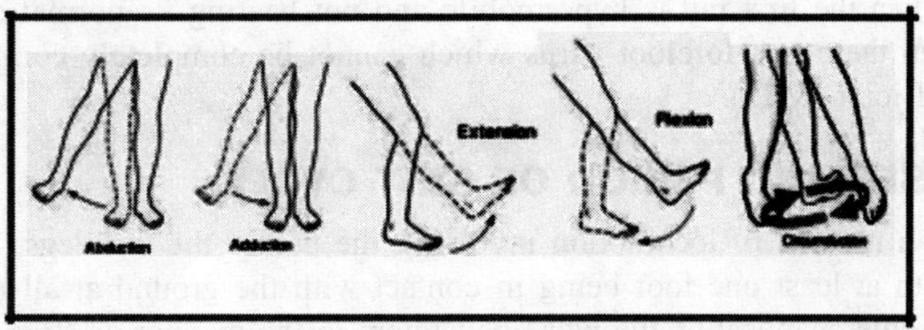

Fig. 9.5 Various motion actions of ankle and foot joint (abduction & adduction, extension & flexion, circumduction

All movements of the ankle during normal gait: dorsiflexion, plantar flexion, supination, and pronation occur at multiple levels and involve several hindfoot and midfoot joints (Fig. 9.2). The motions of plantar flexion and dorsiflexion are the major contributors to overall ankle motion are of particular import during clinical evaluation. The average plantar flexion and dorsiflexion of the ankle has been measured at 40°-56° and 13°-33°, respectively. During most activities of daily living, only a partial ROM is required: walking on an even surface (10°-15° plantar flexion and 10° dorsiflexion), walking upstairs (37° total ROM), walking downstairs (56° total ROM). It is observed in walking up and down hill the movements of the foot and ankle provides dorsiflexion and plantar flexion. However, eversion and inversion are found in standing sideways on a hill. A study of elderly patients found that a total ROM of 15°-20° is normally sufficient to complete their daily routines. It may get altered with age and gender. Limb dominance and asymmetry in the actions of the lower extremities during walking and running is a common phenomenon. The ROM of the ankle may get reduced in patients with systemic diseases (e.g., diabetes mellitus).

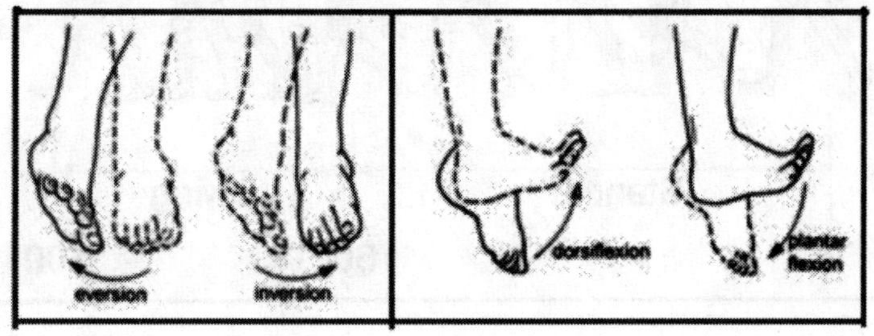

Fig. 9.6 Foot and toe movement (Eversion & inversion, dorsiflexion & plantarflexion, abduction & adduction)

Foot and toe movements are generally capable of motion in two directions:plantar flexion or dorsiflexion. In addition, the joints permit abduction and adduction of the toes. The foot as a whole (excluding the toes) has two movements: inversion and eversion. All the joints of the hindfoot and midfoot contribute to these complex movements that are ordinarily are combined with movements at the ankle joint Fig. 9.6).

9.9 LOAD TRANSMISSION IN THE FOOT AND ANKLE JOINT

It is observed that inversion or eversion of the ankle in the neutral or dorsiflexion position leads to a decrease in total contact areas and an increase in average pressures. In general, the total contact area was higher and the average pressure was lower in dorsiflexion. The patterns of pressure distribution were different in dorsiflexion and plantar flexion: the main force transmission area moved interiorly to posteriorly on the talus during the movement from dorsiflexion to plantar flexion. The talar dome contributed the most to the load transfer with 77% to 90% of the load. After application of full loading, contact strains higher than 15% were observed in more than 40% of the contact area, showing that the ankle cartilage on both sites, tibial and talar, experienced large deformation during the load transmission in the ankle joint. It is observed that most ankle joint loading during the stance phase occurred across the articular surfaces, while the stabilization due to ligament tension played an inferior role. The medial ligaments have a more important part at the ligamental stabilization than the lateral ligaments: the deep deltoid ligament transferred the most force during the stance phase (57.2%), followed by the superficial part of the deltoid ligament (26.1%). Normal pressure flow through the foot starts slightly lateral in the heel and flows forward to between the first and second metarsal and exits through the great toe. A situation which would cause a shift lateral will occur when the first ray is hypermobile and not bearing its normal amount of weight. Extreme lateral shift will occur when there is a forefoot varus which cannot be completely compensated for at the subtalar joint (STJ) or the midtarsal joint (MTJ).

9.10 NORMAL PHASES AND PERIOD OF GAIT CYCLE

Walking can be defined as a method of locomotion involving the use of the two legs, alternately, to provide both support and propulsion, with at least one foot being in contact with the ground at all times. While the terms gait and walking are often used interchangeably, the word gait refers to the manner or style of walking, rather than the actual walking process. The gait cycle is the time interval between the exact same repetitive events of walking. The defined cycle can start at any moment, but it is generally begun when one foot contacts the ground. If it starts with the right foot contacting the ground, then the cycle ends when the right foot makes contact again.

There are two phases of the gait cycle: (Fig. 9.7)

(1) Stance phase is the part of the cycle when the foot is in contact with the ground. It comprises about 60% of the cycle, beginning with initial foot strike and ending with toe-off.

(2) Swing phase occurs when the foot is in the air and comprises 40% of the cycle, beginning with toe-off and ending with second (ipsilateral) foot strike.

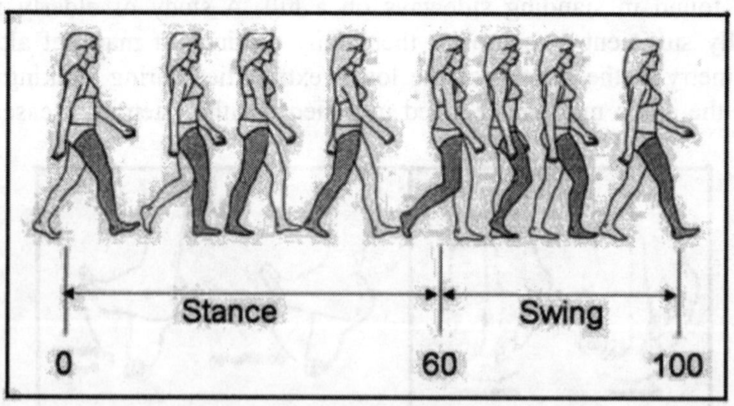

Fig. 9.7 Periods of normal gait cycle with body position for saggital plane motion

The phases of gait cycle are further divided in to 6 subcycles (Sub-phases) consisting of 6 major periods: (1) Initial Double Support, (2) Single Limb Stance (3) Second Double Support (4) Initial swing (5) Mid-swing (6) Terminal swing. Periods 1-3 make up the stance phase, while periods 4-6 make up the swing phase. It can be observed that the opposite limb repeats the same sequence of events, but is displaced by 50%. For example, while left toe off occurs at 12%, right toe off occurs at 62% of the gait cycle. The stance and swing phases (six periods) of both feet and their relative time occurrence and is shown in Figure 9.8.

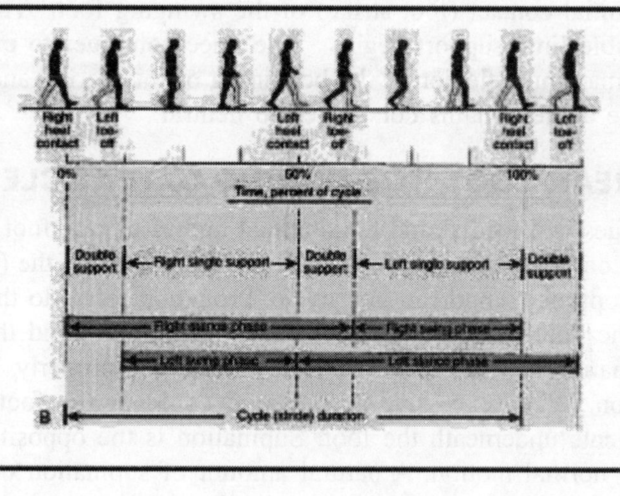

Fig. 9.8 Gait cycle showing major periods of motions

9.11 DETAILS OF PERIODS OF GAIT CYCLE

Period 1 Initial Double Limb Support (0-12% of gait cycle). This period begins with the first foot strike and ends with the opposite foot's toe-off. The center of mass (COM) begins to rise during this period as the shift to a single limb occurs. It is characterized by rapid loading with shock absorption (mostly from the knee) and slowing the body's forward momentum. The hip begins flexed and then extends as the opposite foot comes off the ground. On the contrary, the knee begins fully extended and then flexes during this period. The ankle goes from a neutral position to plantar flexed until the foot is flat on the floor.

Period 2 Single Limb Support (12-50% of gait cycle). This period begins with the opposite toe off and ends with opposite foot strike. As the body passes over the grounded foot, the COM reaches its peak height while both the forward and vertical velocity decreases. The COM then falls until the end of this period at opposite foot strike, as forward and vertical velocity increase again and forward shear reverses to aft shear (at ~30% of cycle). The hip continues to extend throughout this period as the limb progresses from an anterior to a posterior position in relation to the body. The knee extends again until it hits a maximum extension and starts to flex. The ankle dorsiflexes throughout the phase but actively plantarflexes towards the end to resist this dorsiflexion.

Period 3 Second Double Limb Support (50-62% of gait cycle). This period (also called pre-swing) begins with opposite foot strike and ends with toe off. The COM is at its lowest level during the beginning portion of this period as both feet are in contact with the ground. The COM then starts to rise again as the opposite limb is in its initial double limb support. As body weight is transferred to the opposite limb, the trailing limb ends its extension and prepares to swing in front of the body. As a result, the hip and knee flex and the ankle plantarflexes during this time in order to prepare for the limbs lift off.

Period 4 Initial Swing (62-75% of gait cycle). This is the beginning of the second single limb support interval. It begins when the foot is lifted off the ground (toe-off) and ends when the swinging foot clears the ground and is opposite the standing opposite foot (feet adjacent). The COM reaches its peak height towards the end of this period as the opposite foot is in the middle of single limb support. The ankle continues to plantarflex but then begins to dorsiflex in order to clear the ground. The knee continues to flex rapidly to its peak at the end of this period, which largely results from the hip flexing as well.

Period 5 Mid-Swing (75-85% of gait cycle)) This is the middle third of the swing phase. It begins with foot clearance where the feet are adjacent and ends when the swinging foot is in front of the body and the tibia is vertical. The COM falls during this time as the opposite limb is in the middle portion of the single limb support period. The leg's advancement is carried out by continued hip flexion. The knee begins to extend quickly due to momentum and in response to gravity. The ankle continues to dorsiflex to neutral.

Period 6 Terminal Swing (85-100% of gait cycle) This is the last third of the swing phase. It begins when the tibia is vertical and ends with initial contact (foot strike) of the swinging foot. The COM reaches its lowest level at the end of this period as double limb support begins. The knee continues to extend in order to complete limb advancement. The hip reaches maximum flexion at the beginning of this period and tends to extend slightly before initial double limb support. The ankle remains dorsiflexed to neutral.

9.12 RESPONSES OF REAR FOOT/TOE DURING GAIT CYCLE

Bones and surrounding soft tissues and medial and longitudinal arches of rear foot serves as the chief load-bearing function in the foot (Fig. 9.9). Loading properties depend on the kinematics of the (metatarso-cuneiform bones) also called as first ray and provides optimal support in gait cycle. Pronation refers to the inward roll of the foot during normal motion and occurs as the outer edge of the heel strikes the ground and the foot rolls inward and flattens out. A moderate amount of pronation is required for the foot to function properly, however damage and injury can occur during excessive pronation. When excessive pronation does occur the foot arch flattens out and stretches the muscles, tendons and ligaments underneath the foot. Supination is the opposite of pronation and refers to the outward roll of the foot during normal motion. A natural amount of supination occurs during the push-off phase of the running gait as the heel lifts off the ground and the forefoot and toes are used to propel the body forward. However, excessive supination (outward rolling) places a large strain on the muscles and tendons that stabilize the ankle, and can lead to the ankle rolling completely over, resulting in an ankle sprain or total ligament rupture.

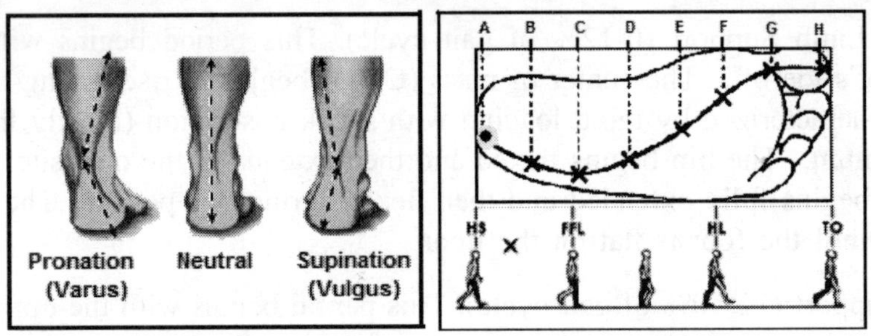

(a) (b)

Fig. 9.9 A & B, Pronation and supination foot and function rear foot during motion

9.13 CONCLUSION

Biomechanical evaluation of ankle and foots joint are necessary for understanding gait balance control system in push-off and propulsion in motion at normal and abnormal situations. The deterioration/degeneration of the human balance and posture control during standing and walking in the elderly and many pathologies has forced researchers and clinicians to understand more about how the system works and how to quantify its status at any point in time. Abnormal gait pattern, such as common limping (secondary other complications), gastrocnemius and soleus muscles weakness and hemi paretic stroke, etc. have been found to alter gait and locomotion in human. Also, lack of propulsion and other important abnormalities include the instability of the knee and the lack of motion at the ankle, associated with muscle imbalance and plasticity have been found to cause lower limb disabilities. The application of motion analysis to sports medicine and functional analysis are very important for optimizing the performance. Sports clinicians, bioengineers, physical therapists and kinesiologists plays significant role and depend upon the level of clinical gait analysis and assisting in expanding clinical remedial methodologies. It is our endeavour that review may provide and encourage research in gait analysis technology. Dynamic knee stability of lower extremities is necessary in locomotion and is of interest of clinicians and scientists. Measuring technology for the analysis of human motion has advanced significantly and provided accurate knowledge about the performance of human locomotion. It has helped in evaluation medical disorders and affects their treatment. Researches are desired to address problems by creating techniques to capture data rapidly, accurately, and efficiently, and to interpret such data by an assortment of modeling, statistical, wave interpretation, and artificial intelligence methodologies. The success of such efforts rests on both our technical abilities and communication between bioengineers and clinicians.

REFERENCES

1. Barnett C., Napier J. (1952) The axis of rotation at the ankle joint in man. Its influence upon the form of the talus and the mobility of the fibula. J Anat, 86: 1-9.
2. Hicks J.H. (1953) The mechanics of the foot: I. the joints. J Anat, 87: 345-537.
3. Smith, L.K., (1996). Brunnstrom's clinical kinesiology. Davis F.A., Philadelphia
4. Hamill, J., Knutzen, K.M., (2006). Biomechanical basis of human movement. Williams & Wilkins, Philadelphia
5. Hootman J.M., Dick R., Agel J. (2007) Epidemiology of collegiate injuries for 15 sports: summary and recommendations for injury prevention initiatives. J. Athl Train. 311–319.
6. Sobhani S., Dekker R., Postema K., Dijkstra P.U. (2013) Epidemiology of ankle and foot overuse injuries in sports: A systematic review. Scandinavian Journal of Medicine & Science in Sports, 23: 669-686.
7. Donatelli R.A. (1985) Normal biomechanics of the foot and ankle. J. Orthop Sports Phys Therapy, 7: 91-95.
8. Nuber G.W. (1989) Biomechanics of the foot and ankle during gait. Clin Podiatr Med Surg. 6: 615-27.
9. Lehmann J.F., de Lateur B.J., Price R. (1992) Biomechanics of normal gait. Phys Med Rehab Clin North Am, 3: 125-138.
10. Ostrosky K.M., Vanswearingen J.M., Burdett R.G., Gee Z. (1994) A comparison of gait characteristics in young and old subjects. Phys Ther, 74: 637-46.
11. Schneck C. (1998) Normal and abnormal gait. Ninth Annual Review Course in Physical Medicine and Rehabilitation, West Orange, N.J: Kessler Institute for Rehabilitation. pp. 2-9.
12. Inman, V.T. (1976) The Joints of the Ankle: Biomechanical studies of the ankle and subtalar joints, Williams & Wilkins, Baltimore.
13. Meinders M., Gitter A, Czerniecki JM. (1998) The role of ankle plantar flexor muscle work during walking. Scand. J. Rehabil. Med. 30: 39–46
14. R.J. Abboud. (2002) Relevant foot biomechanics. Current Orthopaedics. 16: 165–179.
15. Baker, R., (2007). The history of gait analysis before the advent of modern computers. Gait and posture. 26: 331–342
16. Mayich D.J.1, Novak A, Vena D, Daniels TR, Brodsky J.W. (2014) Gait analysis in orthopedic foot and ankle surgery: Topical review, part 1: principles and uses of gait analysis. Foot Ankle Int. 35: 80-90.
17. Pitkin M. (2013) What can normal gait biomechanics teach a designer of lower limb prostheses? Acta Bioeng Biomech. 15: 3-10.
18. Kirtley C. (2005) Clinical gait analysis: theory and practice. Elsevier; Edinburgh, New York.
19. Perry J. (1992) Gait Analysis: normal and pathological function. Thorofare, Slack, Inc.; NJ:.
20. Kuo A.D., Donelan J.M. (2010) Dynamic principles of gait and their clinical implications. Phys Ther. 90: 157–174.
21. Morgan P, McGinley J. (2014) Gait function and decline in adults with cerebral palsy: a systematic review. Disabil Rehabil, 36: 1-9
22. Lu T.W., Chang C.F., Kaohsiung. (2012) Biomechanics of human movement and its clinical applications J Med Sci [2012, 28 (suppl: 2): s13-25
23. Sutherland D.H. (2002) The evolution of clinical gait analysis. Part II kinematics. Gait Posture 16: 159-179.
24. Williams G.N., Chmielewski T., Rudolph K., et al., (2001) Dynamic knee stability: current theory and implications for clinicians and scientists. J. Orthop Sports Phys Ther, 31: 546-566.
25. Jones, F.W. (1949)Structure and Function as Seen in the Foot, 2nd ed., Bailliere, Tindall, and Cox, London.
26. Novak A.C., Mayich D.J., Perry S.D., Daniels T.R., Brodsky J.W. (2014) Gait analysis for foot and ankle surgeons—topical review, part 2: approaches to multisegment modeling of the foot. Foot Ankle Int, 35:178-191.
27. Morgan P., McGinley J. (2014) Gait function and decline in adults with cerebral palsy: a systematic review. Disabil. Rehabil, 36: 1-9
28. Hanson C.J., Jones L.J.. (1989) Gait abnormalities and inhibitive casts in cerebral palsy. Literature review. J Am Podiatr Med Assoc, 79: 53-59.
29. Leardini A, O'Connor J.J., Giannini S. (2014) Biomechanics of the natural, arthritic, and replaced human ankle joint. J Foot Ankle Res, 7: 8-15.
30. Hurmuzlu Y., Basdogan C., Stoianovici D. (1996) Kinematics and dynamic stability of the locomotion of post-polio patients. J. Biomech. Eng. 118: 405–411.
31. Simon S.R. (2004) Quantification of human motion: gait analysis—benefits and limitations to its application to clinical problems. Journal of Biomechanics. 37: 1869–1880

10

Mechanical Advantages of Curvilinearity of Bones

A. K. Dash, J. Sahoo, D. Dash, D. Mohanty, P. Kar, J. Kumar and S. Samal

Centre of Biomechanical Science & Department of Mechanical Engineering
Institute of Technical Education & Research, SOA University, Bhubaneswar, India
Email: bapun.dd@gmail.com

ABSTRACT

The straight bones present in our body do not take any active part in any functioning of the body, so maximum bones are curved in nature and has greater advantages than the straight ones. Probably due to this curvatures present in bone, the bones have a long service period even if subjected to various loading conditions. In the current analysis, the stress distribution on straight and curved bones have been determined and found that the deformation and the stress distribution of bones with more curvature was less as compared to the bones with less curvature under different load applications. To study this mechanical aspect, it would be highly beneficial for the prosthetics designer to design the prosthetics keeping in mind to reduce the deformation and stress distribution and increase the longevity of the prosthetics..

Keywords: Curvature, stress, deformation, bone.

10.1 INTRODUCTION

Biomechanics is the area of support and exercise science where the laws, principles and methods of mechanics are applied to the structure and function of the human body. Bones are rigid organs that constitute part of the endoskeleton of vertebrates. They support and protect the various organs of the body, produce red and white blood cells and store minerals. Bone tissue is a type of dense connective tissue. Bones come in a variety of shapes and have a complex internal and external structure, are lightweight yet strong and hard, and serve multiple functions. The human bone is a material that has the same strength as cast iron, but achieves this while remaining as light as wood. There are two major kinds of bone, trabecular and cortical. Trabecular (spongy) bone gives supporting strength to the ends of the weight-bearing bone. The cortical (solid) bone on the outside forms the shaft of the long bone. The long bones are found in the limbs, and each consists of a body or shaft and two extremities. The body or diaphysis's cylindrical, with a central cavity, the wall consists of dense, compact tissue of considerable thickness in the middle part of the body, but becoming thinner toward the extremities. The extremities are generally expanded, for the purposes of articulation and to afford broad surfaces for muscular attachment.

No bone in the body is straight in true sense. Each bone is curved in different planes. The long bones are more curved than the other bones, the curve generally taking place in two planes, thus affording greater strength to the bone as compared to straight bones. The bones belonging to this class are: the Clavicle, humerus, radius, ulna, femur, tibia, fibula, metacarpals, metatarsals, and phalanges. The curvatures present in bone might be imparting different mechanical advantages for example curvilinearity of bones facilitates the rotational movement of body and many more for which a normal human-being being subjected to different stresses but not getting fractured, i.e. mechanical failure such as fatigue or fractured. The primary loads experienced by most bones are compressive. The mechanical response of bone to compression, tension and other complex loads depends on the complex structure of bones. Compressive stresses are those of the body weight pushing the bone down, and tensile stresses are from the muscles, pulling the bone apart.

10.1.1 Femur Bone

The femur, or thigh bone, is the most proximal (closest to the body) bone of the leg invertebrates capable of walking or jumping, such as most land mammals, birds, many reptiles such as lizards, and amphibians such as frogs. In vertebrates with four legs such as dogs and horses, the femur is found only in the rear legs. The head of the femur articulates with the acetabulum in the pelvic bone forming the hip joint, while the distal part of the femur articulates with the tibia and patella forming the knee joint. By most measures the femur is the strongest bone in the body.

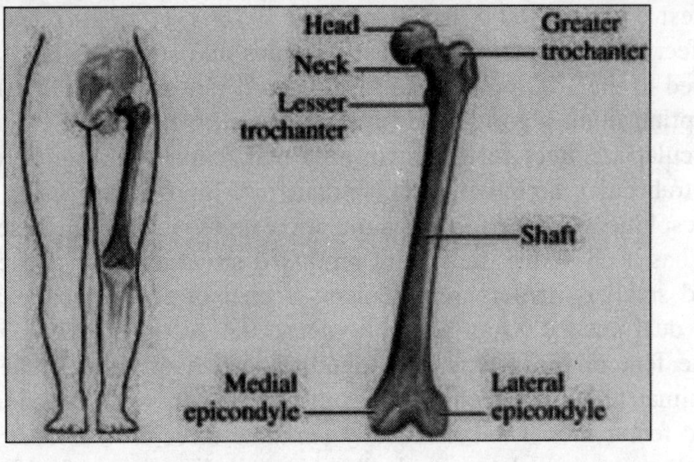

Fig. 10.1 Femour bone anatomy

Currey et al. [1] have studied that the bending strength of a wide variety of bone types is nearly linearly proportional to Young's modulus of elasticity/100. The post-yield deformation of the bone material allows a greater bending moment to be exerted after the yield point has been reached, thereby increasing the strength. Loading in bending results in a much smaller proportion of the volume of the specimens being raised to high stresses than is the case in tension, and this reduces the likelihood of a weak part of the specimen being loaded to failure. Shahar et al. [2] have found out that the stress and strain significantly reduced when all muscles have been considered and bending moment significantly decreased when femur has been loaded primarily in bending. They have experimented on the study for the basis of future studies of canine femur and would provide information regarding hip prortheses and other surgical procedures. Nishiyama et al. [3] have studied and developed a fast and validated method to estimate whole bone structural stiffness and failure load in addition to a sensitivity analysis of varying boundary conditions. They have performed QCT scans on twenty fresh-frozen proximal femurs (age: 77±13 years) and mechanically tested the femurs in a configuration of sideways fall on the hip. They varied the internal rotation angle of the femoral neck from −30° to 45° at 15° intervals and estimated structural stiffness at each angle and found both the FE estimates of structural stiffness and failure load to be nearly equal to the values found in mechanical testing. Juszczyk et al. [4] have tested the failure of 12 cadaveric femurs and demonstrated whether the human proximal femur behaved linearly elastic when loaded to failure. They carried the test at two strain rate covering the physiological range and found that all specimen were broken with a single fracture surface into two pieces and did not showed any sign of permanent deformation. The force-displacement curves have been highly linear upto 99% of fracture force and the remaining 1% has been slightly inclined. They concluded that the proximal femur has a linear-elastic behavior up to fracture, for physiological strain-rates. Zysset et al. [5] have proposed a study to quantify the elastic properties of bone at the lamellar level and compared these properties among osteonal, interstitial and trabecular microstructures from the diaphysis and the neck of the human femur. They concluded that the nanostructure of bone tissue must differ substantially among lamellar types, anatomical sites and individuals. Hoshaw et al. [6] have examined the hypothesis that fatigue loading of the proximal femur results in highly correlated decreases in bone stiffness and strength through the accumulation of bone micro-damage and the damage (micro cracks) in the cortical bone and broken trabecular has been observed in the neck and head region of the femur fatigued until its stiffness was reduced by 40%. Jonkers et al. [7] have established the correlation between calculated stress distributions & changes in bone mineral density and the average values of von mises stress have been calculated for relevant zones of the proximal femur & have found the differences in the stress distribution induced by specific loading conditions & the loading interchanged the patterns of the stress distribution. Antoni et al. [8] have studied the orthotropic properties in a proximal femur FE model using the directions of the principal stresses produced by a physiological load scheme. Two heterogeneous material models, one isotropic and orthotropic has been employed to test their influence on the mechanical behavior of the bone model. They concluded that the mechanical properties have been aligned with the highest principal stress produced from the successive application of a multi load scenario corresponding to 10%, 30% and 45% of the gait cycle. Kasra et al. [9] have determined the effect of strain rate, apparent density, and presence of bone marrow on trabecular bone shear material properties. They concluded that the results have been in bones with and without marrow tested at low (0.002 s_1) and high (0.05 s_1) strain rates, presence of bone marrow did not have any significant effect on trabecular bone shear modulus and strength. The shear strength was proportional to the apparent density raised to the 1.02 power and to the strain rate raised to the 0.13 power. Jang and Kim [10] have developed topology optimization algorithm, design space optimization (DSO), to bone remodeling simulation in order to determine trabecular architecture in human proximal femur with higher computational efficiency. They have represented the full trabecular architecture in human proximal femur using a two-dimensional micro-FE model with 50 mm pixel resolution and performed simulation under three load cases in daily activities. From the quantitative comparison, it has been shown that DSO produced structurally equivalent trabecular architecture with shorter computing time and smaller memory requirement. Scenzi et al. [11] have assessed bone mineral density of the proximal femurs by dual energy X-ray absorptiometry and sectioned the femurs through the center of the femoral head and along the femoral neck axis and found that each of mean trabecular width and percent bone area vary within each proximal femur independently from each other, with dependence on site. Wirtz et al. [12] have develop an orthotropic femur model to realistically simulate three-dimensional bone remodeling & these 3-D models have been implemented into simulation tool, the principal directions of stiffness of cancellous and cortical bone have been determined manually using the orientation of the trabecular structures and the Haversian system & matched with the "CT-density characterized" three-dimensional femur model, the proximal femur has been realized considering orthotropic conditions of bone behaviour & has been formed to realize realistic simulations of the dynamical processes of bone remodelling under different loading conditions. Nazarian et al. [13] have assess whether variations in the primary and secondary tensile and compressive triangular microstructure can account for variations in proximal femur strength in comparison to DXA measured BMD & correlated with their mechanical properties & the work has demonstrated that the mechanical performance of each sub-region is highly dependent on the corresponding trabecular microstructure & improved the assessment of changing bone structure in response to pharmacological interruption. Jang et al. [14] have implemented a surface remodeling technique that performs bone remodeling in the exterior layer of the cortical bone while keeping its interior area unchanged & the femoral bone has normal cortical bone shape and perturbed femoral bone has an artificial bone lump in the inferred cortex & it has been shown that a bone irregularity has some little contribution to bone structural integrity tends to disappear & the trabecular bone supported the total load in the human proximal femur in the cortical bone carried. Antonio et al. [15] have proposed the density distribution results from a given set of loading parameter and found that it has been easy to measure the density distribution of long bones. This improved the robustness of the trained ANN against deviations of the measure density distribution from the predictions of the forward bone tissue adaptation model. They concluded the technique has been capable of predicting loading parameters that resulted in a density distribution close to the measured density distribution.

10.2 METHODOLOGY

Finite element analysis (FEA) is a computerized method for predicting how a product reacts to real-world forces, vibration, heat, fluid flow, and other physical effects. Finite element analysis shows whether a product will break, wear out or work the way it was designed. It is called analysis, but in the product development process, it is used to predict what is going to happen when the product is used.

FEA works by breaking down a real object into a large number (thousands to hundreds of thousands) of finite elements, such as little cubes. Mathematical equation help predict the behaviour of each element. A computer then adds up all the individual behaviours to predict the behaviour of the actual object.

Finite element analysis helps predict the behaviour of products affected by many physical effects, including:

- Mechanical stress
- Mechanical vibration

- Fatigue
- Motion
- Heat transfer
- Fluid flow

10.3 RESULTS AND DISCUSSION

Table 10.1: Data for curvature 1 (Radius of curvature-1310 mm)

Sl. No.	Load (in N)	Maximum Deformation (in mm)	Minimum Deformation (in mm)	Maximum Equivalent Stress (in MPa)	Minimum Equivalent Stress (in MPa)
1.	300	1.6397	0	5.862	0.021977
2.	550	3.0061	0	10.747	0.040291
3.	750	4.0992	0	14.655	0.054942

Results:

- Under 300 N load condition we found out that the total deformation was 1.6397 mm, maximum equivalent stress was 5.862 MPa and minimum equivalent stress was 0.021977 MPa.
- Under 550 N load condition we found out that the total deformation was 3.0061 mm, maximum equivalent stress was 10.747 MPa and minimum equivalent stress was 0.040291 MPa.
- Under 750 N load condition we found out that the total deformation was 4.0992 mm, maximum equivalent stress was 10.747 MPa and minimum equivalent stress was 0.054942 MPa.
- Thus with increase in load conditions, the total deformation and the maximum equivalent stress increases and the minimum equivalent stress decreases correspondingly.

Table 10.2: Data for curvature 2 (Radius of curvature-1300 mm)

Sl. No.	Load (in N)	Maximum Deformation (in mm)	Minimum Deformation (in mm)	Maximum Equivalent Stress (in MPa)	Minimum Equivalent Stress (in MPa)
1.	300	1.6339	0	5.8534	0.0069303
2.	550	2.9954	0	10.731	0.012706
3.	750	4.0846	0	14.634	0.017326

Results:

- Under 300 N load condition we found out that the total deformation was 1.6339 mm, maximum equivalent stress was 5.8534 MPa and minimum equivalent stress was 0.0069303 MPa.
- Under 550 N load condition we found out that the total deformation was 2.9954 mm, maximum equivalent stress was 10.731 MPa and minimum equivalent stress was 0.012706 MPa.
- Under 750 N load condition we found out that the total deformation was 4.0846 mm, maximum equivalent stress was 14.634 MPa and minimum equivalent stress was 0.017326 MPa.
- Thus with increase in load conditions, the total deformation and the maximum equivalent stress increases and the minimum equivalent stress decreases correspondingly.

Table 10.3: Data for curvature 3 (Radius of curvature-1290 mm)

Sl. No.	Load (in N)	Maximum Deformation (in mm)	Minimum Deformation (in mm)	Maximum Equivalent Stress (in MPa)	Minimum Equivalent Stress (in MPa)
1.	300	1.5249	0	5.4632	0.0064683
2.	550	2.8865	0	10.341	0.012244
3.	750	3.9757	0	14.243	0.016864

Results:

- Under 300 N load condition we found out that the total deformation was 1.5249 mm, maximum equivalent stress was 5.4632 MPa and minimum equivalent stress 0.0064683 MPa.
- Under 550 N load condition we found out that the total deformation was 2.8865 mm, maximum equivalent stress was 10.341 MPa and minimum equivalent stress was 0.012244 MPa.
- Under 750 N load condition we found out that the total deformation was 2.8865 mm, maximum equivalent stress was 14.243 MPa and minimum equivalent stress was 0.016864 MPa.
- Thus with increase in load conditions, the total deformation and the maximum equivalent stress increases and the minimum equivalent stress decreases correspondingly.

10.4 ILLUSTRATIONS, DIAGRAMS AND PHOTOGRAPHS

10.4.1 For Curvature–1 (Radius of Curvature:-1310 mm)

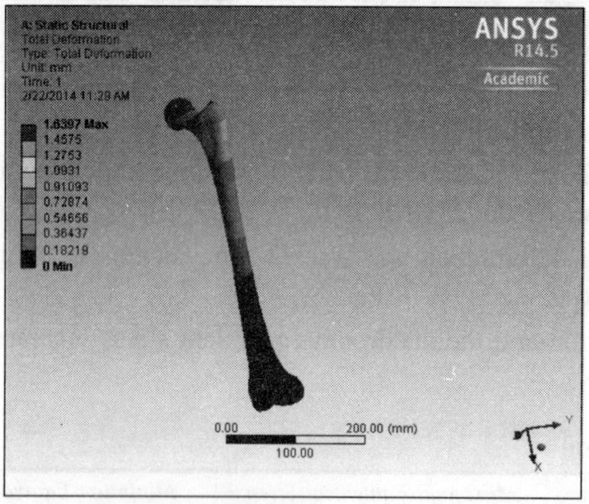

Fig. 10.2 Total deformation at 300 N

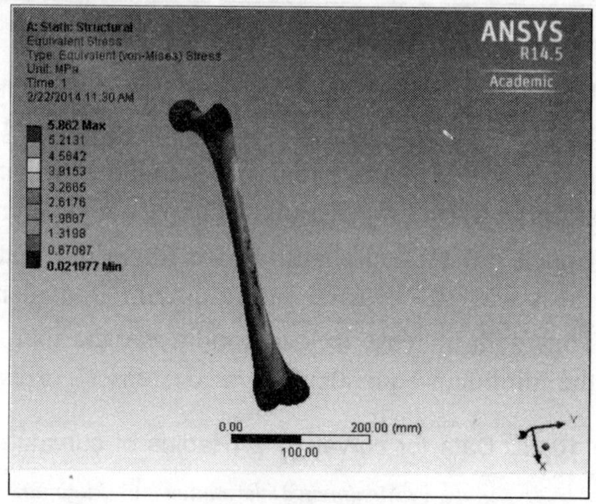

Fig. 10.3 Equivalent stress at 300 N

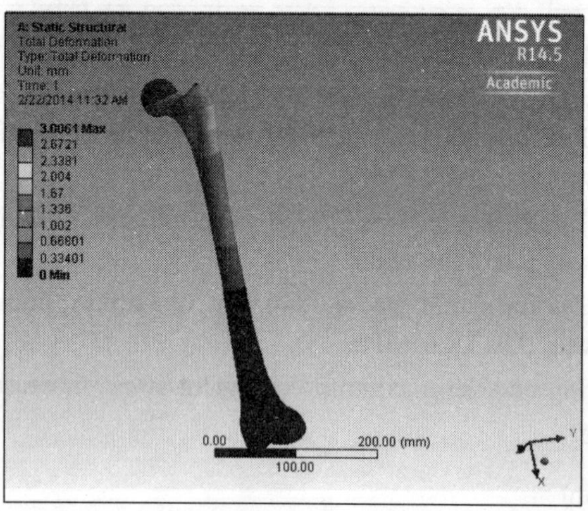

Fig. 10.4 Total deformation at 550 N

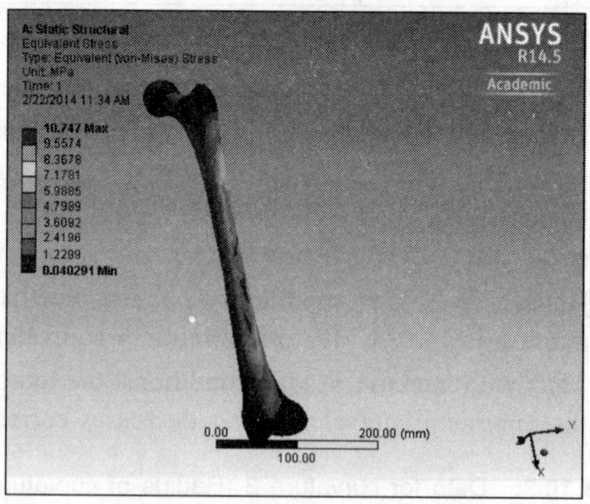

Fig. 10.5 Equivalent stress at 550 N

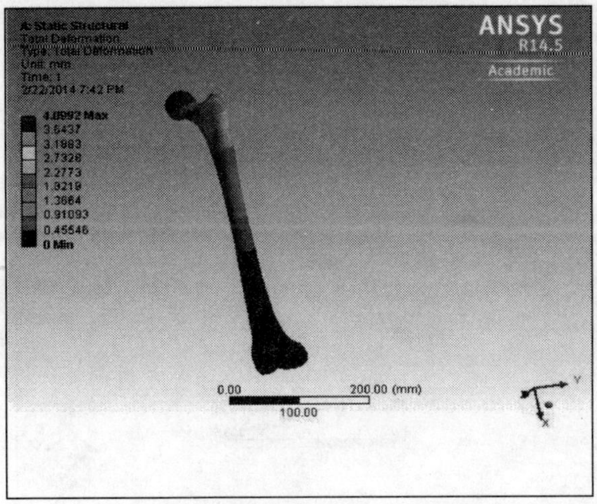

Fig. 10.6 Total deformation at 750 N

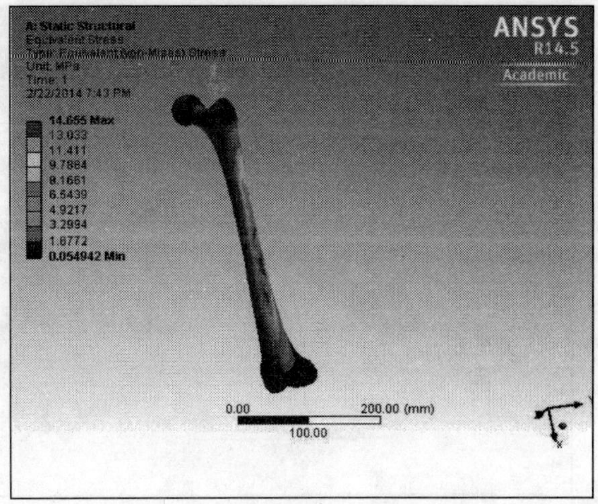

Fig. 10.7 Equivalent stress at 750 N

10.4.2 For Curvature–2 (Radius of Curvature: 1300 mm)

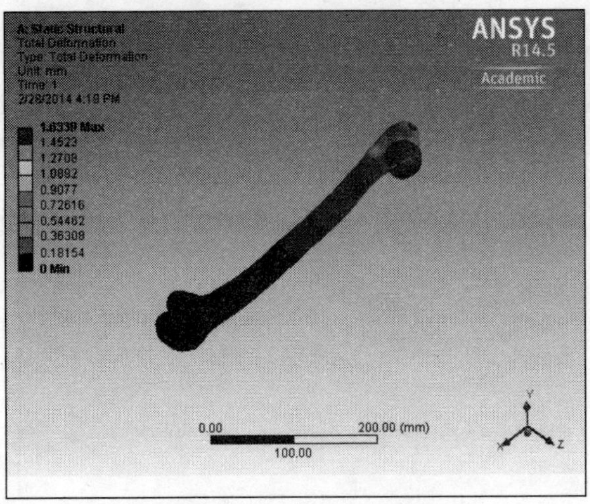

Fig. 10.8 Total deformation at 300 N

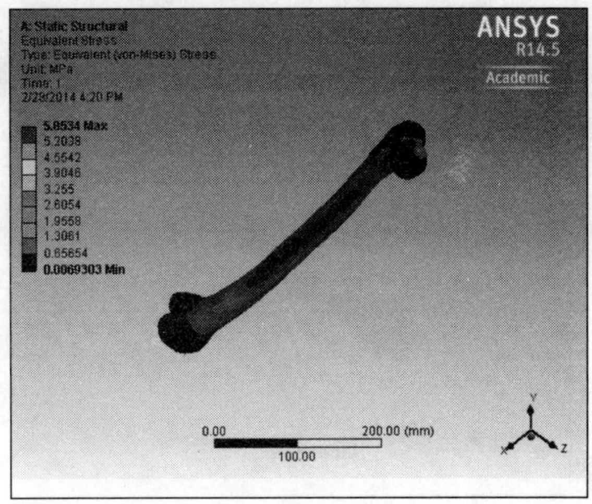

Fig. 10.9 Equivalent stress at 300 N

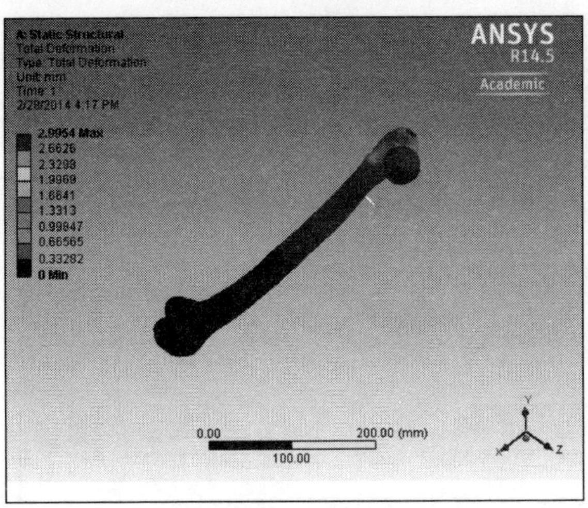

Fig. 10.10 Total deformation at 550 N

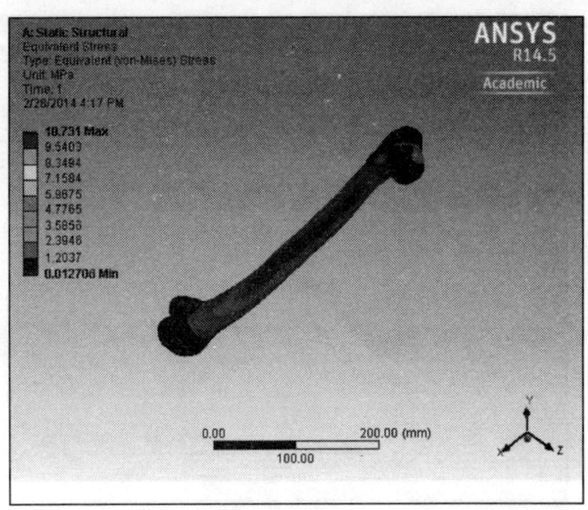

Fig. 10.11 Equivalent stress at 550 N

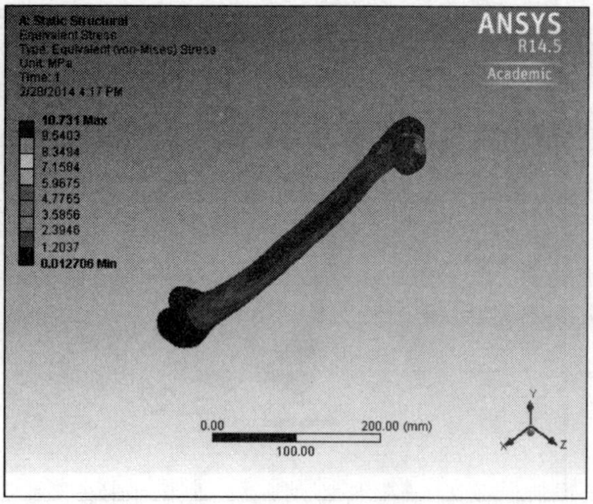

Fig. 10.12 Total deformation at 750 N

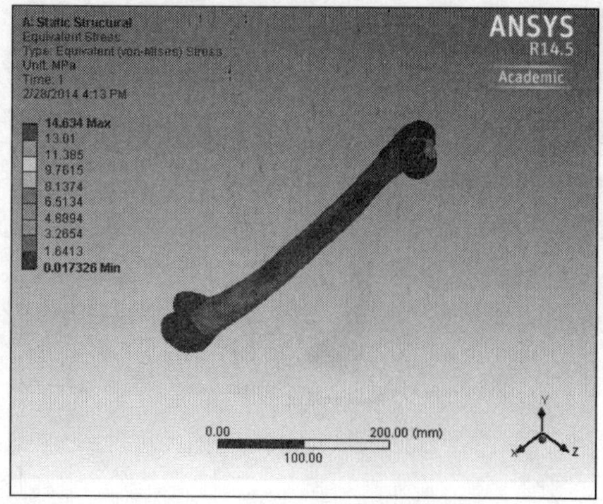

Fig. 10.13 Equivalent stress at 750 N

10.4.3 For Curvature–3 (Radius of Curvature: 1290 mm)

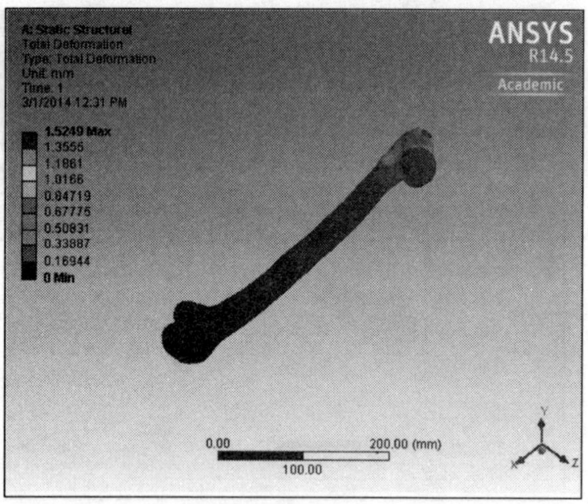

Fig. 10.14 Total deformation at 300 N

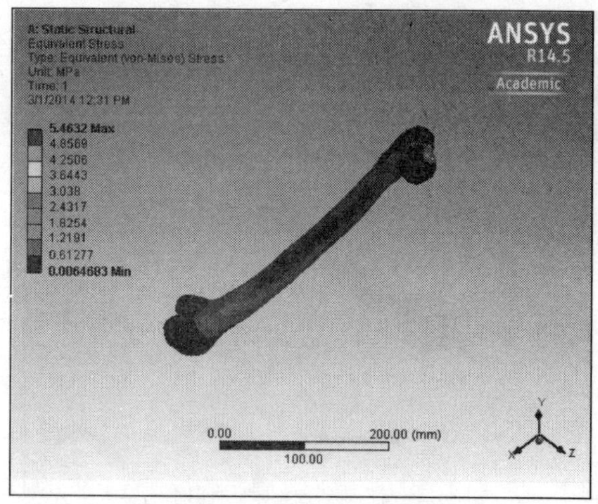

Fig. 10.15 Equivalent stress at 300 N

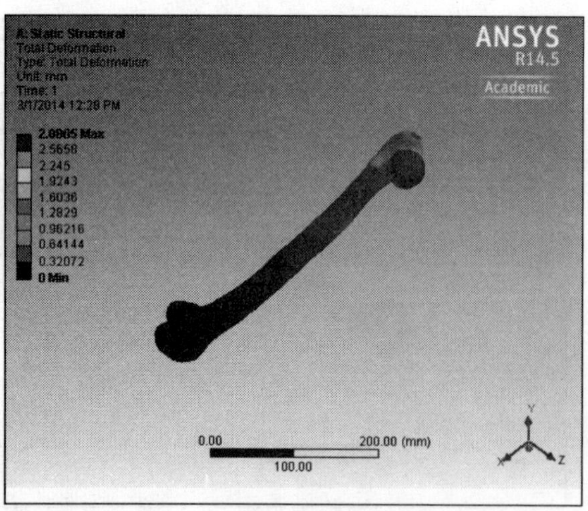

Fig. 10.16 Total deformation at 550 N

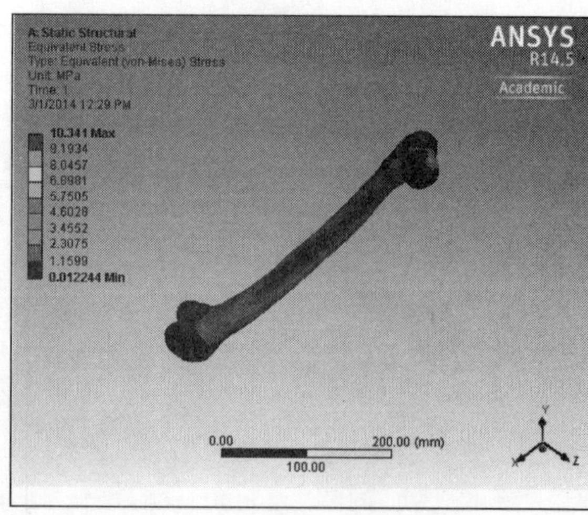

Fig. 10.17 Equivalent stress at 550 N

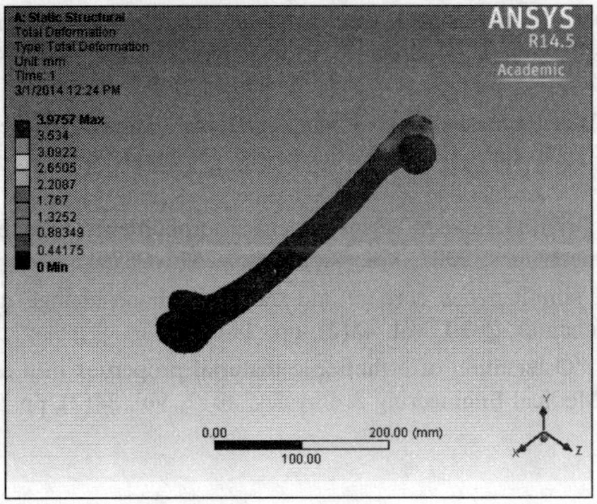

Fig. 10.18 Total deformation at 750 N

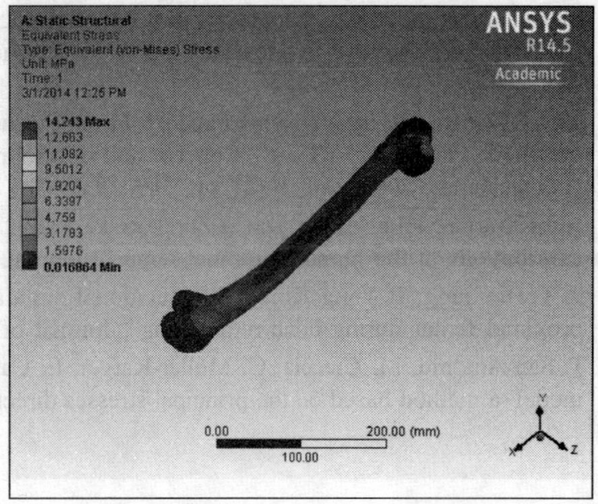

Fig. 10.19 Equivalent stress at 750 N

10.5 CONCLUSION

From the analysis of different radius of curvatures of a bone under different loading condition, we came to a conclusion that for radius of curvature of 1310 mm (i.e. curvature-1), the total deformation and equivalent stress is more as compared to the radius of curvature of 1290 mm (i.e. curvature-2). As the deformation and stress distribution in case of curvature-2 is less than curvature-1, therefore it can be preferred for prosthetics design. Hence this analysis gives us a prominent idea that the curved nature of bone plays an important role in reducing the deformation and stress distribution. Thus, during any accidental case, when the femur breaks, the prosthetics should be designed in such a manner that the curvature of the prosthetics bone should be nearly equal to the natural bone so that the deformation and the stress distribution on the bone would be lowered, thus increasing the longevity of prosthetics.

REFERENCES

1. John D. Currey, Katz, J.L., Keller, T.S., Burstein, A. H., Sasaki, N., Ikawa, T., Wagner, H. D., Weibull, W. "What determines the bending strength of compact bone?", The Journal of Experimental Biology, 1999, Vol. 202, pp. 2495-2503.
2. R Shahar, L Banks-Sills, R Eliasy, "Stress and strain distribution in the intact canine femur: finite element analysis", Medical Engineering & Physics, 2003, Vol. 25(5), pp. 387-395.
3. Kyle K. Nishiyama, Seth Gilchrist, Pierre Guy, Peter Cripton, Steven K. Boyod, "Proximal femur bone strength estimated by a computationally fast finite element analysis in a sideways fall configuration", Journal of Biomechanics, 2013, Vol. 46(7), pp. 1231-1236.
4. Mateusz Maria Juszczyk, Luca Cristofolini, Marco Viceconti, "The human proximal femur behaves linearly elastic up to failure under physiological loading conditions", Journal of Biomechanics, 2011, Vol. 44(12), pp. 2259-2266.
5. Philippe K. Zysset, X. Edward Guo, C. Edward Hoffler, Kristin E. Moore, Steven A. Goldstein, "Elastic modulus and hardness of cortical and trabecular bone lamellae measured by nanoindentation in the human femur", Journal of Biomechanics, 1999, Vol. 32(10), pp. 1005-1012.
6. D.P. Fyhrie, D.D. Cody, A.M. Saad, S.J. Hoshaw, "Decrease in canine proximal femoral ultimate strength and stiffness due to fatigue damage", Journal of Biomechanics, 1997, Vol. 30(4), pp. 323-329.
7. Llse Jonkers, Nicolas Sauwen, Gerlinde Lenaerts, Michiel Mulier, GeorgesVander Perre, Siegfried Jaecques, "Relation between subject-specific hip joint loading, stress distribution in the proximal femur and bone mineral density changes after total hip replacement" , Journal of Biomechanics, 2008, Vol. 41(16), pp. 3405–3413.
8. T. San Antonio, M. Ciaccia, C. Muller-Karger, E. Casanova, "Orientation of orthotropic material properties in a femur FE model: A method based on the principal", Mechanical Engineering & Physics, 2012, Vol. 34, pp. 914-919.
9. Kasra, Marc D. Grynpas, "On shear properties of trabecular bone under torsional loading: effects of bone marrow and strain rate", Journal of Biomechanics, 2007, Vol. 40, pp. 2898-2903.
10. Gwun Jang, Yong Kim, "Application of design space optimization to bone remodeling simulation of trabecular architecture in human proximal femur for higher computational efficiency", Finite Elements in Analysis and Design, 2010, Vol. 46(4), pp. 311-319.

11. Maria-Grazia A scenzi, Nicole Hetzer, Alexandre Lomovtsev, Robert Rude, Aurelia Nattiv, Angela Favia,"Variation of trabecular architecture in proximal femur of postmenopausal women", Journal of Biomechanics, 2011, Vol. 44(2), pp. 248–256.

12. Dieter Christian Wirtz, Thomas Pandorf, Frank Portheine, Klaus Radermacher, Norbert Schiffers, Andreas Prescher, Dieter Weichert, Fritz UweNiethard "Concept and development of an orthotropic FE model of the proximal femur", Journal of Biomechanics, 2003, Vol. 36(2), pp. 289–293.

13. Ara Nazarian, John Muller, David Zurakowski, Ralph Muller, Brian D. Snyder, "Densitometric morphometric and mechanical distributions in the human proximal femur", Journal of Biomechanics, 2007, Vol. 40(11), pp. 2573–2579.

14. In Gwun Jang, Il Yong Kim, "Computational simulation of simultaneous cortical and trabecular bone change in human proximal femur during bone remodeling", Journal of Biomechanics, 2010, Vol. 43(2), pp. 294–301.

15. T. San Antonio, M. Ciaccia, C. Müller-Karger, E. Casanova, "Orientation of orthotropic material properties in a femur FE model: a method based on the principal stresses directions", Medical Engineering & Physics, 2012, Vol. 34(7), pp. 914-919.

11

Biomechanical Analysis and Treatment of Congenital Talipes Equino Varus (CTEV)

E. Vijayaragavan, D. Somasundaram, P. Mohana Krishnan and T. V. Gopal

Department of Mechanical Engineering, SRM University, Kattankulathur, India
Department of Mechanical Engineering, SRM University, Ramapuram, India

Email: vijayaragavan.e@ktr.srmuniv.ac.in

ABSTRACT

Clubfoot deformity also referred as Congenital Talipes Equino Varus (CTEV) is a congenital foot deformity which is well established from birth. The clinical presentation is a child walking on the lateral border of the foot. The foot appears to be club shaped. The current statistics shows that the prevalence of clubfoot is about 1–2 per 1000 live births. The treatment of clubfoot is initiated from birth; severe bone and muscular deformity require surgical correction. But non surgical management is the most preferred way of treatment. The ankle foot orthosis (AFO) have been used to treat a variety of foot deformity, but its application in clubfoot is yet to be understood. In this study, non linear three dimensional finite element model of the clubfoot were developed using the actual geometry of the foot skeleton, which were obtained by 3D reconstruction of MR images. The 3d model is meshed and analyzed using 3d tetrahedral elements with properties E = 7300 MPa, γ = 0.3. The stress distribution in the foot is studied and with this as reference conceptual model of clubfoot orthosis is prepared. The key areas were identified in the foot and the orthotic is designed such a way that the growth pattern of the clubfoot is regulated to attain the normal profile. The FE analysis of the orthosis shows that the continuous usage will regulate the bone growth in the required profile and make the child to walk in the normal posture.

Keywords: CTEV, non-linear FE model, AFO.

11.1 INTRODUCTION

Clubfoot is a congenital condition that affects newborn infants. The medical term for clubfoot is Congenital Talipes Equino Varus. Congenital means that the condition is present at birth and occurred during fetal development. The condition is not rare and the incidence varies widely among different races. In the Caucasian population, about one to two in a thousand infants are born with a clubfoot. The condition affects both feet in about half of the infants born with clubfoot. Clubfoot affects twice as many males as females. Congenital Talipes Equino Varus (CTEV) is a common deformity of the foot in all races. An old Indian prayer book "Yajur Veda" is the oldest account describing it in 10th century B.C. as is the statute of a dwarf with a diastrophic foot seen in the tomb of Tutankhamen. Plaster of Paris casts, introduced by Kite and popularized by Ponseti is the most commonly used method. The more extensive operative procedures, advocated by some authors can be associated with complications and do not always have the desired outcome. Subcutaneous tenotomy of Tendo Achilles is a minor procedure, first described by Lorenz (1782) in Frankfurt and Delpech (1823) in France. Stromeyer of Germany performed the procedure and Little who later brought it to England and led to its popularity [1]. Later experience highlighted its limitations in complete correction of all club foot deformities and more extensive procedures were introduced.

11.1.1 Anatomy of Clubfoot

The ankle is the complex structure in the lower extremity containing tibia, fibula, talus and the calcaneus. In general an ankle contains 26 bones, 33 joints and over 100 muscles, tendons and ligaments. A network of blood vessels, nerves skin and soft tissues are necessary for proper functioning of the ankle [2]. Figure 11.1 represents the anatomy of a normal human foot.

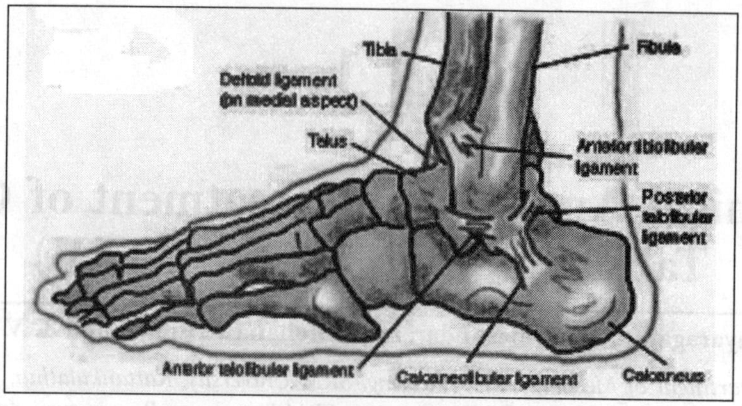

Fig. 11.1 Anatomy of normal human foot

The clubfoot anatomy shown in Fig. 11.2 represents the talar neck which is deviated in the medial and plantar directions. The calcaneus is rotated medially and both the navicular and cuboid is displaced medially.

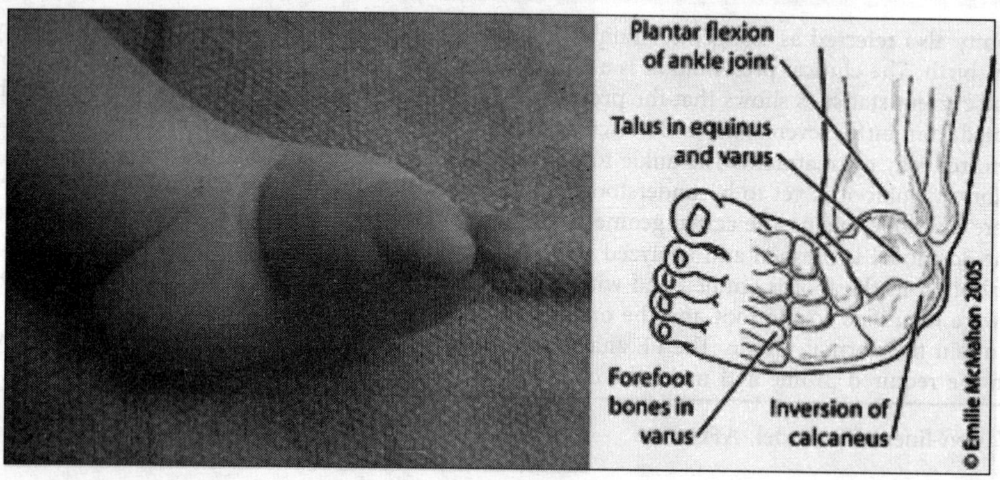

Fig. 11.2 Anatomy of clubfoot

The primary goal of this study was to perform biomechanical analysis of clubfoot, which is done by creating the 3d finite element of the clubfoot using reconstruction of CT images of the foot.

11.2 METHODS

This study consists of (i) Data collection, (ii) Model creation, (iii) Mesh generation, (iv) FE analysis and (v) development of orthosis. The following sections explain the procedures involved.

11.2.1 Data Collection

Twelve subjects (7 male and 5 female) with mean age of 21 years were selected of which one female subject with age 21 years, 5′5″ height and weight 60 kg is selected for further studies of a normal foot. A child of 18 months old with bilateral clubfoot with weight 10 kg is selected for the study. The MRI scans of the foot of both the subjects were taken with the optimal 1 mm thickness using Philips Ingenia 3.0T.

11.2.2 Model Creation

MIMICS 12.1 (software developed by Materialise, Belgium) is the software used for medical image processing. The segmentation of 3D medical images from MRI / CT results in accurate 3D models of the patient's foot. The model from MIMICS is converted to .iges and opened in CATIA for finishing. The model of the normal foot and club foot modified in CATIA is shown in Fig. 11.3.

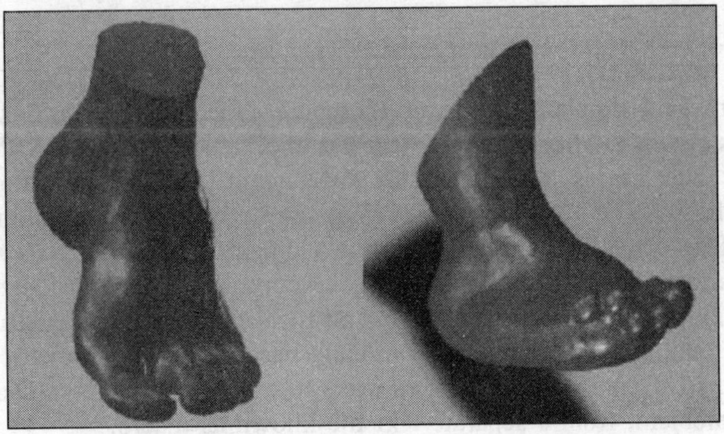

Fig. 11.3 Solid models of normal foot and clubfoot

11.2.3 Mesh Generation

The 3D model of the normal foot and the clubfoot were meshed in Hypermesh using 4 node tetrahedral elements with the element mesh size of 0.1 mm. The meshed models of the foot are shown in Fig. 11.4.

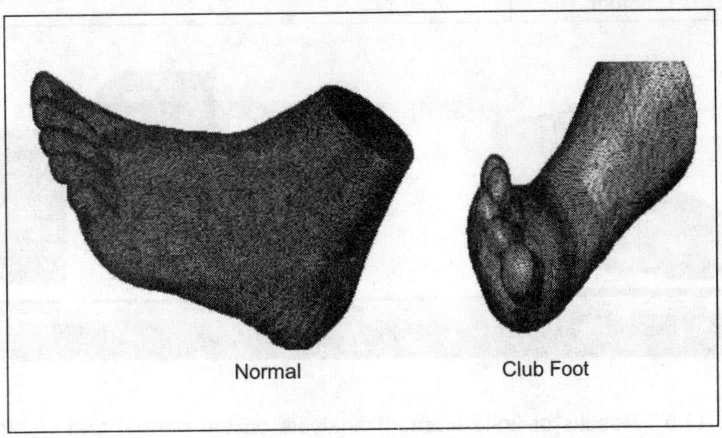

Fig. 11.4 Meshed models of normal foot and clubfoot

11.2.4 FE Analysis

Finite element analysis (FEA) is a computer simulation technique that can predict the deformation of a structure such as a bone when a load is applied, providing a measure of stiffness (N/mm). The structure is divided into a number of discrete, finite elements whose geometry is defined by points on the elements called nodes. Each element has prescribed material properties, such as Young's modulus and Poisson's ratio. Finite element analysis of a human club foot and normal foot is dependent on the overall size which is described by the geometry of the elements, and on the material properties of each element [3-6]. The material properties of the elements are specified in Table 11.1.

Table 11.1: Material properties and element type of the FE model

Component	*Element type*	*Young's modulus E (MPa)*	*Poisson's ratio (n)*	*Cross-sectional area (mm2)*
Bony structures	3D-tetrahedra	7300	0.3	---
Soft tissue	3D-tetrahedra	Hyper elastic	---	---

Contd...

Cartilage	3D-tetrahedra	1	0.4	---
Ligaments	Tension-only truss	260	---	18.7
Fascia	Tension-only truss	350	---	290.7
Ground support	3D-brick	17,000 upper layer, 1,000,000 lower layer	0.1	---

ANSYS, Inc. is an engineering simulation software (computer-aided engineering, or CAE) developer. The core product of Ansys Inc is its ANSYS Multiphysics/ Structure mechanical module. ANSYS, Inc. at Release 12.0, the ANSYS PREP7/POST1 interface is now called the "Mechanical APDL application" within the Workbench environment. In addition, the ANSYS Workbench interface formerly known as "simulation" is now called the "Mechanical application". This code is based on the finite element method and is capable of performing static (stress) analysis, thermal analysis, modal analysis, frequency response analysis, transient simulation and also coupled field analysis. The Ansys multiphysics can couple various physical domains such as structural, thermal and electro magnetics. Many researchers and engineers prefer this module because of its parametric language known as Ansys Parametric Design Language (APDL).The APDL allows users to execute all the commands required to pre process, solve and post process the problem, from a separate text file known as macro.

The results obtained during finite element analysis of the normal foot and club foot for the loading conditions as mentioned in table 2 is shown in Fig. 11.5.

Table 11.2: Load applied to the FE model

Foot	*Load*	*Region of peak stress*
Normal foot	300 N	calcaneus
Clubfoot	50 N	talus

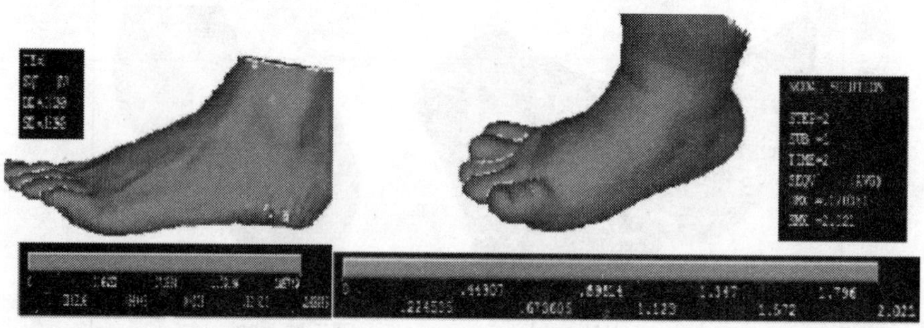

Fig. 11.5 Results of finite element analysis on the normal and clubfoot

11.2.5. Development of orthosis

The FE analysis on clubfoot shows the stress distribution and the stress is more in the talus Portion of the ankle. A comparative study between the normal foot and the clubfoot were made which helps us to identify the key points were the stress is maximum in the clubfoot [7-8]. The stresses in these points are minimised to the desired value in comparison with that of the normal foot.

11.2.5.1 Rapid Prototyping

Rapid Prototyping (RP) can be defined as a group of techniques used to quickly fabricate a scale model of a part or assembly using three-dimensional computer aided design (CAD) data. The 3D model of clubfoot and normal foot are prepared by rapid prototyping using STRATASYS FDM 360 MC. This machine uses the ".stl" file from MIMICS and creates the model using fusion deposition modeling. The optimal orientation for the model is selected and the model is build and post operation methods were performed to clean the part and the final model from the machine is shown in Fig. 11.6. Rapid prototyping decreases development time by allowing corrections to a product to be made early in the process. By giving engineering, manufacturing, marketing, and purchasing a look at the product early in the design process, mistakes can be corrected and changes can be made while they are still inexpensive.

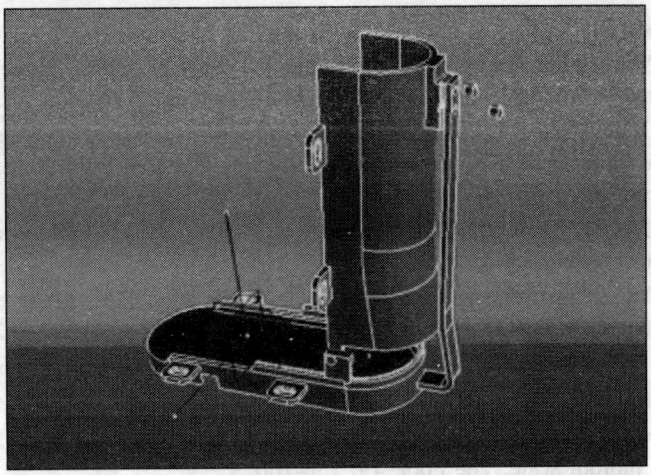

Fig. 11.6 Ankle foot orthosis

11.2.5.2 Design of Orthosis

Figure 11.6 is a model of an Ankle Foot Orthosis (AFO) with a leaf spring and sliding mechanism. This design combines the concept of a hinged AFO with that of a leaf spring AFO, both of which have been shown to assist in generating near-normal gait. The incorporation of the hinge allows for the possibility of a full range of motion in the sagittal plane and the addition of the carbon-fibre leaf spring provides resistance to foot movement both mediolaterally and in the sagittal plane, thus restricting the range of motion as required by the patient. In addition, the carbon fibre strut acting as a leaf spring would have the potential to store and subsequently release a large amount of energy as it elastically deforms due to its shape and material properties [9-10].

The main feature of this design is that as the Sliding Spring AFO is loaded through plantar or dorsiflexion, the leaf spring will elastically deform. Once unloaded, the leaf spring will return to equilibrium, providing energy to the ankle joint and assisting with gait while ascending and descending stairs.A second feature of the Sliding Spring design is that the leaf spring could be adjusted to the unlocked and locked configurations. In the unlocked configuration, the carbon fibre strut and slider block would be able to slide along the length of the guide rail as the user dorsiflexes and plantarflexes. The motion would not be free from resistance however, as there would be forces normal to the back face of the carbon fibre strut forcing the strut against the back of the guide rail. As the individual wearing the AFO dorsiflexes, the carbon fibre strut would slide downwards along the rail until it is either blocked from further motion by the end of the guide rail or until the forces along the line of action parallel to the guide rail balance. At this point, any further application of force would result in a bending of the carbon fibre strut. Upon releasing this loading force, the carbon fibre strut would return to its original position, releasing the energy stored during deflection and propelling the individual forward. As the foot returns to its neutral position, the carbon fibre strut and slider block would return to their original position along the guide rail. The same principle would also apply during plantar flexion, however, the carbon fibre strut and slider block would move upwards along the guide rail rather than downwards.

11.3 CONCLUSION

The Orthotic model for the club foot has been created which has a features of optimum comfort, flexibility, prevent relapses, reduce weight bearing by equal distribution of stress all along the foot which promotes physical growth and prevents psychological distress. It is one of the external support for the clubfoot. So that it may helpful for the infants with clubfoot. For the future scope, making the different kind of Orthotic models for the same clubfoot and analysis of AFO and selecting the best optimized Orthotic model. So that, it may helpful for the clubfoot infants.

REFERENCES

1. F. Herd, M. Macnicol, R.J. Abboud (2004). The need for biomechanical evaluation in the assessment of clubfoot, The Foot, 14, 72–76.
2. Susan J. Hall (2004), Basic Biomechanics, McGraw-Hill, International Edition.
3. Jason Tak-Man Cheung, MingZhang, Aaron Kam-Lun Leung, Yu-Bo Fan (2005). Three-dimensional finite element analysis of the foot during standing—a material sensitivity study, Journal of Biomechanics, 38, 1045–1054.
4. Camacho D.L.A., Ledoux W.R., Rohr E.S., Sangeorzan B.J., Ching R.P. (2002). A three-dimensional, anatomically detailed foot model: a foundation for a finite element simulation and means of quantifyingfoot-bone position, Journal of Rehabilitation Research and Development, 39(3), 401-410.
5. Song HR., Carroll N.C., Neyt J., Carter JM., Han J., D'Amato CR. (1999). Clubfoot analysis with three-dimensional foot models, Journal of Pediatric Orthopaedics, 8(1), 5-11
6. Herzenberg J.E., Carroll NC., Christofersen M.R., Lee E.H., Steve Whit BA., Munroe R. (1998). Clubfoot analysis with three-dimensional computer modeling, Journal of Pediatric Orthopaedics, 8(3), 257-262.
7. Abdul Hadi Abdul Razak, Aladin Zayegh, Rezaul K. Begg, Yufridin Wahab (2012). Foot Plantar Pressure Measurement System: A Review, Sensors, 12, 9884-9912.
8. A.K. Ramanathan, F. Herd, M. Macnicol, R.J. Abboud (2009). A new scoring system for the evaluation of clubfoot: The IMAR-Clubfoot scale, The Foot, 19, 156-160.
9. Lori A. Karol, Kelly A. Jeans (2011). Assessment of Clubfoot Treatment Using Movement Analysis, Journal of Experimental and Clinical Medicine, 3(5), 228-232.
10. Michael P. Dillon, Stefania Fatone, Andrew H. Hansen (2011). Effect of prosthetic design on center of pressure excursion in partial foot prostheses, Journal of Rehabilitation Research & Development, 48(2), 161-178.

12

Relationship of Human Gait Speed with Selected Anthropometric Variables Among Indian Adults

A. Guha Thakurta[1], R. Iqbal[2] and H. V. Bhasin[3]

National Institute of Industrial Engineering Vihar Lake Road, Powai Mumbai – 400087, Maharashtra

Email: guhathakurta@gmail.com

ABSTRACT

The purpose of this study was to investigate the relationship between anthropometric parameters and the gait speed of human locomotion in Indian males and females. Very few studies has been found exploring the relationship between the walking speed and body dimensions like body height, leg length (trochanteric height) cadence and step length. Anthropometric parameters were measured in 83 males and 20 female university level students of average age 24.76±2.45 years and 22.8±1.52 years respectively, average height 172.7±5.83 m and 1.6±0.06 m respectively and average weight 71.18±13.19 kg and 58.4±8.47 kg respectively. The relationship between the gait speed and the anthropometric variables was determined using Pearson multiple correlations matrix and independent t-test. It was observed that correlations exist between height and other parameters though in some cases correlations were moderate. The correlation between gait speed and step length was found to be high (r = 0.719 for males and r = 0.857 for females) and it was significant at $p<0.01$ level. The results of the study suggest that gait speed in humans is related to step length but not significantly dependant on body height and leg length.

Keywords: Gait, anthropometric parameters, leg length (trochanteric height), LKstep length.

12.1 INTRODUCTION

Walking, an important activity of daily living, is a mode of bipedal locomotion in which a period of double support, when both feet are in contact with the ground, is followed by a period in which the body is supported by one lower limb while the other is swung forward (Obaidi et al., 2003). Gait is the manner of walking. This recognizes that although the sequence of foot floor contact meets the above definition, we do not all have an identical walking pattern. Perhaps the most commonly reported measurements of gait are the temporal-spatial parameters. These include walking speed, stride time and length, step time and length, as well as the durations of stance phase and swing phase. These measurements have been referred to as the "basic gait parameters" and the "vital signs of walking" (Wall et al., 1987, 1997; Oberg et al., 1993) This last statement reflects the fact that these measurements are very useful indicators that a gait pattern may be abnormal and, if so, the extent of that abnormality, without necessarily indicating the cause of the problem. The temporal-spatial parameters are relatively easy to measure, requiring little more than a stopwatch, making them practical for clinical use. Humans can walk up to 4 m/s (Borghese et al., 1996) , but natural transition between walking and running is roughly 2.2 m/s (Stefanyshyn et al., 2000; Opila-Correia, 1990). Many authors studied the influence of walking speed on gait parameters to and out their normal ranges (Crosbie and Ko, 2001; Wall et al., 1991). Many investigators have studied the inter subject gait differences in normal samples; some are based on measurements of one or more of the temporal-spatial parameters. Perhaps the most frequently studied differences are those associated with age (Obrien et al., 1983; Hageman, 1995). However, numerous other normative studies have been undertaken to determine the effects of such variables such as walking speed, height or leg length, footwear, and curb clearance (Grieve and Gear, 1966; Rosenrot et al., 1980).

12.1.1 Normal Gait

Series of rhythmical, alternating movements of the trunk & limbs which result in the forward progression of the centre of gravity (Schafer, Kharb et al. 2011)

12.1.1.1 Gait Cycle

Single sequence of functions by one limb.

Begins when reference font contacts the ground.

Ends with subsequent floor contact of the same foot.

12.1.1.2 Cadence

Number of steps per unit time

Normal: 100 – 115 steps/min

12.1.1.3 Step Length

Distance between corresponding successive points of heel contact of the opposite feet.

Rt step length = Lt step length (in normal gait).

12.1.1.4 Stride Length

Distance between successive points of heel contact of the same foot. Double the step length in normal gait (Whittle, Perry)

The primary objectives of the study were to find the correlation between step length height, BMI and average leg length of Indian adults with their gait speed.

12.2 METHODS AND MATERIALS

12.2.1 Sample

Male: Data from a sample of 83 students between the age 20 and 30 were collected

Female: Data from a sample of 20 students between the age 20 and 30 were collected

12.2.2 Procedure

The experiment includes the measurement of height, weight, BMI, leg length and the gate speed. The procedure for measurement of gait speed is described below.

12.2.3 Equipment Required

1. Paper tape (4 pieces)
2. Tape measure (English or metric)
3. Stopwatch.
4. Height measurement-Anthropometric measuring set
5. Weight – Weighing machine
6. Leg length – Anthropometric measuring set.

Setup

Using paper tape a start line and end line on the floor were marked. An speed adjustment zone of 1 step to 3 meters based on available space and place a piece of tape at this distance on the floor was marked, then the test distance of 50 m was measured and a piece of tape was placed on the floor The subjects were asked to walk at a comfortable pace from before the start line to the end line and were made to stand behind the start line. Using a stopwatch, time from when the subject's leading limb (toe) crosses the test distance line until the leading limb crosses the end test distance line were noted.

12.2.4 Calculation

Comfortable gait speed = test distance/time to complete test distance.

Average leg length = (left leg length + right leg length)/2

Body Mass Index (BMI): The body mass index (BMI), or Quetelet index, is a measure for human body shape based on an individual's mass and height. [BMI= Mass in kg/(Height in m)^2]

Step length is the distance between the point of initial contact of one foot and the point of initial contact of the opposite foot. In normal gait, right and left step lengths are similar.

Stride length is the distance between successive points of initial contact of the same foot. Right and left stride lengths are normally equal.

Cadence or walking rate is calculated in steps per minute.(Kirtley et al., 1985; Zajac et al., 2003)

Velocity, the product of cadence and step length, is expressed in units of distance per time. Free speed refers to the individual's comfortable walking speed. Since individuals walk at different speeds depending on the situation, normal velocity values are somewhat arbitrary (Owings and Grabiner, 2004)

12.2.5 Data Analysis

An independent t-test was used to test whether the variables were significantly correlated or not. Relationships between the step length and walking speed were characterized using the Pearson correlation coefficient (r).

For all the variables included in the study, descriptive analyses were conducted. All statistical analyses were per-formed with SPSS.

12.3 RESULTS AND DISCUSSION

Table 12.1: Demographic data parameters age (years)

Height(cm)	*Age (years)*	*Height(cm)*	*Weight(kg)*	*BMI*	*Leg length(m)*
Males (N = 83)	24.76 ± 2.45	172.7 ± 5.83	71.18 ± 13.19	21.7 ± 2.90	0.99 ± 0.05
Females (N = 20)	23.27 ± 1.66	158.97 ± 5.87	57.68 ± 10.05	22.9 ± 3.60	1.0 ± 0.16

Mean ± Standard Deviation

Table 12.2: Walking speed and step length (gait parameters considered)

Parameters	*Walking Speed(m/s)*	*Step Length(m)*
Males (N = 83)	1.4±0.10	0.8±0.05
Females (N = 20)	1.3±0.11	0.7±0.04

Mean ± Standard Deviation

Table 12.3: Correlations between the parameters

Paeameters	*Walking Speed*	*Step Length*	*Leg Length*	*BMI*	*Body Height*
Walking speed (Males) Males (N = 83)	1	0.719**	–0.048	0.070	0.070
Walking speed (Females) Females (N = 20)	1	0.857**	–0.171	0.304	–0.467*

**. Correlation is significant at the 0.01 level (2-tailed).
*. Correlation is significant at the 0.05 level (2-tailed).

In Table 12.1, the mean age in years, body height in metres, and body weight in kilograms were measured. The mean age of the male subjects were 24.76±2.45 years and females were 23.2±1.66 years. The mean height for males were 172.7±5.83 cm and for females 158.97±5.87cm.The mean body weight for males were 71.18±13.19 kg and for females 57.68 ± 10.05 kg. The basal metabolic rate (BMI) calculated for males are 21.7±2.90 and for females are 22.9±3.60 which are within normal range of 18.50 and 24.99 (Adapted from WHO, 1995, WHO, 2000 and WHO 2004). The mean leg length of males is 0.99±0.05 m and females is 1.0±0.16m. In Table 12.2, the mean gait speed (walking speed) and step length for males are 1.4±0.10 m/s and 0.8±0.05m respectively and for females 1.3±0.11m/s and 0.7±0.04 m respectively. In Table 12.3, walking speed (gait speed) was found to be significantly highly correlated with step length in both males and females. The Pearson coefficient (r) were 0.719 for males and 0.857 for females at $p < 0.01$ level. Most of the power parameters were found to have a quadratic relationship with gait speed. Of the moment parameters, four had a linear relationship with gait speed; while four had a quadratic one (Lelas et al., 2002). Step length asymmetry is related to propulsive force generation during hemiparetic walking (Balasubhramanium et al., 2007). All these studies found a relationship between step step length and walking speed.

12.4 CONCLUSION

It was observed that correlations exist between height and other parameters though in some cases correlations were moderate. The correlation between gait speed and step length was found to be significantly high and it was significant at $p<0.01$ level. The results of the study suggest that gait speed in humans is related to step length but not significantly dependant on body height and leg length. We were able to provide some insights into the relationship between step lengths and walking speed in Indian context.

REFERENCES

1. R.C. Schafer. Book: "Clinical Biomechanics: Musculoskeletal Actions and Reactions, DC, Chapter 4, Ficc Body Alignment, Posture and Gait.
2. Kharb A, Saini V, Jain YK, Dhiman S.A Review of Gait Cycle and Its Parameters. IJCEM International Journal of Computational Engineering & Management 2011; 13.
3. M. Whittle.Gait analysis-An introduction.
4. Jacquelin Perry. Gait Analysis: Normal and pathological function.
5. C. Kirtley, M.W. Whittle, R.J. Jefferson. Influence of walking on gait parameters. Journal of biomedical engineering 1985; 7(4): 282-288.
6. Tammy M Owings, Mark D Grabiner. Variability of step kinematics in young and older adults. Gait and Posture 2004; 20(1): 26-29.
7. Zajac F.E., Neptune R.R., Kautz S.A. Biomechanics and muscle coordination of human walking: part II: lessons from dynamical simulations and clinical implications. Gait Posture 2003; 17: 1-17.
8. Balasubramanian C.K., Bowden M.G., Neptune R.R., Kautz S.A. Relationship between Step Length Asymmetry and Walking Performance in Subjects with Chronic Hemiparesis. Arch Phys Med Rehabil 2007; 88: 43-49.
9. Borghese N.A., Bianchi L., Lacquaniti F. Kinematic determinants of human locomotion. Journal of Physiology 1996:494 (3): 863-879.
10. Lelas J.L.,. Merriman GJ, P.O. Riley P.O., D.C. Kerrigan D.C. Predicting peak kinematic and kinetic parameters from gait speed. Gait & Posture 2002.
11. Obaidi S.A., Wall JC, Yaqoub A.A., Ghanim M.A., Basic gait parameters: A comparison of reference data for normal subjects 20 to 29 years of age from Kuwait and Scandinavia.Journal of Rehabilitation Research and Development 2003; 40 (4).
12. Wall J.C., Charteris J., Turnbull G.I. Two steps equals one stride equals what? The applicability of normal gait nomenclature to abnormal walking patterns. Clin Biomech 1987; 2:119-25.
13. Öberg T., Karsznia A, Öberg K. Basic gait parameters: reference data for normal subjects, 10-79 years of age. J Rehabil Res Dev 1993; 30(2): 210-23.
14. Wall J.C., Brunt D. Clinical gait analysis: Temporal and distance parameters. In: Van Deusen J, Brunt D, editors. Assessment in occupational therapy and physical therapy. Philadelphia: W.B. Saunders; 1997. p. 435-47.
15. Wall J.C., Scarbrough J. Use of a multimemory stopwatch to measure the temporal gait parameters. J Orthop Sports Phys Ther 1997; 25(4):277-81.
16. O'Brien M., Power K., Sanford S., Smith K., Wall J. Temporal gait patterns in healthy young and elderly females. Physiother Can 1983; 35(6):323-26.
17. Hageman P.A. Gait characteristics of healthy elderly: a literature review. Issues Aging 1995; 18(2):14-18.

18. Grieve D.W., Gear R.J. The relationships between length of stride, step frequency, time of swing and speed of walking for children and adults. Ergonomics 1966; 9(5): 379-99.
19. Rosenrot P., Wall J.C., Charteris J. The relationship between velocity, stride time, support time and swing time during normal walking. J Hum Mov Stud 1980; 6: 325-35.
20. Stefanyshyn D.J., Nigg B.M., Fisher V, O'Flynn B., Liu W. The influence of high heeled shoes on kinematics, kinetics, and muscle EMG of normal female gait. J Appl Biomech 2000; 16(3):309-19.
21. Opila-Correia K.A. Kinematics of high-heeled gait. Arch Phys Med Rehabil 1990; 71(5):304-9.
12. Crosbie J, Ko V. Changes in the temporal and distance parameters of gait evoked by negotiation of curbs. Aust J Physiother 2000; 46(2):103-12.
23. Wall J.C., Hogan D.B., Turnbull G.I., Fox R.A. The kinematics of idiopathic gait disorder. A comparison with healthy young and elderly females. Scand J Rehabil Med 1991; 23(3): 159-64.

13

Development of an EMG Based Smart Functional Electric Stimulator

Dinesh Bhatia

Department of Biomedical Engineering
North Eastern Hill University, Shillong–793022, Meghalaya, India

Email: *dbhatia@nehu.ac.in*

ABSTRACT

Musculoskeletal disorders are common ailments in human and cause various disabilities. Smart Functional Electrical Stimulation (FES) is a musculoskeletal rehabilitation therapeutic system provides effective and safe activation of neurmuscular tissue and generate physiological stable and controllable muscle contractions. The technology is based on stimulation of skeletal muscles through FES-evoked-electromyography (EMG) instrumentation, in order to help for regeneration and rejuvenation of malfunctioning / recurrent activated muscle state. We have developed an excellent feedback based Integrated Electrical Stimulater+Electromyographic System for optimal excitation of malfunctioning/damaged skeletal muscles, so tonicity as well as performance of disabled muscles can restored, also it can facilitate the early prediction of onset of muscular disorder. FES+EMG system may provide relief to paraplegic and physically challenged persons who needs and desires to restore their normal gait patterns for independent and dignified life in society. Device may improve quality of life (QOL) and daily activities of living (DAL) of physically challenged people without any fear or favour.

Keywords: Electromyography, stimulator, feedback, rehabilitation, musculoskeletal system.

Functional Electrical Stimulation is a rehabilitation technique where electrical current is applied to an individual with a disability to enhance that person's ability to function and live independently.

13.1 INTRODUCTION

Neuromuscular functional disabilities are developed in extremities affected by paralysis, resulting from spinal cord injury, head injury, stroke and other neurological disorders. Functional Electrical Stimulation (FES) is an electrical stimulation rehabilitation technology provide improvement and regeneration of mal/non functioning of neuro-musculoskeletal system, in people who suffered injury or disease due to spinal cord, stroke or any neurological malfunctioning. It is a rehabilitation technology through which paralysed muscles can be activated by applying small electrical current and voltages in order to restore wholly or partially body functions [9]. Despite the significant technological advancements occurred in two decade, but almost all FES systems works on "open-loop" manner and the stimulator's controller receives no feedback about the actual state of the system. Also, FES devices are adjusted by means of buttons which are mounted on walking frame. It requires constant pressing of buttons and slight error may results could result in a fall and great discomfort to patient and caused loss of confidence in the system [1]. Electromyography is well known skeletal muscles assessment system. EMG recordings can provide good assessment and monitor the functioning of different FES applications and many researchers have started adopting EMG as a trigger or feedback for designing a FES system [1]. Current commercial FES systems are not able to measure EMG activity during an active stimulation phase of the device, due to possibility of damage to the EMG amplifiers caused by the high voltage electrical pulses, measurement artifacts caused by the stimulation and the suppression of EMG signals due to electrode polarization [4]. In order to overcome these problems, proposed EMG based smart FES systems is developed in BME Laboratory for the lower extremities, may help in recording the electrical activities associated during stroke and in incomplete spinal cord-injured (iSCI) individuals. Also, it would be possible that muscle contractions during FES or muscle activity and which is independent of FES could observe the residual motor functions in disable individuals and may help in online detection of volitional muscle coordination or unwanted spastic reactions [11]. These smart systems would benefit gait rehabilitation in the individuals by controlling the timing and intensity of stimulation accurately that will allow flexibility for adapting to different walking speeds, stairs climbing, avoiding obstacles during gait and to a variety of movements unrelated to locomotion. These can easily be used to support physiotherapists in performing the repetition-based rehabilitation tasks more easily on these subjects. Moreover, these systems will be portable, fail-safe, easy to set up and operate, and would possess sufficient processing power to compute large quantities of data in real time mode.

Recently (2009), Chen et al. have used the intensity of the residual EMG to modulate the stimulation intensity during a FES-assisted cycling on an ergometer in stroke patients [2]. The recorded EMG signal of each muscle was divided into four suitable ranges. According to the range of EMG the stimulation patterns determine when to stimulate and which muscle should be stimulated. The proposed method simply distinguishes between correct and incorrect muscle activation by comparing the activities of both legs to avoid activation of electrical stimulation due to incorrect muscle activation. For the EMG measurement, recording electrodes different from the stimulation electrodes were used [2]. In 1989, Graupe discussed about development of a patient-responsive EMG control for FES [5]. The EMG was measured from the trunk muscles above the level of lesion in paraplegics. The EMG signals measured from different upper trunk muscles were analyzed by an EMG signature discriminator to recognize the different intended walking functions such as right step, left step, standing-up, or sitting-down. The below-lesion FES-induced EMG from the stimulated muscles was used to regulate stimulation intensity to counteract the muscle fatigue. They were using the surface EMG signals measured at the patient's erector spinae muscles (a muscle group of the back region) above the lesion level to provide control commands to the neuromuscular stimulator attached to the paralyzed lower limb muscles [5]. The potential of using sEMG as a natural trigger signal during FES-assisted gait was investigated by Dutta et al. [3]. They tried to detect residual motor activity below the level of injury in two volunteers with iSCI, which was integrated and then used to trigger the FES-assisted stepping during locomotion task. The quality of FES-assisted gait initiation using EMG-triggered control was compared with the auto-triggered FES (in which stimulation was triggered automatically at a fixed rate) and with a hand-switch-triggered control if required [3]. Four neurologically intact volunteers provided normative data for comparison. The gait initiation using EMG-triggered FES was found to be more coordinated and dynamically more stable than auto-triggered and hand-switch-triggered techniques [3, 5].

Yamaguchi et al. presented a sEMG-triggered stimulation system to help hemiplegics in standing-up tasks by means of FES [10]. By exploiting the synchronization of both lower limbs, the EMG detected at the intact quadriceps, hamstrings, rectus abdominis and Erector spine during standing-up was used to predict the action that the patient intends to do and to trigger the stimulation of the impaired side. The experiment was performed using one hemiplegic subject having a paretic left leg. The sampled raw EMG signal was detected, rectified, and then filtered by using a moving average filter. For each muscle, the integrated EMG is compared to a threshold value. If the integrated EMG from all muscles is simultaneously above the corresponding threshold levels, then the stimulation is switched on [10]. Peasgood et al., 2000, introduced a single channel closed-loop system for EMG-controlled FES in drop foot patients [7]. Ineffective dorsiflexion during swing phase (drop foot) is present in 20% of the stroke population with partial recovery. Residual EMG activity of tibialis anterior was detected and processed to control the stimulation intensity of this muscle giving the patient ability to control the stimulation levels. To apply EMG-proportional control the authors had to low-pass filter the volitional EMG signal to obtain a smooth controlling action. However, the system was found to be slow in response to the patient's intention due to time delay introduced by above filtering technique. A delay of 700 ms was reported. In walking tests, it was shown, that the time delay forced patients with drop foot to walk more slowly to obtain the support at the right time and this was a severe drawback of this system. Yeom et al., 2010, presented a similar system to continuously control the lower extremity paretic muscles by means of EMG-controlled FES. They employed separate stimulation and EMG recording electrodes [11]. Compared to the work by Peasgood et al., an adaptive Gram-Schmidt filtering algorithm was applied in real-time to cancel the stimulation artifacts from the recorded electromyograms. The feasibility of using vEMG to continuously control the FES-assist dorsiflexion was tested using one healthy subject. The results showed the feasibility of using EMG-controlled FES with adaptive stimulation-artifact removing filter.

Futami et al., 2005, proposed a locally EMG-controlled FES system for the knee extension of partially paralyzed patients [4]. The intensity of the applied FES was controlled in proportion to the voluntary EMG detected from the paretic knee extensor muscle. The authors reported that this EMG-driven FES is applicable to daily activities such as transfer from bed to wheelchair and standing-up without any control-mode selection for each consecutive activity to aid, since EMG reflects the neural commands from the brain. A switching circuit was used to protect the input stage of the EMG amplifier from any damage by stimulation pulses, and to reduce the transient artifacts. Kamono et al., 2001 investigated an EMG-driven FES system with EMG recording from the stimulation electrodes for correction of drop foot [6]. Electrodes were placed on the belly of tibialis anterior. The stimulation took place at a low frequency of 15 Hz and intensity of the stimulation was set again proportionally to the recorded volitional EMG at the same muscle. From the low stimulation frequency it can be concluded that stimulation artifacts last longer. The device was experimentally evaluated in a 24 years old intact subject.

The majority of these published articles show drawback of not discussing the tuning of FES controller parameters in detail, regardless if EMG was measured from stimulation electrodes or not. If these parameters could be possibly controlled, then the resultant system would show remarkable improvements in performing patient tasks when initiated. These facts caught our attention to develop a FES system which could be tuned as per requirements on individual subjects and make it more customized.

13.2 METHODOLOGY

The number of patients participating in this study as approved by ethical committee at All India Institute of Medical Sciences (AIIMS), Delhi is 30 in number based on literature studies and consensus of the research team. All patients should show willingness to participate and provide consent to undergo this study. EMG will be acquired from six different positions, i.e. flexor and extensor of hip and knee, plantarflexor and dorsiflexor of ankle using surface electrodes. This EMG data would then be analysed in a specially developed graphical user interface (GUI) software developed by the research team in Matlab to derive significant signal parameters. Based on these results, the FES device would be developed as shown in Fig. 13.1. The hardware of the system consists of an EMG amplifier (Fig. 13.2) and FES stimulation controller that are controlled by processing unit. The FES controller could be tuned as per the requirements of individual subjects through intensity and duration. The processing unit samples the amplified and processed EMG signals from the identified muscles with a preselected sampling frequency and then sends out the control signals through the stimulation controller to provide the required stimulation to the individual based on its need using FES electrodes. The software of the system was developed by MATLAB program which processed the incoming signals and decides the amplitude of voltage, magnitude of current and duration of application. Currently, the study is ongoing at AIIMS, Delhi.

13.3 CONCLUSION

Although FES devices to restore upright mobility have undergone dramatic changes since their beginning in the early 1960s, they have not yet reached a technological potential sufficient to benefit a great number of FES users worldwide. Particularly, the products available in International market do not address the need of Indian masses. They are designed based on studies carried out in European and US markets with significant high costs of these products available in the country. The available systems lack providing requisite feedback to the user on performance and also do not address issue of kinematic redundancy and how to generate the optimal excitation patterns to drive the skeletal system. To overcome these problems, there is a need for development of a reliable smart system that is simple to set up, to operate and more importantly, enable a satisfactory quality of gait. To achieve this, the use of externally mounted EMG based smart FES systems seems to be the most suitable approach at present. In the near future, further advancements in this technology should overcome the major practical obstacles often described in the donning and doffing of EMG based FES systems. The development of such a system would be of great benefit to significant number of paraplegic and physically challenged persons who frequently visit physiotherapy clinics, hospitals and other health care centres around the country to help them restore their normal gait patterns, making them functionally independent in carrying out their activities of daily living (ADL) tasks thereby improving their quality of life (QOL).

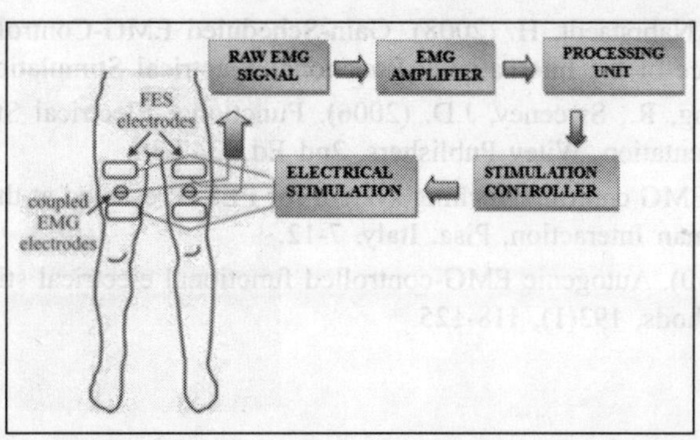

Fig. 13.1 Block diagram of smart EMG based FES system

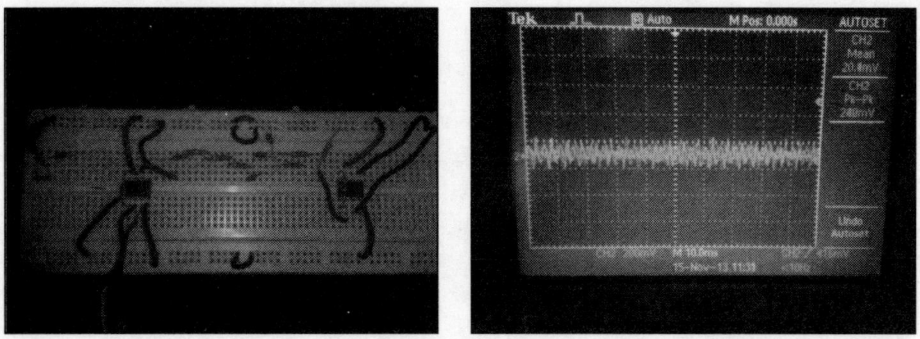

Design of EMG Amplifier | Output of EMG Amplifier

Fig. 13.2 EMG amplifier for the FES based system developed in laboratory

Acknowledgements

This work is supported by funding received (IDP/MED/27/2010) from the Department of Science and Technology, Government of India, New Delhi. The author acknowledges the support from the Department of Physical Medicine and Rehabilitation, All India Institute of Medical Sciences (AIIMS), Delhi and Centre for Biomedical Engineering, Indian Institute of Technology, Delhi.

REFERENCES

1. Braz, G.P., Russold, M. and Davis G.M. (2009). Functional Electrical Stimulation Control of standing and stepping after Spinal Cord Injury: A Review of technical characteristics. Neuromodulation: Technology at the Neural Interface, 12(3), 180-190.
2. Chen, C., He, Z. and Hsueh, Y. (2009). An EMG feedback control Functional Electrical Stimulation cycling system. Journal of Signal Processing Systems, 64(2), 195-203.
3. Dutta, A., Kobetic, R. and Triolo, R. (2009). Gait Initiation with Electromyographically triggered electrical Stimulation in people with partial paralysis. Journal of Biomechanical Engineering, 131(8), 2009.
4. Futami, R., Seki, K. and Kawanishi, T. (2005). Application of Local EMG-Driven FES to Incompletely Paralyzed Lower Extremities. Presented at the Proceedings of 10th Annual Conference of the International FES Society, Montreal, Canada.
5. Graupe, D. (1989). EMG pattern analysis for patient-responsive control of FES in paraplegics for walker-supported walking. Biomedical Engineering, IEEE Transactions, 36 (7), 711-719.
6. Kamono, A. and Kotaka, Y. (2001). Development of Gait Assist Device for Foot Drop Patients, Presented at the Proceedings of 6th Annual Conference of the International FES Society, Cleveland, Ohio, USA, 178-180.
7. Peasgood, W., Whitlock, T., Bateman, A., Fry, M., Jones, R. and Davis-Smith, A. (2000). EMG-controlled closed loop electrical stimulation using a digital signal processor. Electronics Letters, 36(22), 1832.

8. Shalaby, R., Schauer, T. and Nahrstaedt, H. (2008). Gain-Scheduled EMG-Controlled Electrical Stimulation. Proceedings of the 13th Annual Conference of the International Functional Electrical Stimulation Society, Freiburg, Germany.
9. Venkatasubramaniam, G., Jung, R., Sweeney, J.D. (2006). Functional Electrical Stimulation. The Wiley Encyclopedia of Medical devices and Instrumentation, Wiley Publishers, 2nd Ed., 347-36.
10. Yamaguchi, H. et al. (1999). EMG control switching system for FES. Presented at the Proceedings of the IEEE International Workshop on Robot and Human Interaction, Pisa, Italy, 7-12.
11. Yeom, H. and Chang, Y. (2010). Autogenic EMG-controlled functional electrical stimulation for ankle dorsiflexion control. Journal of Neuroscience Methods, 193(1), 118-125.

14

Effect of Vibration on Implant Used for Fracture Fixation of Bone

A. K. Dash, D. K. Agarwalla, D. Sahoo, C. K. Swain, S. R. Behera, A. K. Roul and D. P. Senapati

Department of Mechanical Engineering
Institute of Technical Education & Research, Bhubaneswar, India

Email: diptiprasadsenapati@gmail.com

ABSTRACT

Implants are generally used for fixation of fracture in bones. But during its service life it is found that the implant fails to fulfil its objective. Different parameters are responsible for failure of implants, vibration is one of them. In the current study vibration, analysis has been carried out on the bone plate assembly for both natural and forced vibration condition to identify the magnitude of frequency of vibration responsible for maximum deformation using finite element analysis. From the analysis of the results, it is observed that the different modes of vibration have a significant effect on the deformation of implant. The second mode of vibration during natural vibration analysis creates the maximum deformation. Hence, the patients should be advised to avoid to expose to the above method of vibration for early recovery.

Keywords: Vibration, fracture, implant., femur.

14.1 INTRODUCTION

Biomechanics is the science where the laws, principles and methods of mechanics are applied to the structure and function of the human body as shown in Fig. 14.1(a), (b). The human bone that has higher strength as that of different material but achieves this while remaining as light as wood. There are two major kinds of bone, trabecular (spongy) and cortical. Trabecular bone gives supporting strength to the ends of the weight-bearing bone. The cortical (solid) bone on the outside forms the shaft of the long bone. The long bones are found in the limbs, and each consists of a body or shaft and two extremities. The body or diaphysis is cylindrical, with a central cavity, the wall consists of dense, compact tissue of considerable thickness in the middle part of the body, but becoming thinner toward the extremities; the extremities are generally expanded, for the purposes of articulation and to afford broad surfaces for muscular attachment. No bones in body are straight in true sense. Each bone is curved in different planes. The long bones are more curved than the other bones, the curve generally taking place in two planes, thus affording greater strength to the bone as compared to straight bones. The bones belonging to this class are the clavicle, hummers, radius, ulna, femur, tibia, fibula, metacarpals, metatarsals, and phalanges.

14.1.1 Femur Bone

The femur, or thigh bone, is the most proximal (closest to the body) bone of the leg invertebrates capable of walking or jumping, such as most land mammals, birds, many reptiles such as lizards, and amphibians such as frogs. In vertebrates with four legs such as dogs and horses, the femur is found only in the rear legs.

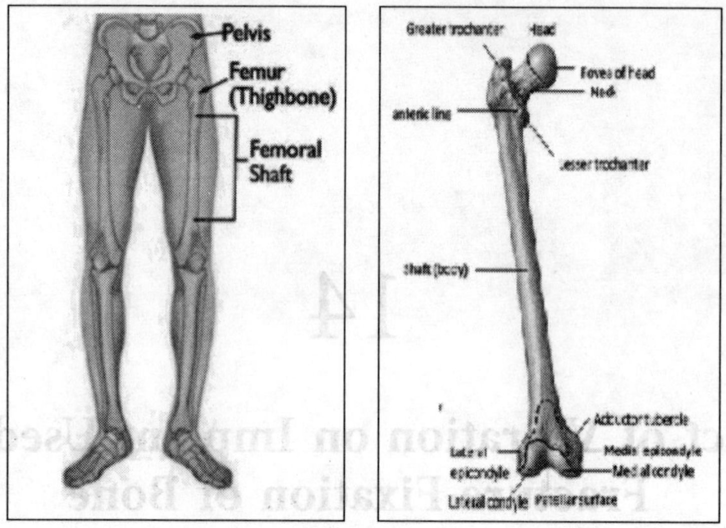

(a) (b)

Fig. 14.1 (a) Femur bone human anatomy (b) Femur bone

14.1.1.1 Femur Shaft Fractures (Broken Thighbone)

The thigh bone (femur) is the longest and strongest bone in human body. Because the femur is so strong, it usually takes a lot of force to break it. Car crashes, for example, are the number one cause of femur fractures. The long, straight part of the femur is called the femoral shaft. When there is a break anywhere along this length of bone, it is called a femoral shaft fracture.

14.1.1.2 Types of Femoral Shaft Fractures

Femur fractures vary greatly, depending on the force that causes the break. The pieces of bone may line up correctly or be out of alignment (displaced), and the fracture may be closed (skin intact) or open (the bone has punctured the skin).

Doctors describe fractures to each other using classification systems. Femur fractures are classified depending on:

- The location of the fracture (the femoral shaft is divided into thirds: distal, middle, proximal)
- The pattern of the fracture (for example, the bone can break in different directions, such as cross-wise, length-wise, or in the middle)
- Whether the skin and muscle above the bone is torn by the injury

The most common types of femoral shaft fractures include:

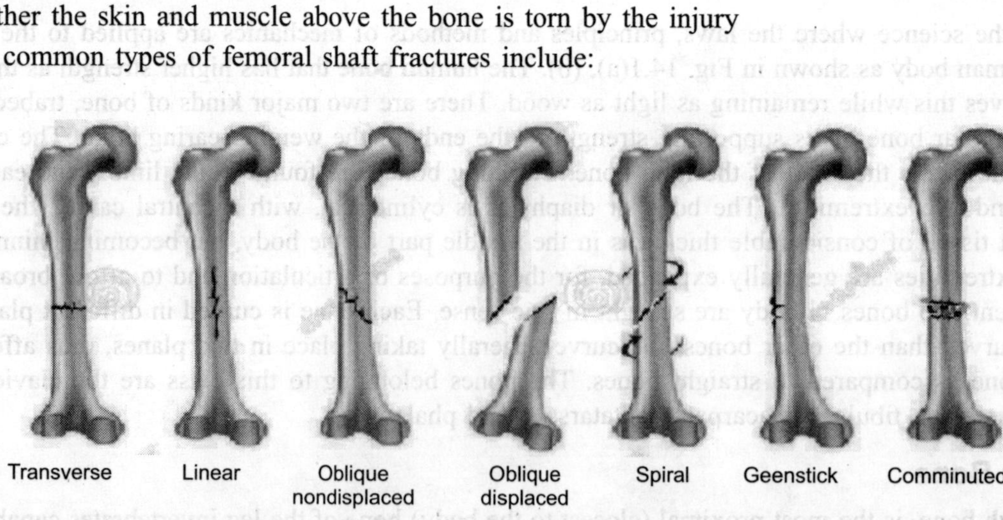

Fig. 14.2 Type of bone facture

Transverse fracture is a type of fracture, the break is a straight horizontal line going across the femoral shaft where as Oblique fracture has an angled line across the shaft and Spiral fracture line encircles the shaft like the stripes on a candy cane. A twisting force to the thigh causes this type of fracture. In comminute fracture the bone has broken into three or more pieces. In most cases, the number of bone fragments corresponds with the amount of force required to break the bone. In open fracture if a bone breaks in such a way that bone fragments stick out through the skin or a wound penetrates down to the broken bone, the fracture is called an open or compound fracture. Open fractures often involve much more damage to the surrounding muscles, tendons, and ligaments. They have a higher risk for complications especially infections and take a longer time to heal.

Pastrav et al. [1] have analysed that after the insertion the implant stability increases and also resonance frequencies increase, allowing the assessment of the implant fixation by vibration analysis while changing contact tolerance & keeping other system parameters constant. The higher modes are more sensitive to the contact change. Ezenwa et al. [2] have obtained that single-frequency methods has been reported to improve bone mineral density. He developed a model of femur bone by using computational software and analysed that with and without MVDMVF (multiple vibration displacements at multiple vibration frequencies input) and load applied to the femur model, then the MVDMVF generate stress level equivalent to the level during walking and stair climbing. Jaecques et al. [3] have found that the vibration modes of the human femur and of selected total hip replacement prosthetic stems were studied by experimental modal analysis using computational software. Two hypotheses were the differences between the vibrational behaviour of an intact femur and of the system formed by a femur and an implanted prosthesis will increase with increasing complexity of the mode shapes & when a prosthesis is implanted in a femur, the vibration modes of the prosthesis will be more affected by this interaction than the modes of the femur. Jones et al. [4] have identified that the experiments which seek to obtain evidence of the usefulness of vibration in the diagnosis of implant loosening by utilizing the experimental technique involved the application of a sinusoidal force to the distal end of a femur containing an implanted prosthesis and collecting the output signal using an accelerometer placed at the proximal end. The output signal was stored on a computer with a digital signal processing board and subsequently analyzed using the computational software. Lannocca et al. [5] have found out that good initial stability is crucial to avoid stem loosening and bone cracking; therefore, the surgeon must achieve optimal press-fitting. The present study aimed to design and test a prototype device able to evaluate the primary mechanical stability of cement less prosthesis, based on vibration analysis. Pastrav et al. [6] have identified that the vibrational technique designed to assess the initial stability of spherical implants. Vibration analysis technique has been successfully applied in biomechanics to determine bone mechanical properties, to monitor fracture healing, and to assess the stability of femoral stems in vitro and in vivo. Gefen [7] have come up with a finite element computer simulations to test the effect of various engineering designs of fixation screws on the predicted screw bone stress transfer, and consequently, on the biomechanical conditions for osteosynthesis for design and evaluation of bone screws, including geometrical features, material characteristics and even coatings. Pastrav et al. [8] have found that the frequency response function change provides reliable information regarding the stability evolution of the stem-femur system during the insertion. The protocol described in this paper can be used to accurately detect the insertion end point and to reduce the risk for intraoperative fracture. A protocol using a vibration analysis technique for the characterization of the primary bone-prosthesis stability. Pastrav et al. [9] success rate of orthopedic dependent on the initial fixation . The protocol described in this paper can be used to accurately detect the insertion end point and to reduce the risk for intra-operative fracture in total hip arthroplasty. The vibrational method based on frequency response function (FRF) analysis can be adapted and extended to other kinds of orthopedic implants. Rowlands et al. [10] have found out that Hip prosthesis loosening can be determined in vivo using a vibration-based technique called vibrometry. In this technique, a low frequency Sinusoidal vibration is applied to the femoral condyles and the resulting vibration is measured at the greater trochanter. the use of an ultrasound probe to detect mechanically induced vibration through significant amounts of soft tissue appears to be viable and could lead to enhanced detection of prosthesis loosening using this technique. Perez et al. [11] have observed that vibration analysis methods have been successfully used to assess implant stability of a bone, to monitor fracture healing and to measure bone mechanical properties. The utility of the combination of a vibrational technique and the finite element (FE) method to detect implant loosening and simulate the change in resonance frequency during the osseointegration process of a cementless Total Hip Arthroplasty (THA). Bolland et al. [12] have found that the peak loads and hoop strains transmitted to the femoral cortex during graft compaction and subsidence of the stem in subsequent mechanical testing were reduced. This innovative technique has the potential to reduce the risk of intra-operative fracture and to improve graft compaction and therefore prosthetic stability.

From the literature cited above, it is quite evident that, vibration plays a vital role in the management of fractures in bone. Researchers have explored various techniques in connection to vibration for analysing the patterns of load and magnitude of vibration responsible for implant failure. In the current quest for research, effort has been made to analyse the effect of vibration on a plate fixed to the cortical zone of the femur bone and it is observed that on the higher modes of vibration the failure chances of implants increases.

14.2 VIBRATION ANALYSIS USING FINITE ELEMENT METHOD

The finite element method (FEM) is considered one of the well established and convenient technique for the computer solution of complex problems in different fields of engineering: mechanical engineering, biomedical engineering, hydrodynamics, heat conduction, etc. However, it has been suggested that it may be possible to distinguish between a secure and a loose prosthesis using a vibration technique. In fact, vibration analysis methods have been successfully used to assess femur implant stability to monitor fracture healing and to measure bone mechanical properties. The user constructs a model of the part to be analyzed in which the geometry is divided into a number of discrete subregions, or "elements," connected at discrete points called "nodes." Certain of these nodes will have fixed displacements, and others will have prescribed loads. Finite element analysis can be done conveniently as part of the computerized drafting-and-design process. In the present study, the main goal is to simulate the change in the resonance frequency during the osseointegration process under different type of load like free load and forced load. The finite element method was used and a computational modal analysis was conducted to obtain the natural frequencies and mode shapes under vibration. The effects were studied on femur bone and different contact conditions at the bone-implant interface. The results were in agreement with in experimental and computational observations, and differences among the different cases studied were detected. As the osseointegration process at the bone-implant interface evolved, the resonance frequency values of the femur-prosthesis system also increased. In summary, vibration analysis combined with the finite element method was able to detect boundary conditions at the bone-implant interface in cases of both osseointegration and loosening.

14.2.1 Steps Adopted in Current Analysis

- Modeling of the cortical zone of the bone.
- Modeling of the plate and screw
- Assembly of the implant with the bone
- Importing the assembly to the computational software
- Taking material properties
- Mesh the model with relevance center fine.
- For the natural frequency the boundary condition is given at the bottom face of the model and applies the free load.
- Compute the solution model and found the analyses data.
- For forced frequency the boundary is fixed at the bottom part of the model.
- Giving the magnitude value 500 N and select required direction of applied load. The load is given at the top of the model and the direction is towards the bottom.
- Then we found the analyzed data.

14.3 RESULTS AND DISCUSSION

Table 14.1: Statistics mesh

Nodes	11925
Elements	4962

Contd.

Table 14.2: Structural steel constants

Density	7850 kg m^{-3}

Table 14.3: Natural frequency

Modes	*Natural frequency (Hz)*	*Maximum deformation (mm)*
1	339.1	0.15098
2	1530.6	0.2067
3	2515	0.19262
4	3267.8	0.095667

Table 14.4: Forced frequencies (At 500 N load)

Modes	*Frequency(Hz)*	*Maximum deformation (mm)*
1	1550	6.1534×10-6
2	3100	
3	4650	
4	6200	

From the natural frequency of vibration it has been observed that the maximum deformation increases from mode 1 and mode 2 but the deformation again decreases from mode 3 to mode 4. From the forced vibration for different modes of vibration that means from mode 1 to mode 4 the maximum deformation remains constant with respect to increasing frequency range. When the fractured bone with implant is subjected to natural frequency of vibration of magnitude 1530.6 Hz .The maximum deformation occurs i.e. 0.2067 mm which should be avoided the patient. So as to avoid failure of implant.

14.4 ILLUSTRATIONS, DIAGRAMS AND PHOTOGRAPHS

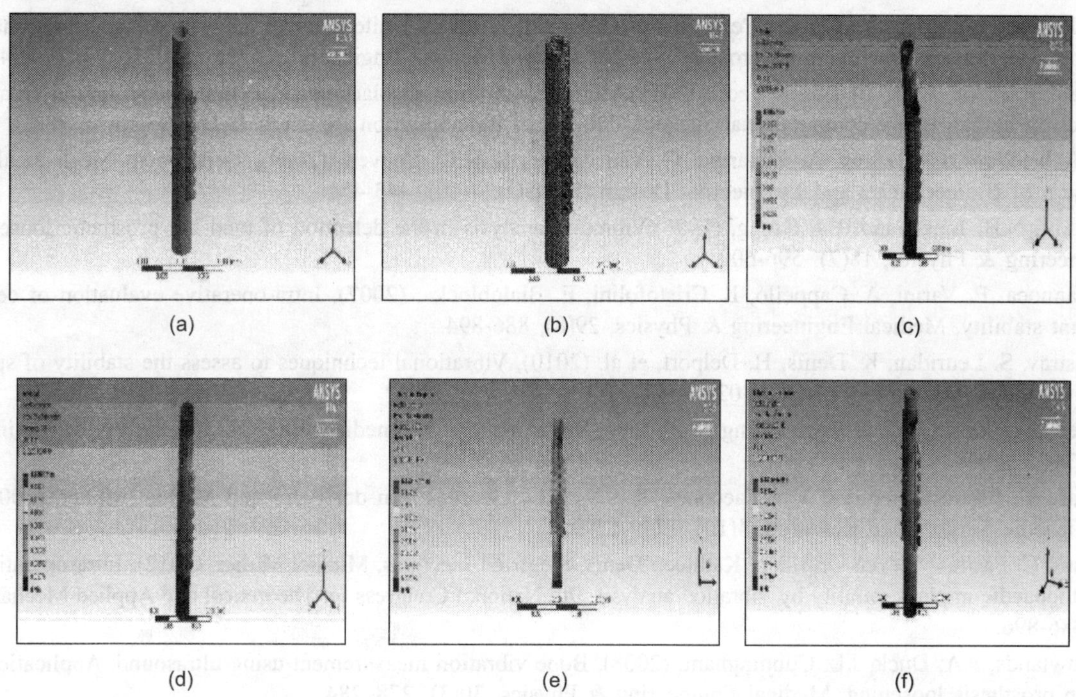

Fig. 14.3 (a) Pylon in complete assembled mode (b) Image after meshing (c) Showing deformation mode-1 (d) Showing deformation mode-2 (e) At 500 N mode (f) Deformation at 500 N

14.4.1 Vibration Analysis (Natural Vibration)

From the computational model as shown in Fig. 14.3(b), 4(c), 4(d), 4(e) and Fig.4 (f), it is observed that, the natural frequency of vibration for mode 1, mode 2, mode 3, mode 4 are 339.1 Hz, 1530.6 Hz, 2515 Hz 3267.8 Hz respectively and the corresponding maximum deformations are 0.15098 mm, 0.2067 mm, 0.19262 mm, 0.095667 mm respectively as shown in Table 14.3.

14.4.2 Vibration Analysis (Forced Vibration)

From the computational model, it has been analyzed that, in case of forced frequency the applied load of 500 N for boundary condition, the frequency of mode 1, mode 2, mode 3, mode 4 are 1550 Hz, 3100 Hz, 4650 Hz, 6200 Hz then all the corresponding maximum deformation is 6.1534 x 10-6 respectively as shown in Table 14.4.

14.5 CONCLUSION

It is observed from the result data analysis

The natural frequency of vibration for the bone and implant are 339.1 Hz, 1530.6 Hz, 2515 Hz and 3267.8 Hz for the 1st, 2nd, 3rd and 4th mode of vibration.

From the natural frequency of vibration it has been observed that the maximum deformation increases from mode 1 and mode 2 but the deformation again decreases from mode 3 to mode 4.

From the forced vibration for different modes of vibration that means from mode 1 to mode 4 the maximum deformation remains constant with respect to increasing frequency range.

When the fractured bone with implant is subjected to natural frequency of vibration of magnitude 1530.6 Hz. The maximum deformation occurs, i.e. 0.2060 mm which should be avoided by the patient. so as to avoid failure of implant.

REFERENCES

1. L.C. Pastrav, J. Devos, G. Van der Perre, S.V.N. Jaecques, (2009),A finite element analysis of the vibrational behavior of the intra-operatively manufactured prosthesis–femur system, Medical Engineering & Physics,31(4), 489–494 .
2. Bertram Ezenwa, PhD; Han Teik Yeoh, (2011),Multiple vibration displacements at multiple vibration frequencies stress impact on human femur computational analysis, Journal of Rehabilitation Research & Development, 48(2), 179-190.
3. S.V.N. Jaecques, C. Pastrav, A. Zahariuc, G. Van der Perre, K.U. Leuven, (2009), Department Mechanical Engineering, Division of Biomechanics and Engineering Design (BMGO), 4(10), 443-456.
4. P.L.S. Li, N.B. Jones, and P. J. Gregg, (1996),Vibration analysis in the detection of total hip prosthetic loosening, Medical Engineering & Physics, 18(7), 596-600.
5. M. Lannoca, E. Varini, A. Cappello, L. Cristofolini, E. Bialoblocka, (2007), Intra-operative evaluation of cement less hip implant stability, Medical Engineering & Physics, 29(8), 886-894.
6. L. Pastrav, S. Leuridan, K. Denis, H. Delport, et al. (2010), Vibrational techniques to assess the stability of spherical press-fitted implants: preliminary results, 1029-1043.
7. A. Gefen, (2002), Medical Engineering & Physics, Department of Biomedical Engineering, Faculty of Engineering, 24(5), 337–347.
8. Leonard C. Pastrav, Siegfried V.N. Jaecques, Ilse Jonkers, Georges Van der Perre and Michiel Mulier, (2009), Journal of Orthopaedic Surgery and Research, 4(10), 1749-1799.
9. Leonard C. Pastrav, Steven Leuridan, Kathleen Denis, Siegfried Jaecques, Michiel Mulier, (2012), Intra-operative monitoring of orthopaedic implant stability by vibration analysis, 9th National Congress on Theoretical and Applied Mechanics, Brussels, 28: 886-896.
10. A. Rowlands, F.A. Duck, J.L. Cunningham, (2008), Bone vibration measurement using ultrasound: Application to detection of hip prosthesis loosening, Medical Engineering & Physics, 30(3), 278–284.
11. Ma Angeles Perez, Belen Seral-Garcia, (2013) Aragón Institute of Engineering Research (I3A), University of Zaragoza, Spain, Orthopedic Surgery. 16: 1022-1031.
12. B.J.R.F. Bolland, A.M.R. New, S.P.G. Madabhushi, R.O.C. Oreffo, D.G. Dunlop. (2007), University of Southampton, Southampton, England, 89: 686-92.

15

Ergonomic Advice for Physical Exercises to Indian Women During Kitchen Work for the Prevention of Musculoskeletal Disorders

C. S. Ram and Bhavani Ramachandran

I. T. S. Paramedical College
Murad Nagar, Ghaziabad, India

Email: csram@its.edu.in

ABSTRACT

Implants are generally used for fixation of fracture in bones. But during its service life it is found that the implant fails to fulfil its objective. Different parameters are responsible for failure of implants, vibration is one of them. In the current study vibration, analysis has been carried out on the bone plate assembly for both natural and forced vibration condition to identify the magnitude of frequency of vibration responsible for maximum deformation using finite element analysis. From the analysis of the results, it is observed that the different modes of vibration have a significant effect on the deformation of implant. The second mode of vibration during natural vibration analysis creates the maximum deformation. Hence, the patients should be advised to avoid to expose to the above method of vibration for early recovery.

Keywords: Vibration, fracture, implant, femur.

15.1 INTRODUCTION

Musculo Skeletal Disorders (MSDs) are significant public health problems due to their high impact on the disability, personal suffering, absence from work, and their direct, indirect costs to the health care system. Several risk factors have been shown to relate to developing musculoskeletal disorders, including repetition, vibration, awkward posture, sustained posture, high loading, and forceful movement. In completing household chores, many factors contribute to increased discomfort and pain in the musculoskeletal system. The exact frequency of these risk factors and their burden on the health of home makers is yet to be fully ascertained[1].

Unpaid physical activity consists of volunteerism and household work; however, the largest proportion of unpaid activity is household work. Apart from personal care, leisure, sleep/rest, and socialization, unpaid work can be divided into household work, parenting, and caretaking. The homemaker is the predominant person in the home. But, physical amenities and designing of her house are not always planned as per her needs of health, security and work place comfort. Ergonomic applications relative to the prevention of MSDs have focused on primarily industrial and office settings. Such applications have sought not only to decrease the risk of injury but also to maximize efficiency, eliminate waste, and improve productivity and comfort. However, ergonomic intervention based studies did not fully address the home and leisure activities of house makers. Many reasons exist why the home environment is an area ripe for ergonomic intervention. The performance of domestic work is still one of the world's primary occupations, with a significant number of accidents occurring inside the home as compared to outside[2].

Household work involves numerous physical (e.g., lifting and carrying, poor work postures, use of muscular strength) and psychosocial (e.g., time pressure, lack of influence on own work, lack of professional development) factors which accelerate the reduction in the work ability with ageing [3]. Among many, one of the main problems in home work arises from ergonomically incorrect working conditions and equipment in private homes.

The quality of housing can enhance or diminish the well being of individual and families as that of the entire community. Hence, this is a major field where much improvement could be achieved by considering more fully the human factors involved. Therefore, home ergonomics in general and kitchen ergonomics in particular is becoming very important amongst home scientist, ergonomists, industrialists, builders and interior designers[2].

The branch of ergonomics dealing with work-related musculoskeletal disorders has in the last few years, seen a flurry of activities pertaining to the scientific basis of the profession, in particular studies to identify risk factors and intervention studies. Many signs of a maturing research field in Physical Therapy are observed, including peer scientific reviews and papers that reflect critically on methods and paradigms. Active surveillance can help to identify the risk factors of daily household chores in a standardized manner so that women homemakers can prevent these disorders before they appear.

It is suggested that investigations to determine the amount of physiological effort, the optimum dimensional accord between equipment and user, and the provision of satisfactory environmental conditions, are needed in order to ensure that the housewife performs her tasks with the minimum of unnecessary effort and maximum satisfaction[3].

This study aspires to examine the effects and feasibility of an ergonomic advice and exercise program on the work content and load of home care workers with Musculo skeletal disorders[4, 5]. This study focuses on the participatory ergonomic intervention with various work-site measures which were found to be effective and feasible for preventing disability among ageing female house wives performing activities, involving muscular locomotion. Health ergonomists need a clearer concept of the role of psychosocial factors in the development of musculoskeletal disorders and a best approach to reduce such risks [6]. In particular, a better understanding of the injury process and thereby the identification of risk indicators of this process is needed.

Broadly the study was undertaken to address the

1. Subjective measurement of pain due to Musculoskeletal disorders in homemakers using VAS scale and Neck Disability Index.
2. To implement appropriate ergonomic intervention program together with Physical Therapy Exercises and study its effects in pain relief in self reported musculoskeletal disorders.
3. Analyze the efficacy of Ergonomic Intervention program and compare with the available data generated from other work related MSDs.

15.2 MATERIALS & METHODS

15.2.1 Study Design

Four experimental design groups, with one control group consisting of 10 members, one experimental group receiving only exercise program, one experimental group receiving exercise program and ergonomic advice and one experimental group receiving only ergonomic advice.

15.2.1.1 Samples Definition of Study Subjects

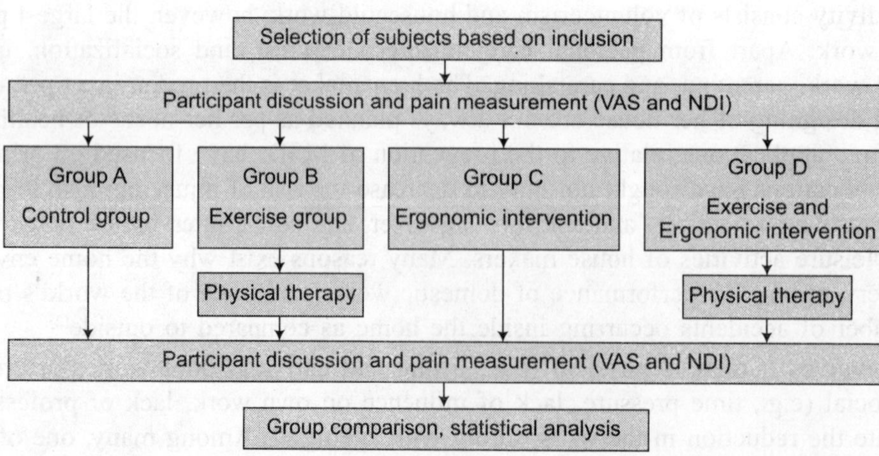

Female homemakers from different locations in southern India aged between 25-40 years, married, non-working, non-menopausal, not previously diagnosed for pain in the upper limb.

Female homemakers aged between 25 and 40 years, Married, and having children, Housewives; Regular physique maintaining women were included in the study. Physically handicapped women; Female homemakers assisted by maid/servant; Women who had undergone major surgeries in spine, upper limb and heart; Women with malnutrition; Women undergoing radiation or chemotherapy for cancer were excluded from experimental research design was planned to study the effectiveness of ergonomic advice and exercises in homemakers with pain in CANS (Cervical, Arm, Neck Shoulder) regions. The subjects were selected based on the inclusion criteria and divided into 4 groups and were educated through didactic lecture and counselling [9]. The contents of the counselling were described in detail below. A Questionnaire was prepared and given to them.

Subjective measurement of pain was done using Visual Analog Scale (VAS) and Neck Disability Index (NDI) studies. Also appropriate physical therapy exercises for pain control were suggested. Two week window period was given to the patients for regular practise and follow up. A follow up session was conducted after 15 days of first counselling and the differences in the Pre and Post Therapy sessions were noted and statistically analysed.

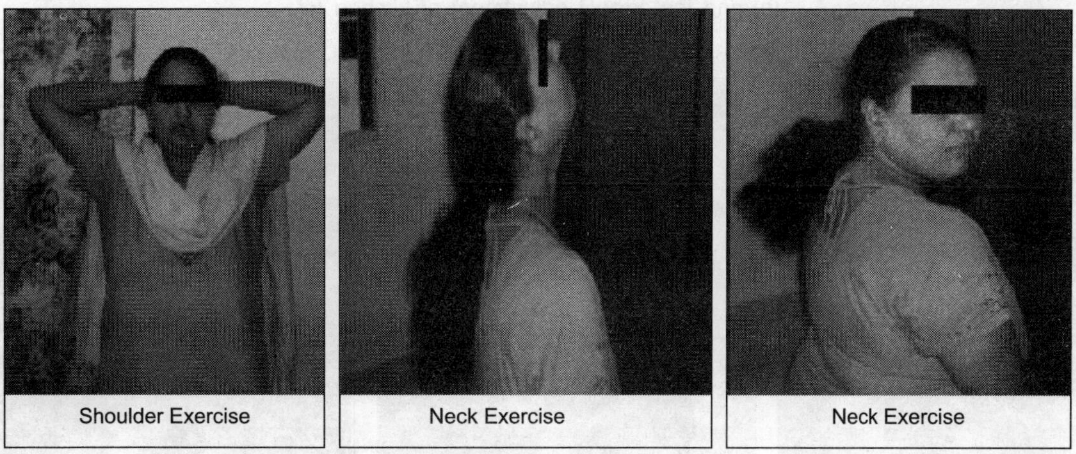

Fig. 15.1 Subject performing shoulder and neck exercises

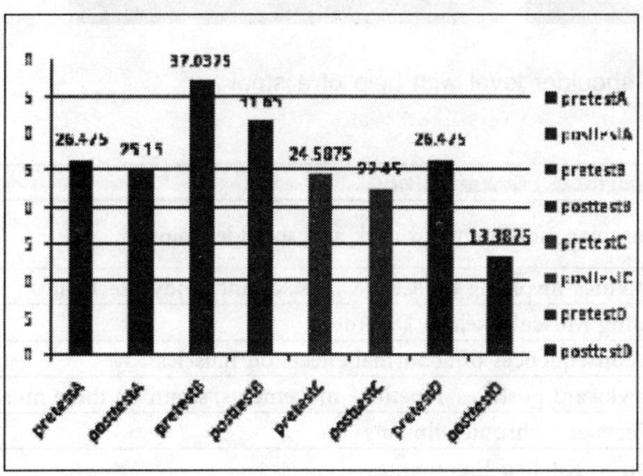

Fig. 15.2 Mean pre test and post test scores

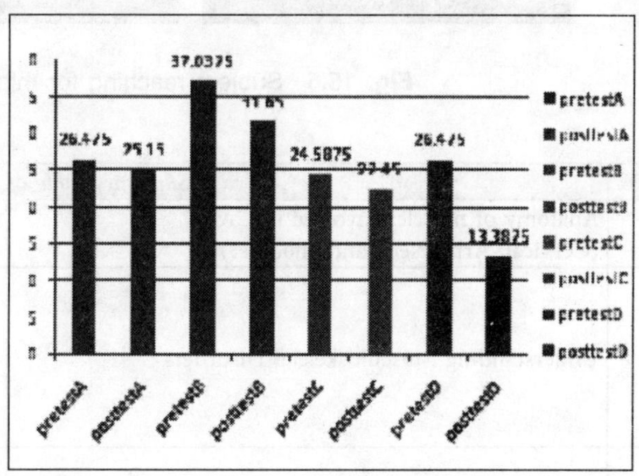

Fig. 15.3 Mean pre test and post test scores of group A-to-D on VAS scale each group on NDI scale

(a) (b) (d)

Fig. 15.4 Subject demonstrating in Home kitchen (a), Upper back exercise (b) and the correct adjustment of kitchen (c)

Fig. 15.5 Subject reaching for things above shoulder level with help of a stool

S. No.	*Summary table of ergonomic advice & exercise activity*	
1.	Anatomy of muscles involved in CANS (Cervical, Arm, Neck and Shoulder)	Normal muscular movements in neck and shoulder joints
2.	Understanding Musculoskeletal Disorders	Normal activities involving the CANS muscles and Body Mechanics
		Understanding Musculoskeletal Disorders
		Cause and consequences of abnormal effects on muscles
		Impact of awkward postures, repetitive movements, strain on these muscles
		Adverse effects and chronic ailments
3.	Concept of Ergonomics	Introduction to Kitchen Ergonomics
		Concept of Kitchen Design and functional layout of kitchen storage
		Kitchen work triangle, Work surface dimensions
4.	Ergonomic Intervention in homemakers	Functional Kitchen Plan and Storage
		Environmental Conditions
		Working Postures in Kitchen
		Proper use of body muscles in different postures (Standing, Bending etc)
5.	Physical Therapy Exercises and Recommendations	Simple neck exercises to relieve pain
		Simple upper back exercises to relieve pain

15.3 RESULTS AND DISCUSSION

The subjective measurement of pain before and after treatment was compiled as shown below and analysed using standard statistics methods. The participants were clustered in to 4 groups, Group A (Control Group who were not monitored with ergonomics and exercise therapy); Group B (Exercise Group who were recommended exercises for the pain reduction but no ergonomic advise was given); Group C (Ergonomic Group participated in the kitchen making advise and were involved in ergonomic intervention, were not suggested of any physical therapy exercises for pain management); and Group D (subjects received both ergonomic intervention program and exercise therapy for the recommended course of time). The mean difference in the pain reduction was derived from the variation in the pain levels after the window period. The results indicate that the levels in pain reduced notably with exercise activities, but were very significant when combined with ergonomic advice.

15.4 CONCLUSION

The results infer that ergonomic intervention together with proper physical therapy exercises plays pivotal function in treatment of musculoskeletal disorders with respect to CANS muscles. The present study results are limited to a window period of 15 days giving rise to hope that with much longer treatment sessions shall be an excellent support therapy.

REFERENCES

1. Joan S. Ward (1970) Ergonomics in the home. Applied Ergonomics,. 1: 223-227.
2. E. Grandjean (1970) Ergonomics of the home. pp. 344. Applied Ergonomics, 1: 223-227.
3. Tiina Pohjonen, Anne Punakallio, Veikko Louhevaara. (1998) Participatory ergonomics for reducing load and strain in home care work. Intern J Indus Erg, 21: 345-352.
4. R.H. Westgaard. (2000) Work-related musculoskeletal complaints: some ergonomics challenges upon the start of a new century. Applied Ergonomics 31: 569–580.
5. Strazdins L., Bammer G. (2003) Women, work and musculoskeletal health. 2003. Elsevier Science Ltd.
6. Gangopadhyay S; Bandyopadhyay T. (1999) Ergonomics study on analysis of normal activities of Indian women in kitchen. Biomedicine. 19: 123-8.
7. Kenneth N.K. Fong A.E. Clara Y.T. Law. (2008) Self-Perceived Musculoskeletal Complaints: Relationship to Time: Use in Women Homemakers in Hong Kong. J Occup Rehabil 18: 273–281.
8. Michael Feuerstein, Cherise B. Harrington. (2006) Secondary prevention of work-related upper extremity disorders: recommendations from the annapolis conference. J Occup Rehab, 16: 401–409.
9. Marut, Michele; Hedge Alan. (2000) Ergonomic Survey of Household Tasks and Products. Human Factors and Ergonomics Society Ann. M. Proc. Cons Products, Human Factors and Ergonomics Society, pp. 506-510.
10. Fadi A. Fathallah, Rima R. Habib. (2010) Full-Time Homemakers: Workers Who Cannot Go Home and Relax, Intern J Occupational Safety and Ergonomics. 16: 113–128.
11. Lyndall Strazdins, Gabriele Bammer. (2004) Women, work and musculoskeletal health. Social Science & Medicine 58: 997–1005.
12. Bernacki E.J., Guidera J.A., Schaefer J.A., Eal. (1999) An Ergonomics program designed to reduce the incidence of upper extremity work related Musculoskeletal disorders. J Occup Environ Med. 41: 1021-1172.

16

Modulatory Effects of Lead on Morphology, Mineralogy and Mechanical Properties in Weight Bearing Bones of Rat

D. V. Rai, Jayanand Manjhi, Payal Bhardwaj, Harkiran Kaur and R. K. Saxena

Centre for Biomedical Engineering, Shobhit University, Meerut, Uttar Pradesh

Panjab University, Chandigarh–160014, India

Email: dv.rai@shobhituniversity.ac.in

ABSTRACT

Lead is the most commonly used metal and therefore commonly encountered in the environment. Its continuous release into the environment can cause serious threat to human health. Lead has long been known to alter the hematological system but there is a lacuna in the study regarding skeletal system. So the study was aimed at assessing the effect of lead (Pb) intoxication on biophysical changes of younger rat bone (30 days, group I) and older rat bone (80 days, group II). Forty wistar strain male rats were divided into two age groups, i.e. 30days and 80 days. Each group has been subdivided into 2 groups, i.e. control and lead treated rats. The animals of experimental subgroup were supplemented with lead in drinking water at 250 mg/l for five weeks and provided ad libitum, to produce a subclinical toxicity. Lead treatment resulted in significant lower femoral head width as compared to the control ones. 34.91% (Lead I) and 53.71% (Lead II) decrease was found in the cortical thickness in the lead treated groups as compared to their respective controls. Ash content analysis showed a decrease in mineral content in the lead treated groups. Biomechanical measurements were performed on the lead and control groups. A significant decrease ($p<0.001$) in hardness, breaking strength, toughness and brittleness of tibia was observed in all lead groups as compared to their respective controls. The lead triggers weakening of bone mechanical properties due to geometry, cortical thickness as well as mineral loss. The results of the study infer that administration of lead affects biomechanical properties thus reducing the strength related properties of long bones.

Keywords: Lead intoxication, mineral content, breaking strength, toughness and brittleness.

16.1 INTRODUCTION

Lead from years of use in paint, gas, and industrial sources can be found in soil and dust. Today, little lead is used in the production of gas and paint. Because it does not decay or biodegrade, lead in soil and dust stays around for many years. The U. S. Environmental Protection Agency (EPA) has set a drinking water limit for lead at 15 ug/l (Andrews, S. L., 1992). Bone is a dynamic organ involved in the architectural plan of human body for mechanical behaviour and mineral metabolism. Lead modulates the antioxidant defence system. The mechanical function provides the structural framework for the organism that permits support, locomotion, and protection of organs. It undergoes constant growth, resorption, and remodeling. Specialized cells called osteoblasts initiate bone formation, osteocytes monitor bone mechanical stresses, and osteoclasts resorb bone. Bone density is determined by the dynamic balance between bone formation and bone resorption. Its formation starts with the deposition of organic matrix by osteoblasts followed by the process of mineralization [Rai, D.V., et al., 1986]. The mechanical properties of bone result from the combined properties of the components of its extra cellular matrix, which is composed of organic osteoid (primarily Collagen type I) and the mineral phase, largely in the form of a poorly crystalline hydroxyapatite (HA). The concentrations of ions of mineral phase like calcium, phosphorous in plasma and extracellular fluid determines the rate at which hydroxyapatite is formed and deposited. Below a critical concentration of these ions, mineralization of organic matrix does not occur at normal rate. Therefore, knowledge of mechanical properties of collagenous component of bone is required for composite modeling of bone tissue and for understanding age-related reduction in ductility and strength of bone [Catanese, J., 1999]. Mechanical strength of bone tissue depends mainly on the nanosize apatite crystals. It is generally considered that the mineral resists compression and the collagen fibres withstand torsion and tension. The orientation of these fibres seems to be load related [Singh, S., et al, 1981]. However, it is considered that the most important determinant of mechanical properties of bones is the determination of its strength which depends on the mineral content. The relationship between the mineral content and strength of the bone has been studied extensively [Vose, G. P., Kubala, A. L., 1959; Marcus, R., Wong, M., 2002.].

Parameters that have been investigated as determinants of the mechanical properties of cortical bone are apparent density (proportional to porosity), ash density (total mineral divided by bulk volume), number of osteons, collagen composition and content, orientation of collagen fibres and mineral, composition of cement lines, bonding between mineral and collagen phase, and accumulation of micro-cracks in the bone matrix and around osteons [Schaffler, M.B., Burr, D.B., 1988; Martin, R.B., Ishida, J., 1989]. The most important determinant of modulus and strength is the ash or mineral content. Recent evidences also suggest that bone mineral density is correlated with bone strength and fracture risk [Veenland, J. F., et al., 1997]. Mechanical properties of cortical bone are also governed by water content of bone. With ageing, the amount of water in bone tissue decreases while the amount in the pores increases. The rate of bone loss varies from bone to bone, with sex, age, race, nutritional conditions [Turner, C. H., 1993] and with environmental conditions. Hardness and toughness had been found to increase till a peak value is reached, and after that it decreases with increasing age [Ding, M., 2000].

Longitudinal bone growth and bone strength both tends to decrease in lead toxicity [Ronis, M. J., et al., 2001]. Decrease in bone strength is due to the effect of lead on mineral bone matrix, where it displaces calcium [Pounds, J. G., Rosen, J. F., 1986]. Lead toxicity has direct effects on the osteoblast function including inhibition of the vitamin D3-stimulated synthesis of osteocalcin, a major non-collagenous protein required for mineralization [Hicks, D. G., et al, 1995]. It also interferes with calcium homeostasis and calcium regulated secondary messenger system in bone cells and induces inhibition of growth plate development [González-Riola, J., et al., 1997] that has been reported earlier. Although the effects of lead toxicity do not start early in adults but with increase in age, in pathological conditions like osteoporosis, osteopenia; lead starts getting mobilized from bones and because of high turnover its concentration get on increasing in plasma, affecting almost every organ. All these mechanisms of action of lead toxicity on bone matrix and cells indicate its ability to disrupt the mineralization process or normal growth.

In order to understand the bone as multiphase composite material, its physical parameters such as femoral head width, cortical thickness, ash content and the mechanical parameters such as bone strength, toughness, brittleness and hardness which gives overall information about the bone growth under the influence of lead toxicity are investigated. Thus, a correlation has been between the mechanical properties and morphometric parameters of bone.

16.2 MATERIALS AND METHODS

Chemicals Used: Lead acetate (~99% pure) was purchased from Qualigens Fine Chemicals: A division of Glaxo Smith Kline Pharmaceuticals Limited, Mumbai, India.

Animal Model and Experimental Conditions: To carry out the present investigation, male Wistar rats were procured from the Central Animal House of Panjab University, Chandigarh and were housed in polypropylene cages bedded with sterilized rice husk. They were given free access to clean drinking water (tap water) and standard animal pellet diet (Ashirwad Industries, Kharar, Punjab, India), throughout the experiment. The temperature of the animal room was maintained at 21±1°C, humidity 50-60% and a 12 h dark and light cycle. Animals were acclimatized for one week to laboratory conditions before the start of the experiment. The experimental protocols were approved by the Institutional Ethics Committee and conducted according to Indian National Science Academy Guidelines for the use and care of experimental animals.

Experimental Design: Forty wistar strain male rats were taken and divided into two age groups (30 and 80 days). These were further subdivided into total of four groups, i.e. Control I (CI), Lead I, Control II (CII), Lead II. Lead (as acetate) was administered in drinking water at 250 mg/l for five weeks and provided ad libitum, to produce a subclinical toxicity. Glacial acetic acid was added to the drinking water solution of control and lead treated groups at a concentration of 12.5 μl/l to prevent the precipitation of lead acetate.

Femoral head width: At the end of study, rats from all the groups were sacrificed and their long bones, i.e. femur and tibia were excised, cleaned of soft tissue. Femoral head width was noted using a vernier caliper to the nearest 0.01 mm.

Histological studies: Tibia bones were rapidly cleared off of all the muscles, tissues and ligament attachments and were then immediately placed in 10% neutral buffered formalin for one day. Then samples were placed in 6-7% HNO_3 solution for a week with daily changes of the decalcifying solution and then embedded in wax [Pearse, A. G. E., 1968]. Transverse section of bones of 7 μm thickness were cut using manual hand driven microtome, then dewaxed in xylene, rehydrated in descending series of alcohol and stained with H/E. After dehydration in ascending series of alcohol, stained sections were mounted in DPX mountant and viewed under a light microscope for measurement of cortical thickness and photographed at 40X.

Ash content analysis: The ash content of control and treated bone were done to get the percentage of organic and inorganic matrix in bone. The samples were placed in a crucible and heated to 850°C for 24 hrs in a muffle furnace (Laboratory model, manufactured by Narang Scientific works Pvt Ltd, New Delhi). Taking the initial and the final weights of the samples, the percentage weight loss for each group gave the ash content.

Bone Mechanical Properties: Tibia bone samples cleaned of from soft tissue were taken and kept in normal saline at 4ºC for 24 hours. After this hardness, breaking strength, toughness and brittleness of the tibial bones were measured using Texture Analyzer (Model TA.HD. Stable Microsystems UK) using HDP/3PB Three Point Bend Rig. In this assembly, the bone samples were placed horizontally on the stationary probe (consisting of two plates which were separated by very few centimeters apart). Another probe which was movable (shape like a knife) accelerated in a constant velocity to the bone sample from upward to downward and broke the bone at mid diaphysis into two pieces. The instrument was attached to a computer system, which recorded breaking force, hardness, toughness and brittleness. The following settings were made for sample measurements: Pre test speed 1.0mm/sec, test speed 1.0 mm/sec, post-test speed 5.0 mm/sec, rupture test distance 4.0mm, force 0.98067 N, time 0.10 S and count 5. The force required to break the bone sample from the fixed distance is expressed as breaking strength and signifies the firmness of bone.

Statistical analysis: Statistics analysis of the data were performed by analysis of variances (one way ANOVA). Following one way ANOVA post Hoc test using least significance difference (LSD) and by student 't' test at $p = 0.05$ using SPSS statistical software data for individual parameters represents average value calculated from three parallels.

16.3 RESULTS AND DISCUSSION

Body weights in terms of weight gain and gain percentage (%) was calculated (Table 16.1). Decrease in femoral head width of lead treated groups with increase in age was observed (Table 16.2). The cortical thickness decreased significantly ($p < 0.001$) in the lead treated group as compared to the control group (Table 16.3). Ash analysis showed a decrease in mineral content in both the lead treated groups (Table 16.4). A significant decrease ($p < 0.001$) in hardness, breaking strength, toughness and brittleness of tibia was observed in all lead treated groups as compared to their respective controls (Fig. 16.1, Table 16.5).

Table 16.1: Body weight gain

Groups	*Week 0*	*Week 1*	*Week 2*	*Week 3*	*Week 5*	*Week 6*	*Weight gain*	*Gain %*
CI	64.99 ±0.9	78.93 ±0.7	87.12 ±1.5	101.13 ±0.96	113.3 ±1.7	127.00 ±1.6	62.01	95.41
Lead I	65.36 ±1.4	68.06 ±1.3[a]	76.06 ±1.3[a]	81.12 ±1.2[a]	90.75 ±1.9[a]	105.25 ±1.4[a]	38.89	61.03
CII	132.96 ±0.9	144.57 ±1.1	148.57 ±1.8	157.37 ±4.1	167.37 ±4.1	177.37 ±4.1	44.41	33.40
Lead II	132.79 ±1.6	138.87 ±1.3[a]	144.87 ±1.5[a]	151.00 ±2.9 b	160.42 ±3.9 b	167.00 ±3.8 a	34.21	25.76

Data: Mean ± Standard Deviation (N = 8)

Statistically significant differences are represented by:

(c) $p \leq 0.05$, (b) $p \leq 0.01$, (a) $p \leq 0.001$

Table 16.2: Femoral head width

Groups	Femoral head width (mm)	
CI	28.67±0.35	25.05± 0.53
Lead I	26.54±0.38***	23.02±0.12***
CII	31.78±0.17###	27.41±1.22###
Lead II	28.45±0.42***	25.81±0.24**

Data: Mean ± Standard Deviation (N = 10)

Statistically significant differences are represented by:

$p \leq 0.05$*, $p \leq 0.01$**, $p \leq 0.001$*** lead treated group w.r.t. control

$p \leq 0.05$ #, $p \leq 0.01$ ##, $p \leq 0.001$ ### CII w.r.t lead II

Table 16.3: Cortical thickness of bone

Sample	Cortical Thickness (μm)
CI	647.95±1.70
LeadI	480.29±1.61*** (-34.91%)
CII	677.93±2.02 ###
LeadII	441.04±1.62*** (-53.71%)

Values are mean ± S. D. (N = 10)

*** $p \leq 0.001$ Values in bracket are % change in lead treated groups w.r.t. control

$p \leq 0.001$ Values in bracket are % change in control groups with increase in age

Table 16.4: Ash content analysis

Groups	Dry weight (gms)	Ash wt (gms)	Total mineral content (%)	Organic content (%)
CI	0.2328	0.1198	51.45	48.55
Lead I	0.2023	0.0877	43.40*	56.57
CII	0.2483	0.1545	62.22	37.78
Lead II	0.2345	0.1272	54.25*	45.75

$p \leq 0.05$*, $p \leq 0.01$**, $p \leq 0.001$*** lead treated group w.r.t. control

Table 16.5: Mechanical parameters of bone

Groups	Hardness (N)	Rupture Energy (N.Sec)	Gradient (N/Sec)
CI	62.40+ 2.60	89.87+2.60	26.03+0.72
Lead I	33.66+0.70***	71.86+4.60***	11.82+0.33***
CII	66.08+0.51###	110.07+1.43###	27.76+0.95###
Lead II	35.13+0.52***	95.35+1.16***	22.50+0.91***

Values are mean ± S. D. (N = 10)

*** $p \leq 0.001$ Values in bracket are % change in lead treated groups w.r.t. control

$p \leq 0.001$ Values in bracket are % change in control groups with increase in age

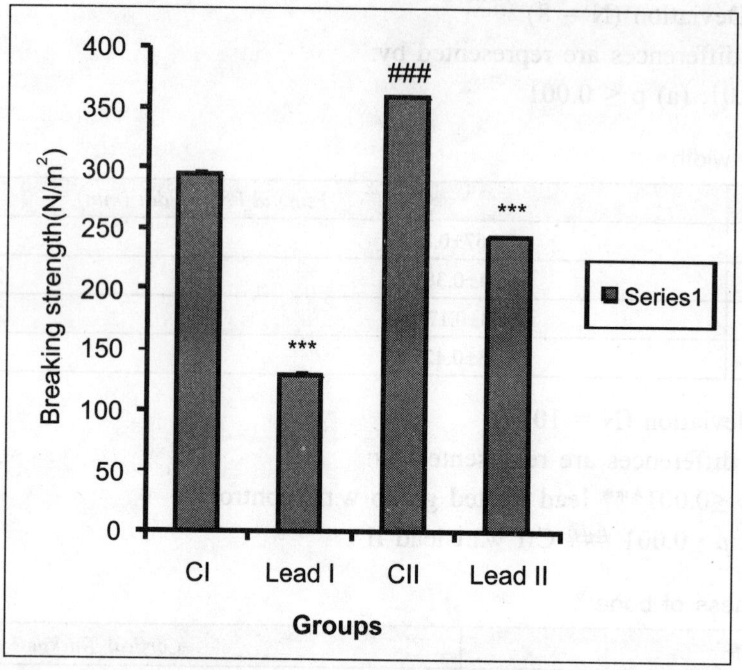

Fig. 16.1 Breaking strength parameter of bone

Values are mean ± S. D. (N = 10)

$p \leq 0.05$*, $p \leq 0.01$**, $p \leq 0.001$*** lead treated group w.r.t. control

$p \leq 0.05$ #, $p \leq 0.01$ ##, $p \leq 0.001$ ### CII w.r.t. lead II

In the present study evidence is provided to show that plumbism in rats induces changes in femoral head width, cortical thickness, ash content and mechanical strength. Lead is toxic metal that may interact metabolically with nutritionally essential metals. Lead interacts with calcium in the skeletal system to produce osteodystrophies and also replaces zinc on heme enzymes and various antioxidant enzymes. Thus lead directly acts on the bone cells, replacing zinc and influencing calcium homeostasis. Lead is a divalent action, and it binds strongly to sulfhydryl groups on proteins. Lead's toxicity can be attributed to distortion of enzymes and structural proteins. Many of lead's toxic properties are due to its ability to mimic or compete with calcium. At picomolar concentrations, lead competes successfully with calcium for binding sites on cerebellar phosphokinase C and thereby affects neuronal signaling [Markovac, J., Goldstein, G.W., 1988] and inhibits calcium entry into cells [Simons, T., 1993]. A study from our lab showing effects of lead on bone antioxidant defense system [Bhardwaj, P., et al., 2009].

It is suggested that lead treatment resulted in retardation of skeletal growth, as indicated by lower femoral head width, which is in conformity with the earlier reports [Campbell, J.R., et al., 2004]. Literature also supports that developing skeletal system has been found to be more sensitive to toxicity than that of adult bone. Femoral head width is observed to decrease from 7.20 mm to 6.76 mm and 7.27 mm to 6.69 mm in both the experimental groups respectively, i.e. Lead I and Lead II. The smaller femoral head width in lead exposed rats is an important finding in the present study, as the active growth plate is found in this region. This is consistent with data of previous experiments given by Jamieson [Jamieson, A. J., et al., 2006]. The cortical thickness decreased significantly ($p < 0.001$) in the lead treated group as compared to the control group (Table 16.2). Histomorphometric studies suggest the decrease in bone formation and thickness of growth cartilage due to lead toxicity in rats. These results demonstrate that the normal physiological functions of bone are being affected by lead treatment. These results are supported by earlier studies done on demineralized and pathological bone [Lewandrowski, K. U., et al., 1996]. The results are further supported by the studies which indicate bone loss occurence in both compact and trabecular bone regions in lead treated rats and humans [Gerhardsson, L., et al., 2005].

Ash content analysis showed the decrease in mineral content of 18.54% and 14.69% for Lead I and Lead II groups respectively when compared to their respective controls. A significant decrease ($p < 0.001$) in hardness, breaking strength, toughness and brittleness of tibia was observed in all lead treated groups as compared to their respective controls (fig II, table III). Till attainment of skeletal and sexual maturity bone strength increases but it starts decreasing after attainment of peak bone mass. Results show the same trend in mature group CII where breaking strength, hardness, brittleness and toughness are found to be significantly increased with respect to CI group. The decrease in breaking strength can be due to the replacement of Ca^{2+} by Pb^{2+} ions which affects the calcium metabolism and mineralization of bone. Lead has also been shown to inhibit vitamin D3- stimulated synthesis of osteocalcin, a major non-collagenous constituent of bone mineralization [Hicks, D.G., et al., 1995; Long, G.J., et al., 1991]. Therefore, it can be concluded that lead treatment affects breaking strength and hardness of bone by altering mineral matrix interactions. Toughness can be defined as the energy required to break bone with time. Energy absorbed is higher in control groups as compared to lead treated groups. Gradient or brittleness is the carrying capacity of force per unit second. Therefore, a loss of toughness reflect decrease in water bound to collagen due to mineralization, or increase in collagen cross-links that displaces water in the collagen [Nyman, J.S., et al., 2006]. In literature it is well supported that with increased mineralization, brittleness increases [Turner, C.H., 2001]. The important thing to be noted is that with lead treatment the time needed for breaking of bone gets decreased, which clearly depicts low strength and altered mineralization of bone Thus with lead treatment the homogeneity of the bone matrix is being altered because of mineral loss as lead replaces calcium in bone and make it weak. Therefore, it is concluded that lead intoxication impairs collagen and mineral phase interactions leading to bone loss. Plumbism is known to derange bone metabolism and causes osteoporosis. It impairs osteoblastic activity that results in reduced bone formation and mineralization [Rai, D. V., et al., 2008].

The results provide evidence that lead intoxication in rats induces profound changes in bone morphological and mechanical properties. Lead intoxication has also been related to osteoporosis, leading to decrease in bone mineral content, thereby, reducing the bone weight and resulting in inferior micro-architectural and mechanical properties of bone. Therefore in the present study, we intend to relate lead toxicity as a predisposing factor to bone loss which leads to the weakening of bone. This study will further help in ameliorating our understanding of action of lead toxicity on bone metabolism in different age groups by ultra structural, histological and mechanical studies.

REFERENCES

1. Andrews, S.L., 1992. Lead and our environment. Food science and human nutrition. Michigan state university.
2. Bhardwaj, P., Kaur, H., Rai, D.V., 2009. New insight into the effects of lead modulation on antioxidant defense mechanism and trace element concentration in rat bone Interdisc Toxicol Vol. 2(1): 18–23.
3. Campbell, J.R., Rosier, R.N., Novotny, L., Puzas, J.E., 2004. The association between environmental lead exposure and bone density in children. Environmental Health Perspective 112: 1200-1203.
4. Catanese, J., 3rd, Iverson, E.P., Ng RK, Keaveny, T.M., 1999. Heterogeneity of the mechanical properties of demineralized bone. J Biomech 32(12):1365-9.
5. Ding, M., 2000. Age variations in the properties of human tibial trabecular bone and cartilage. Acta Orthop Sc and Suppl 292: 1-45.
6. Gerhardsson, L., Akantis, A., Lundström, N.G., Nordberg, G.F., Schütz, A., Skerfving, S., 2005. Lead concentrations in cortical and trabecular bones in deceased smelter workers. J Trace Elem Med Bio 19(2-3): 209-15.
7. González-Riola, J., Hernández, E.R., Escribano, A., Revilla, M., Ca-Seco Villa, L.F., Rico, H., 1997. Effect of lead on bone and cartilage in sexually mature rats: A morphometric and histomorphometry study. Environ Res 74: 91–93.
8. Hicks, D.G., O' Keefe, R.J., Reynolds, K.J., Puzas, J.E., Rosier, R.N. 1995. Effects of lead on autocrine/paracrine regulation of growth plate chondrocytes. J Toxicol Appl Pharmacol 140: 164-172.
9. Jamieson, A.J., Taylor, C.G., Weiler, H.A., 2006. Marginal zinc deficiency exacerbates bone lead accumulation and high zinc attenuates lead accumulation at the expense of bone density in growing rats. Toxicol Sci 92(1): 286-294.
10. Lewandrowski, K.U., Venugopalan, V., Tomford, W.W., Schomaker, K.T., Mankin, H.J., Dentesch, T.F., 1996. Kinetics of Cortical Bone Demineralisation–A New Method For Modifying Cortical Bone Allografts. J Biomed Mater Res 31: 365-372.
11. Long, G.J., Pounds, J.G., Rosen, J.F., 1991. Lead intoxication alters basal and parathyroid hormone-related cellular calcium homeostasis inrat osteosarcoma (ROS 17/2.8) cells. Calcif. Tissue Int 50: 451-458.
12. Marcus, R., Wong, M., Heath, H., 3rd, Stock, J.L., 2002. Antiresorptive treatment of postmenopausal osteoporosis: comparison of study designs and outcomes in large in clinical trials with fracture as an end point. Endrocr Rev 23: 16-37.
13. Markovac, J., Goldstein, G.W., 1988. Picomolar concentrations of lead stimulate brain protein kinase C. Nature 334: 71–73.
14. Martin, R.B., Ishida, J., 1989. The relative effects of collagen fiber orientation, porosity, density and mineralization on bone strength. J Biomech 22: 419-426.

15. Nyman, J.S., Roy, A., Shen, X., Acuna, R.L., Tyler, J.H., Wang, X., 2006. The influence of water removal on the strength and toughness of cortical bone. J Biomech 39(5): 931-8.
16. Pearse, A.G.E., 1968. Histochemistry, theoretical and applied, (J. and A., Churchill, London).
17. Pounds, J.G., Rosen, J.F., 1986. Cellular metabolism of lead: A kinetic analysis in cultured osteoclastic bone cells. Toxic Appl Pharmacol 83: 531-545.
18. Rai, D.V., Kumar, G., Tewari, P., Saxena, D.C., 2008. Acute and chronic dose of alcohol affect the load carrying capacity of long bone in rats. Journal of Biomechanics 4120-24.
19. Rai, D.V., Behari, J., Saha, S., 1986. The effect of mineral deficient diet on the structural and mechanical properties of long bones. Biomedical Engineering V Recent Development (Ed.) by Saha S, Pergmon Press NY. pp. 456-460.
20. Ronis, M.J., Aronson, J., Gao, G.G., Hougue, W., Skinner, R.A., Badger, T.M., Lumpkin, C.K., 2001. Skeletal effects of developmental lead exposure in rats. Toxicol Sci 62: 321-329.
21. Schaffler, M.B., Burr, D.B., 1988. Stiffness of compact bone: effects of porosity and density. J Biomechanics 21: 13-16.
22. Simons, T., 1993. Lead-calcium interactions in cellular lead toxicity. Neurotoxicology 14: 77–85.
23. Singh, S., Behari, J., Rai, D.V., 1981. Elastic Constants Relating to two Phase Bone System. Ultrason. Int June 30-June 2, Brighton, UK.
24. Turner, C.H., 1993. Age, bone material properties, and bone strength. Calcified Tissue International 53(1): S32-S33.
25. Turner, C.H., 2001. Biomechanics of Bone: Determinants of Skeletal Fragility and Bone Quality. Osteoporosis International 13(2): 97-104.
26. Veenland, J.F., Link, T.M., Konermann, W., Meier, N., Gelsema, E.S., 1997. Unraveling the role of structure and density in determining vertebral bone strength. Calcif Tissue Int 61: 474-479.
27. Vose, G.P., Kubala, A.L., 1959. Bone Strength-Its Relationship to X-Ray Determined Ash Content. Hum Biol 31: 262-270.

17

Solid Model Generation of Knee Joint and Alignment Analysis in Total Knee Arthroplasty (TKA) in Obese Patient: A Finite Element Study

Nishant Kumar Singh, Rati and Sanjay Kumar Rai

School of Biomedical Engineering, Indian Institute of Technology
Banaras Hindu University, Varanasi, India

Email: nishant.singh.bme09@iitbhu.ac.in

ABSTRACT

Three-dimensional FE model of the healthy human knee was reconstructed by CT-DICOM images with image processing algorithms that included the main structures of the knee joint bones. Total Knee Arthroplasty (TKA) is the resurfacing of the damaged surfaces of the knee and replacing the lost cartilage and diseased bone with metal and plastic components. Knees wear out for a variety of reasons, including inflammation from arthritis, injury, or simple wear and tear. Longevity of TKA depends to a greater extend on its alignment during surgery. The current alignment for the knee surfaces looks into the morphology of knee in average patients. In normal knee, the femur is orientated inwards and the alignment of the tibia is directed outwards to provide a foot alignment that is parallel to the ground, this alignment is known as Genu Valgum. In this case an even distribution of the subject's weight is inserted in the knee. For obese patients, the thighs are naturally abducted (directed outwards), which creates a case of Genu Varum. Even with this variation in the alignment of the femur and tibia, these patients receive the same alignment of average patients. With the help of finite element analysis on a knee joint we have regenerated the anatomical alignment of an obese patient and examined that there is linear increase in stress levels with an increase in varum alignment.

Keywords: Knee joint, finite element analysis, total knee arthroplasty, knee alignment, varus & valgus.

17.1 INTRODUCTION

The knee is one of the most complex and vastly load bearing joint of the human body [1]. With many pathological disorders and several factors like increasing age, injuries, waning of bone and muscles occurs. This tends to wear joint. This wear results in the development of Osteoarthritis which causes pain in the knee joint and cumulative failure of joint function [2]. Widely accepted solution of this problem is total knee replacement (TKR) or total knee arthroplasty (TKA) [3]. In this surgical procedure the damaged or osteoarthritic surfaces of bone are replaced with artificial knee component [4]. Alignment of femur and tibia plays important role in success and longevity of TKA. In average patient alignment of femur and tibia is in a manner such that the femur is directed inwards and the tibia is directed outwards [5] [6]. This alignment, which is also called as Genu Valgum, helps in a transfer of the weight of the body through the feet with balanced ground reaction forces [7].

In the case of obese patients however, their femur is oriented outwards and their tibia inwards, known as Genu Varum [8]. It has been shown that this abnormality affects the location of centre of pressure and the transfer of forces through the lower extremity limbs in a manner such that the patient will be able to keep full balance and have their feet parallel to the ground. It is therefore important to help achieve this alignment when TKA is performed, but it is not the case. TKA done for obese patients are fitted with TKA that has a Valgum alignment and designed for average patients in general practice. This improper alignment of femerotibial angle for obese patients tend to reduce the life of TKA [9].

This work is focused on generation of finite element model of knee joint in Simpleware ScanIP image processing software and simulating the knee joint with Genu Varum deformity with angles ranging from 9-17 degrees in ANSYS.14.5 and results will be compared from average subject data with Genu Valgum deformity. By performing these simulations a better perception will be achieved in viewing the stress levels and how they are distributed on the articulating surfaces. From these results it is expected that surgeons improve the methods in which they insert knee replacements for obese patients and for knee component design industry to optimize its design.

17.2 GEOMETRICAL MODELING OF THE KNEE JOINT

17.2.1 CT Data Analysis, Segmentation and Smoothing

CT scan images [10] of lower extremity of a subject (age 55 years, female) with healthy femur and tibia bones were acquired from Institute of Medical Sciences, BHU, Varanasi. The images acquired have an arterial phase and were obtained in helical mode from a 64 slice CT scanner, GE Medical System by aligning the lower extremity of the patient to the CT scanner rotating in clockwise direction, apparatus power condition were at 120 kV/350 mA. The images obtained were in DICOM (Digital Imaging and Communication in Medicine) format and were imported directly to trial licensed copy of Scan IP 6.0 for further 3D Image data visualization, solid model generation, analysis & meshing. After visualizing the images in the preview, a histogram [11] (Grayscale range -2000 to 2000) was evaluated and set to a customized range of window: 400, level: 40 (Fig. 17.1.).

The area of interest was cropped and resampled from entire stack of images in order to lower the memory requirements from 257.50 MB to 17.90 MB thereby making the segmentation and processing steps easier. Cropping reduces the total number of slices from 1030 to 510. However, there are a wide range of techniques through which segmentation can be performed [12], in this project we have applied two methods, firstly, by region growing [13] and secondly, applying manual segmentation on each slice. In order to clearly visualize the cortical and cancellous bone of both tibia and femur, each slice was segmented into two regions [14], namely a cortical region (segmented with a mask of red color) and a cancellous region (segmented with a mask of blue color at the distal end and green at the proximal end) (Fig. 17.2). The region growing was applied with number of iterations to be 3, multiplier value of 2.0 and initial neighborhood radius (pixels) of 2.

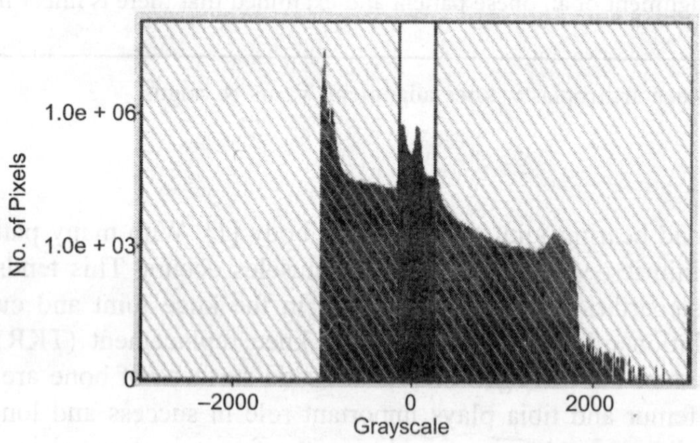

Fig. 17.1 Histogram of the acquired DICOM images

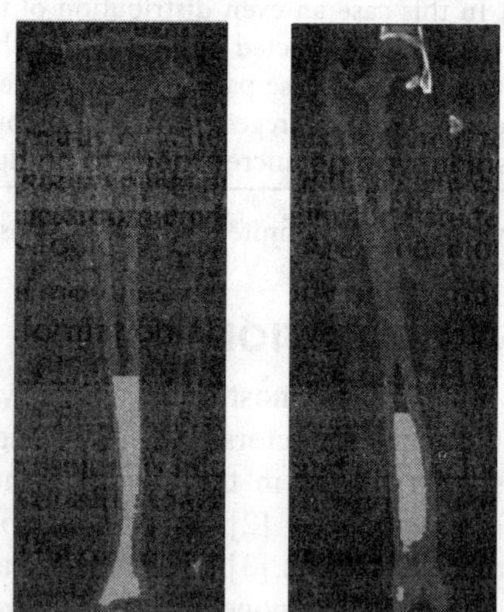

Fig. 17. 2 Coronal view of the segmented portion of tibia and femur

The purpose of applying region growing segmentation was that, though, it does not provide complete segmentation in one go but presents a clear view of parts near the complex regions of proximal and distal ends of tibia and femur with a new mask so that the complete region of interest can be visualized properly. This makes the next step of manual segmentation much easier.

Each slice was manually segmented using paint with threshold by adjusting the lower and upper threshold value at 40 and 255, respectively as shown in Fig. 17.3 (e, f, g, h). A 3D preview of the complete segmented portion of tibia and femur was then visualized to locate any holes or patches that might be present in the geometry. Smoothing of each mask was carried out using Recursive Gaussian filter which reduces noise and reduces detail levels [15]. It has a blurring effect on the active mask. Mathematically, it has the effect of low pass filtering the image. The application of Recursive Gaussian results in data loss of cortical region at proximal and distal portions of both tibia as well as femur (Fig. 17.3 (a, b, c, d)).

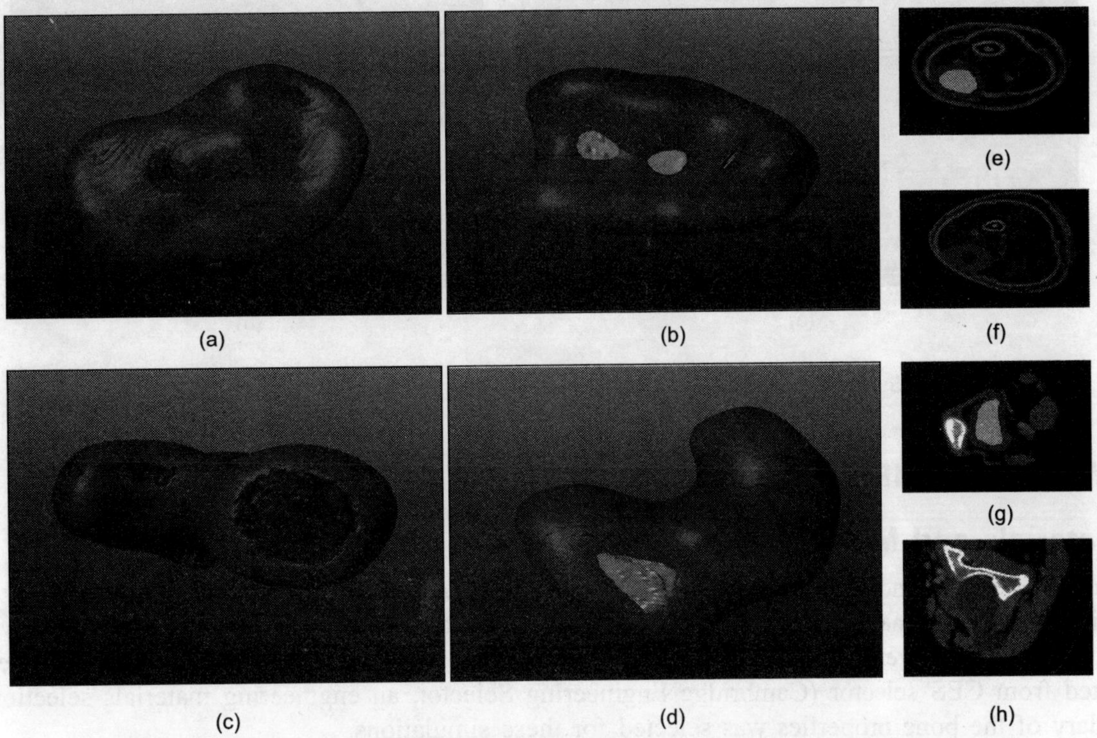

Fig. 17.3 Result of smoothing operation showing data loss in cortical regions at (a) proximal portion of tibia (b) distal portion of tibia (c) proximal portion of femur (d) distal portion of femur (e, f, g, h) Illustration of manual segmentation of tibia & femur

In view of the significance of data loss, morphological dilate filter of 1 (cubic values) was applied before performing smoothing operation so as to grow the mask of cortical region. Hence, Gaussian sigma of 2.5 (cubic values) was applied on cortical region mask and of 2.0 on the mask of cancellous region (Fig. 17.4).

Fig. 17.4 Image smoothing by applying recursive Gaussian filter on (a) tibia and (b) femur

17.2.2 Solid model generation

The processed region of interest was then converted to IGES format with NURBS (Non-Uniform Rational B-Splines) meshing (Fig. 17.5). However, the generated solid model of required region has number of irregular faces so, NURBS meshing is best suited regardless of tetrahedral meshing or other meshing technique, which was further imported into CAD packages [16] for assembling and regeneration of knee joint.

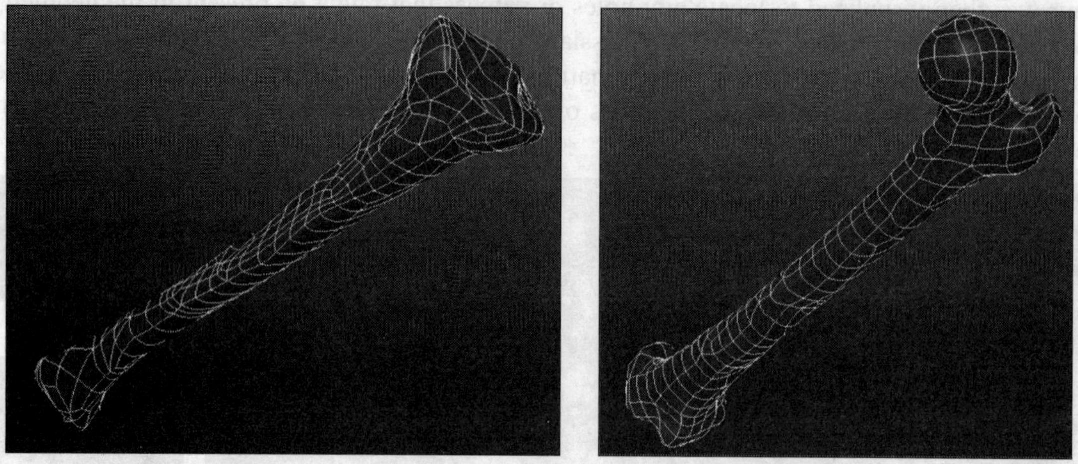

(a) (b)

Fig. 17.5 Solid model of (a) Tibia (b) Femur with NURBS mesh in IGES format

17.3 SIMULATION METHODOLOGY

17.3.1 Materials and Methods

3D solid model of femur and tibia was imported in CREO 2.0 to reconstruct knee joint in assembly (Fig. 17.6). Bone behavior was approximated by a homogeneous isotropic linear elastic material, distinguishing between cortical and cancellous bone. The average mechanical properties of each type of bone tissue are shown in Table 17.1 which were extracted from CES selector (Cambridge Engineering Selector, an engineering materials selection tool). The lower boundary of the bone properties was selected for these simulations.

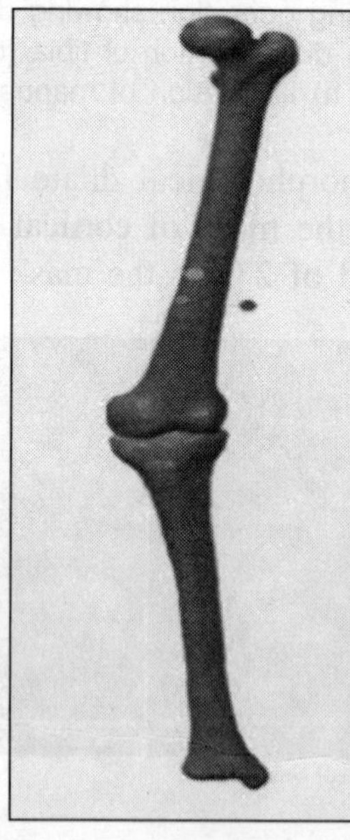

Figure 17.6 Illustration of the Knee joint constructed in CREO-2.0 in assembly mode

Table 17.1: Table of the material properties of bone used for the simulations (from CES selector).

Structural Properties of Bone	*Value*
Young Modulus	1.7e10 Pa
Poisson's Ratio	0.31
Density	1.8e3 kg/m^3
Thermal Expansion	1e-5/°C
Tensile Yield Strength (elastic limit)	1.2e8 Pa
Compressive Yiels Strength	1.14 e8 Pa
Bulk Modulus	* 1.8e10 Pa
Values marked * are estimates	

Table 17.2: A table of the statistics of the patients needed for FEA simulations

	Virus angle (degrees)			*BMI*
	Mean	Min	Max	
Non-Obese Patient	9	2	21	25
Obese Patient	14	2	25	34

Constructed knee joint was imported in Ansys Work bench 14.5 (FEA simulation software) as an STEP file where simulation was done.

A data set of patients was received (Table 17.2) from IMS BHU and values extracted from this data set were used for the simulations. Weight level used was set to 95 kg approximately equaling 930 N. This load was applied to the joint at the two following locations: 1) where the femur joins the hip assuming the direction of the force parallel to the direction of gravity. 2) at the ankle joint of the tibia assuming the feet are parallel to the ground hence a normal GRF (opposite to the direction of gravity). It is assumed that a subject is standing straight with one leg supporting the full weight of the body.

On the femerotibial structure of Fig. 17.1 the forces were applied at the medial point of the femur head socket and the middle of the tibia/ankle joint location. The coordinates of these points came out to be as such:Femur Load Location: (0.26, 0.32, 0.75) Tibia Load Location: (0.14, 0.32, -0.08).

Since rotation is about the Y-axis the rotation matrix is used in order to change this alignment to 9° varum it is necessary to rotate the femur 9° clockwise and the tibia 9° counterclockwise.

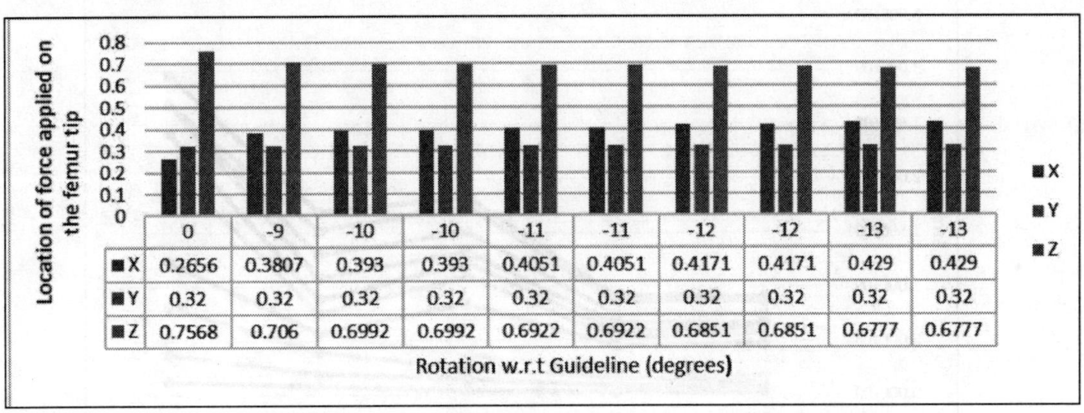

Fig. 17.7 Location of force applied on the femur tip for each different femorotibial alignment (Guideline to 17°)

$$R_y(\theta) = \begin{bmatrix} \cos\theta & 0 & -\sin\theta \\ 0 & 1 & 0 \\ \sin\theta & 0 & \cos\theta \end{bmatrix} \qquad \begin{bmatrix} X' \\ Y' \\ Z' \end{bmatrix} = \begin{bmatrix} \cos\theta & 0 & -\sin\theta \\ 0 & 1 & 0 \\ \sin\theta & 0 & \cos\theta \end{bmatrix} \begin{bmatrix} X \\ Y \\ Z \end{bmatrix}$$

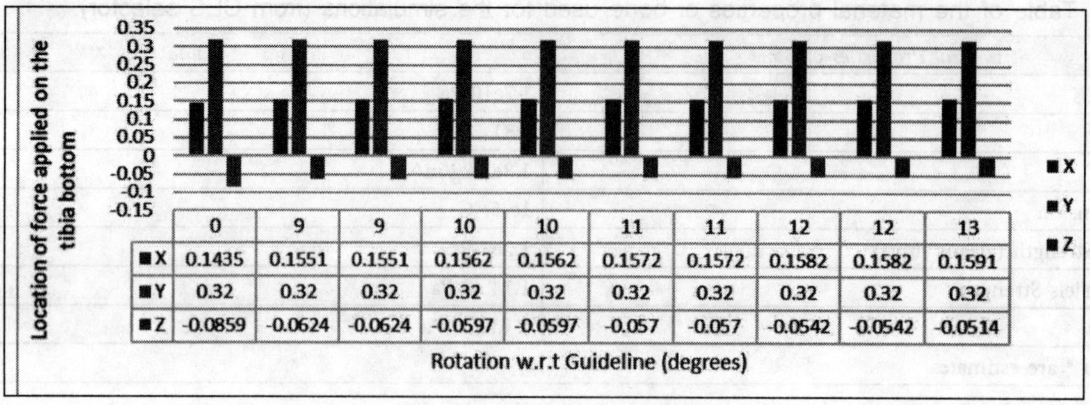

Fig. 17.8 Location of force applied on the tibia bottom for each different femorotibial alignment (Guideline to 17°)

where; $\begin{bmatrix} X \\ Y \\ Z \end{bmatrix}$ is the coordinates in valgum and θ the value of rotation is (counter clockwise as the positive direction of rotation). The calculated coordinates for each case is given below (Figs. 17.7 & 17.8).

17.3.2 Meshing and Convergence

For the meshing of the parts a simple triangular mesh was used with the elements sizes varying. The mesh size was decreased to finer levels starting from 0.005 m to 0.002 m. This refinement is important since it is essential in determining a convergent solution for stresses on the finite element bodies. By keeping all conditions uniform with each refinement of the mesh the values for stress should converge to a finite value.

17.4 RESULTS

At the end of each simulation a data file was extracted which held the values of Von-Mises Stress (Equivalent Stress) for each node. In the case of each femerotibial alignment there were 5 meshing refinements and each refinement encompassed 5 time steps. From the stress results for each meshing refinement the maximum, median and average Von-Mises stress was tabulated by averaging the maximum values of the 5 time steps.

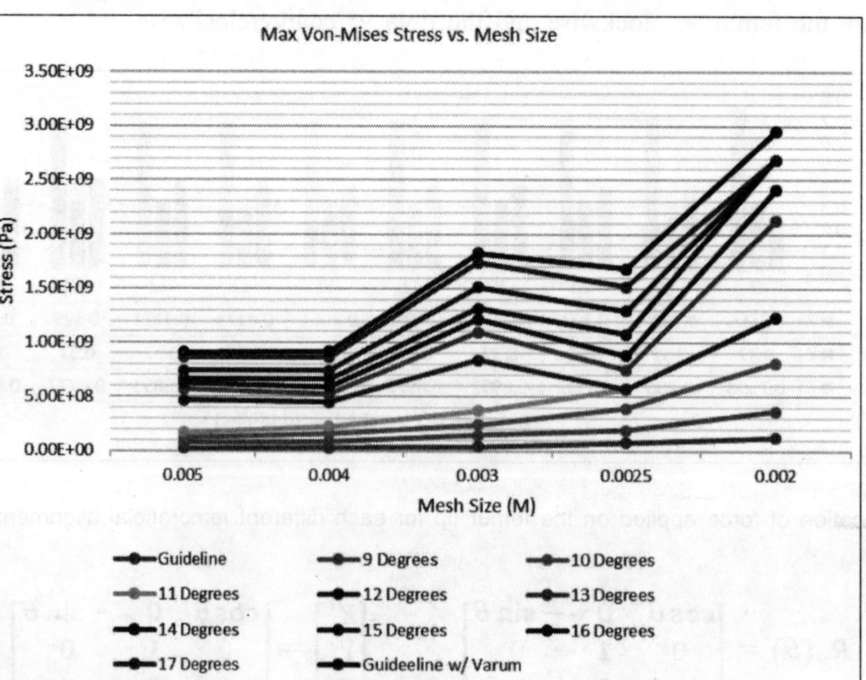

Fig. 17.9 A graph displaying the curves of maximum Von-Mises stress

In the first case the response in the knee joint for normal alignment is seen. In the second case the results encompass the Genu Varum deformity for angles ranging between 9°-17°. And for the final case the scenario was simulated for an obese patient where the alignment is 9° valgum. From Fig. 17.9 it is conclusive that with an increase in varum angle the maximum stress value increases.

17.5 DISCUSSION

From above results it was observed that average Von-Mises stress on the articulating surfaces increases linearly with the increase in varum femerotibial angle. In the case of obese patients it was also observed that by fitting them with a valgum alignment for their TKA average stress increases by 10% compared to a varum alignment with less uniform distribution of Von-Mises stress.

17.6 CONCLUSION

By successful FEA simulation in ANSYS 14.5 there is a very important conclusion which shows that valgum alignment could substantially decrease average stress values in the articulating surfaces when an obese patient is fitted with a valgum alignment for their TKA they will naturally tend to reverse this alignment and exert a varum alignment in their knee so for obese patient they must have a varum alignment to have their feet parallel to the ground. Therefore the fitted valgum alignment will be old-fashioned which supports the assumption made by us that TKA for obese patients with valgum alignment is a faulty practice.

REFERENCES

1. Walker, P.S. and Hajek, J.V. (1972). The load-bearing area in the knee joint. Journal of Biomechanics, 5(6), 581–584.
2. Ralphs, J.R. and Benjamin, M. (1994). The joint capsule: structure, composition, ageing and disease. J. Anat 1994, 184 (Pt 3), 503-509.
3. www.orthoinfo.aaos.org
4. www.aboutstryker.com/pdfs/knee_replacement.pdff
5. Siston, R.A.; Patel, J.J.; Goodman, S.B.; Delp, S.L. and Giori, N.J. (2005). The variability of femoral rotational alignment in total knee. J Bone Joint Surg Am, 87(10), 2276-2280.
6. Siston, R.A.; Patel, J.J.; Goodman, S.B.; Delp, S.L. and Giori, N.J. (2006). The high variability of tibial rotational alignment in total knee arthroplasty. Clinical Orthopaedics and Related Research 2006, 452, 65-69.
7. Cooke, T. D.V.; Sled, E.A. and Scudamore, R.A. (2007). Frontal plane knee alignment: a call for standardized measurement. The Journal of Rheumatology, 34(9), 1796-1801.
8. Lozano, L.M.; Núñez, M.; Sastre, S. and Popescu, D. (2012). Total knee arthroplasty in the context of severe and morbid obesity in adults. The open obesity journal, 4, 1-10.
9. Moreland, J.R. (1988). Mechanisms of failure of total knee arthroplasty. Clin Orthop, 226, 49-64.
10. Ribeiro, N.S.; Fernandes, P.C.; Lopes, D.S.; Folgado, J.O. and Fernandes, P.R. (2009). 3-D Solid and Finite Element Modeling of Biomechanical Structures, A software pipeline. 7th EUROMECH Solid Mechanics Conference, J. Ambrosio et.al. (eds.), Lisbon, Portugal.
11. Ramesh, N.; Yoo, J.H. and Sethi I.K. (1995). Thresholding based on histogram approximation. IEEE Proc Vision Image Signal Proc; 142, 271-9.
12. Sharma, N. and Aggarwal, L.M. (2009). Automated medical image segmentation techniques. Journal of Medical Physics, Vol. 35, No. 1, 3-14.
13. Fan, j.; Yau, D.K.Y.; Elmagarmid, A.K.; and Aref, W.G. (2001). Automatic Image Segmentation by Integrating Color-Edge Extraction and Seeded Region Growing. IEEE transactions on Image Processing, Vol. 10, No. 10.
14. Wang, L.I.; Greenspan, M.; and Ellis, R. (2006). Validation of bone segmentation and improved 3-d registration using contour coherency in CT data. IEEE Trans. on Medical Imaging, 25(3), 324-334.
15. Young, I.T. and Vliet, L.J.V. (1995). Recursive implementation of the Gaussian filter, Signal Processing. Vol. 44, pp. 139–151.
16. Starlya, B.; Laua, A.; Suna, W.; Laub, W.; Bradbury T. (2005). Direct slicing of STEP based NURBS models for layered manufacturing. J. Computer Aided Design, 37, 387-397.

18

Biomechanical Analysis of Successful and Unsuccessful Penalty Corner Push-in

S. Viswanath[1] and R. Kalidasan[2]

Dept. of Physical Education
Bharathidasan University, Tiruchirappalli, India

Email: viswabdu@gmail.com

ABSTRACT

The aim of this study was to investigate the differences in successful and unsuccessful Penalty corner push- in among, men university hockey players (n = 10), who were randomly selected from Trichy city colleges and their age ranged from 18 to 28 years. The high definition camera (DCR SR65) was used to record the performances that were approximately transverse to the trajectory of the ball of the push in. The collected data by using video camera were analysed through the "Max Traq" Software. Video footage was digitized for data analysis of ball speed and drag time. The data was analyzed by use of Paired sample t-test using SPSS. The finding reveals that ball speed and drag time differs significant between successful and unsuccessful push-in.

Keywords: Hockey, penalty corner, push-in, Max Traq software.

18.1 INTRODUCTION

Hockey is a popular sport played commonly in the artificial surface throughout the world. In modern hockey, there is increased in the number of set plays performed during the game[5]. The penalty corner is one of the most important tactical situations in field hockey[4,6]. The penalty corner was introduced in 1908. Penalty corner is awarded for foul committed by the defending team in its own entire 23 meters area. Depending upon the nature of foul penalty corner is awarded, with penalty corner one has greater scoring opportunities, because during penalty corner only five defenders will be permitted within the circle but all the attackers are permitted. The champions of today are seen perfect in the conversion of penalty corners. It is seen that different variations in penalty corner are being adopted and executed successfully. This requires lot of understanding among the specialized players[8].

According to Bhangu1 stated that highest 35 percentage of the goal were scored through the penalty corner by the Netherland team in Atlanta Olympics, which was the highest by a nation. In 1994, world cup the Australian team scored only 4 goals from 28 opportunities when compared to Dutch they have higher conversion of 7 goals form 17 opportunity[5]. Penalty corner execution can be separated into three progressive phases: the push-in, the trap and the strike. Kerr and Kevin[3] opined that all over the entire penalty corner takes about 1.9 seconds and 2.3 seconds respectively for male and female players of national standards of Australia. It is essential that it is performed precisely as it offers an excellent scoring opportunity during the game.

In penalty corner success rate largely depends on various factors such as speed, accuracy, power, technique and coordination during the execution. Push-in part in penalty corner plays critical role in conversion of penalty corner. If push-in is with great speed, the striker has extra time before defender reach the penalty circle. Push-in technique is briefly here under. The push starts with an attacker standing close to the goal line with at least one foot outside the field of play. The left shoulder points in the direction of the push. The hook of the stick rest against the ball. The push-in movements involves a rapid rotation of the hip, shoulders and arms in the direction of the trapper while the body weight is being transferred from the back foot to the front foot. The ball is dragged or pushed over the playing surface by the hockey stick for some distance and then released in the direction of the trapper. In the artificial surface dragging action is used frequently. The trap phase follows when the ball reaches top of the circles and is trapped by another attacking player just outside the circle the trapper propels the ball back into the circles for the phase three to commence, the phase three consists of a third attacker striking the moving ball towards the goal or another attacking player. Some researchers have focused on penalty corner push-in techniques in field hockey[3,8].

There is a paucity of research on the kinematics of the successful and unsuccessful penalty corner push-in execution, for this purpose successful execution of the push-in is defined as pushing the ball accurately with in the 0.50 meters line in front of the first post and the unsuccessful execution is defined as ball which deviate beyond the 0.50 meters and limited to 2 meters on either side is referred as a unsuccessful penalty corner push-in. The objective of the study is to document the ball speed and drag time in penalty corner push-in and also to compare the successful and unsuccessful penalty corner push-in. The purpose of the study was to investigate the differences in successful and unsuccessful penalty corner push-in.

18.2 METHODS

18.2.1 Subjects

Ten men university hockey players, who were randomly selected from Trichy city colleges and their age ranged from 18 to 28 years. The subjects had past playing experience of at least four years and further delimited to the players those who play forward position were only taken as subjects.

18.2.2 Data Acquisition

To acquire biomechanical data, with the assistance of technical person the high definition camera (DCR SR65) was used to capture movements of push-in in artificial surface. The camcorder mounted at the height of one meter, placed 6 meters away perpendicular to the trajectory of the ball of the push in. The shutter speed of the camcorder was adjusted at (1/1000 of a second) in order to eliminate the blurring effects while processing the recordings. A cage with the dimensions of 1.0 × 1.0 m at 4 control points was used to calibrate the space, in which the push-in was performed.

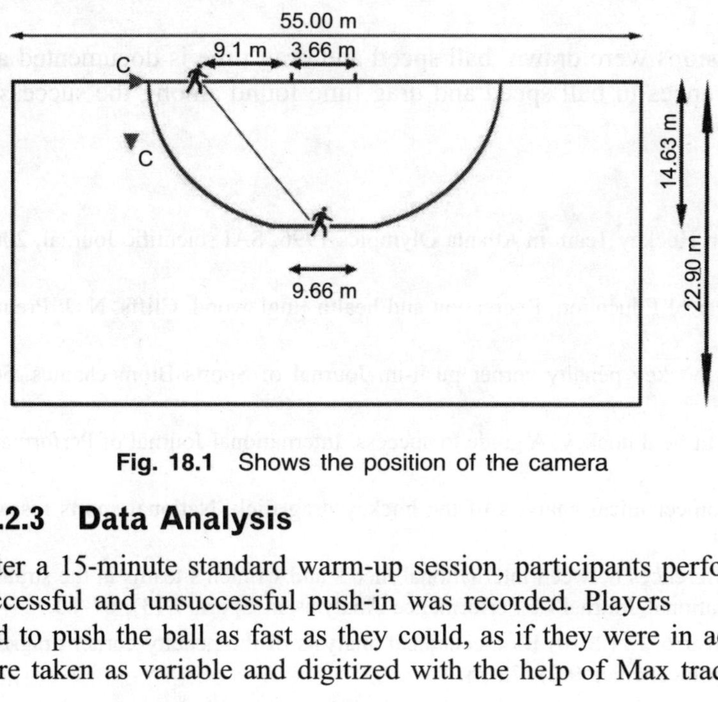

Fig. 18.1 Shows the position of the camera

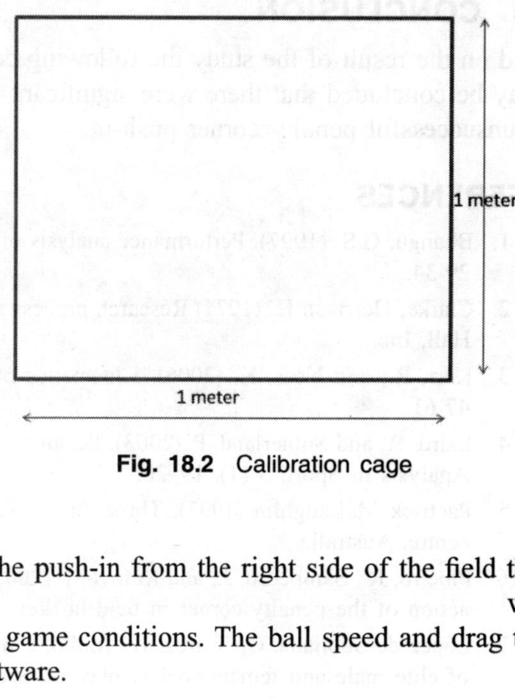

Fig. 18.2 Calibration cage

18.2.3 Data Analysis

After a 15-minute standard warm-up session, participants perform the push-in from the right side of the field three successful and unsuccessful push-in was recorded. Players were told to push the ball as fast as they could, as if they were in actual game conditions. The ball speed and drag time were taken as variable and digitized with the help of Max traq software.

18.2.4 Statistical Technique

Paired Sample t-test was used to compare the ball speed & drag time between successful and unsuccessful push-in was analysed by using SPSS version 15.0.1. The level of significance was fixed at 0.05.

18.3 RESULTS AND DISCUSSION

The ball speed & drag time changes in push-in are shown in Table 18.1. Paired sample t-test was used to test the differences of ball speed & drag time changes between successful and unsuccessful push-in of men hockey players. The results showed significant differences in successful and unsuccessful of penalty corner push-in. The data relevant to the ball speed indicates the significant difference between successful and unsuccessful push-in. The analysis on drag time revealed significant difference between successful and unsuccessful push-in.

Table 18.1: Successful and unsuccessful penalty corner push-in differences of ball speed & drag time

Ball Speed	*Mean*	*Standard Deviation*	*t*	*df*	*sig*
Successful	0.42	± 0.033	-2.751	9	.022*
Unsuccessful	0.48	± 0.056			
Drag time	*Mean*	*Standard Deviation*	*t*	*df*	*sig*
Successful	0.21	± 0.042	-3.127	9	.012*
Unsuccessful	0.25	± 0.024			

Ball speed and drag time in penalty corner push-in play an important role in the success of the penalty corner execution. During the push-in, the body should be moved in a well developed kinematic chain in order to generate the necessary power. Even small changes in the ball speed and drag time can cause the ball to deviate beyond the accurate area. A wrong technique may also play an important role in increasing the risk factors for the injury.

The results of the statistical analyses showed significant differences between the successful and unsuccessful penalty corner push-in. The findings were found to be consistent with the previous results of a study conducted by Kerr and Kevin 3 in which significant contribution was found for the ball speed and drag time in execution of penalty corner push-in. When interpreting the results and planning future research, it should be considered that the present study has certain limitations, like the limited number of the participants and the lack of some other kinematic variables.

18.4 CONCLUSION

Based on the result of the study the following conclusions were drawn, ball speed and drag time is documented and it may be concluded that there were significant differences in ball speed and drag time found among the successful and unsuccessful penalty corner push-in.

REFERENCES

1. Bhangu, G.S. (1997). Performance analysis of Indian Hockey Team in Atlanta Olympics 1996. SAI scientific Journal, 20(3), 29-34.
2. Clarke, Harrison H. (1974) Research process in Physical Education, Recreation and health Englewood. Cliffs, N. J: Prentice Hall, Inc.
3. Kerr, R., and Ness, K. (2006). Kinematics of field hockey penalty corner push-in. Journal of Sports Biomechanics, 5(1), 47-61.
4. Laird, P., and Sutherland, P. (2003). Penalty corners in field hockey: A guide to success. International Journal of Performance Analysis in Sport, 3 (1), 19–26.
5. Pactrick McLaughlin (1997). Three dimensional biomechanical analysis of the hockey drag flick. National sports research centre. Australia.
6. Pineiro, R., Sampedro, J., and Refoyo, I. (2007). Differences between international men's and women's teams in the strategic action of the penalty corner in field hockey. International Journal of Performance Analysis in Sport, 7 (3), 67–83.
7. López de Subijana, C., Juarez, D., Mallo, J. and Navarro, E. (2010) Biomechanical analysis of the penalty-corner drag-flick of elite male and female hockey players. Sports Biomechanics 9(2), 72-78.
8. Viswanath. S and Kalidasan. (2012) Selected biomechanical and kinanthropometrical factors in relation to penalty corner performance in field hockey. ISBS conference proceedings, Taiwan.
9. Wein, Horst. (1981). The advance science of hockey. London: Pelham Books Ltd. (p. 20).

19

Using CFD as a Tool for Analysis of Endodontic Irrigation: Maxillary and Mandibular Teeth Cases

Z. A. Quadri[1], M. F. Khan[1] and K. Zeya[2]

[1]*Dept. of Aerospace Engineering, University of Petroleum and Energy Studies, Dehradun, India*
[2]*Mansarovar Dental College, Bhopal, India*

Email: zeya.quadri@stu.upes.ac.in

ABSTRACT

This paper presents analysis of endodontic irrigation through a 30G needle in apical tooth canal using Computational Fluid Dynamics technique. Comparison of the detailed flow field resulting from CFD and PIV was performed to assess the validity of the CFD model. A simple root canal geometry resembling the prepared root canal of a central incisor was used with 30G irrigation needle. Maxillary and mandibular teeth cases were compared to analyze the effect of flow from needle. It has been exhibited that flow inside the upper teeth need increased inlet velocity or it need deeper insertion of needle.

Keywords: Computational fluid dynamics, endodontic irrigation, root canal therapy.

19.1 INTRODUCTION

Endodontic irrigation is used to eradicate microbes from the root-canal system and to prevent reinfection. The root canal is shaped with hand and rotary instruments under constant irrigation to remove the inflamed and necrotic tissue, microbes/biofilms, and other debris from the root-canal space. The hydrodynamics of the root canal are different than in macro environments. Computational fluid dynamics is an area of research usually associated with large-scale phenomena and related problems but fluid flow problems exist at the microscopic level of the root canal systems as well.

Computational fluid dynamics (CFD) is a new approach in endodontic research to improve our understanding of fluid dynamics in the special anatomic environment of the root canal [2]. Fluid flow is commonly studied in 1 of 3 ways: experimental fluid dynamics; theoretic fluid dynamics; and computational fluid dynamics. CFD gives an insight into flow patterns that are difficult, expensive or impossible to study using traditional (experimental) techniques. Computational Fluid Dynamics (CFD) provides a qualitative and sometimes even quantitative prediction of fluid flows. CFD enables scientists and engineers to perform 'numerical experiments in a 'virtual flow laboratory'. CFD can be used to evaluate and predict specific parameters, such as the streamline, velocity distribution of irrigant flow in the root canal, wall flow pressure, and wall shear stress on the root-canal wall, which are difficult to measure in vivo because of the microscopic size of the root canals.

The development of CFD as a tool in aerospace and automotive industry has been reached to a mature stage. Now CFD is achieving new high in the field of Biomedical Engineering. In recent years use of CFD as a tool in endodontic research is overwhelming. Boutsioukis C et al. [1] monitored ex vivo intra-canal irrigation with three endodontic needles (25, 27 and 30 gauge) and compared them in terms of irrigant flow rate, intra-barrel pressure, duration of irrigation and volume of irrigant delivered. Ertbruggen CV et al. [2] performed computational fluid dynamics (CFD) simulations in a 3D scaled up model of an alveolated bend with rigid walls that incorporated essential geometrical characteristics of human alveolar structures and compared numerical predictions with experimental flow measurements made in the same model by Particle Image Velocimetry (PIV). Boutsioukis C et al. [3] compared the results of a Computational Fluid Dynamics (CFD) simulation of the irrigant flow within a prepared root canal, during final irrigation with a syringe and a needle, with experimental high speed visualizations and theoretical calculations of an identical geometry and to evaluate the effect of off-centre positioning of the needle inside the root canal. Damir S et al. [4] analyzed the influence of the needle type, insertion depth, and irrigant flow rate on irrigant flow pattern, flow velocity, and apical pressure by ex-vivo based endodontic irrigation computational fluid dynamics (CFD) analysis. Li P et al. [5] proposed two novel needles based on a prototype by means of the dimple and protrusion, and flow patterns within a root canal during final irrigation with these needles were numerically investigated.

19.2 PROBLEM DEFINITION AND METHODOLOGY

Two different cases of apical root canal in maxillary and mandibular teeth are simulated and there flow behavior is compared. Velocity profile of irrigant in the root canal is compared with previous experimental result.

19.2.1 Geometry Creation

CAD modeling software Pro/Engineer Wildfire is being used for the creation of flow domain. The root canal was simulated as a geometrical frustum of a cone. The apical constriction and apical foramen were not simulated because they were not within the focus of this study. The apical terminus of the root canal was simulated as an impermeable wall. The length of the root canal model was 18 mm, the diameter was 1.57 mm at the canal orifice and 0.45 mm (ISO size 45) at the most apical point (6.2% taper) (Fig. 19.1).

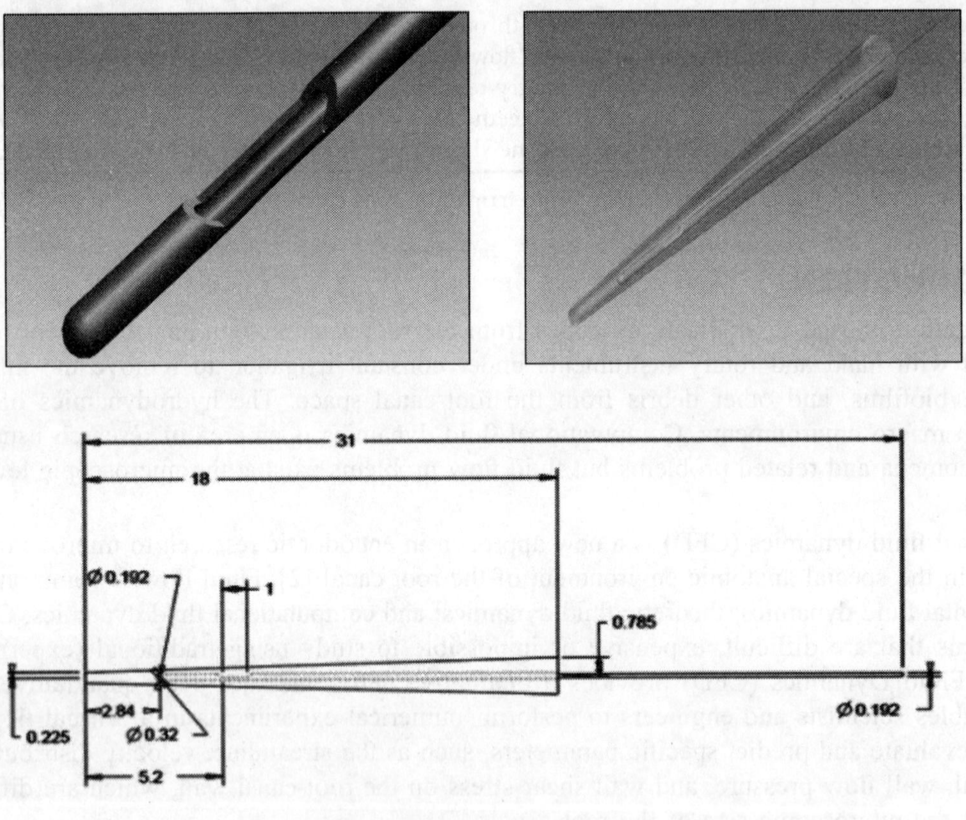

Fig. 19.1 3D CAD model

The needle was modeled using the 30G KerrHawe Irrigation Probe (KerrHawe SA, Bioggio, Switzerland) as a reference, similarly to a previous study (Boutsioukis et al. [10]). The external and internal diameter and the length were defined as Dext = 320 μm, Dint = 196 μm, l = 31 mm, respectively. The needle was fixed and centred within the canal, 3 mm short of the working length. This setup allows for the evaluation of irrigant flow apically to the needle tip, the most challenging area for irrigation.

19.2.2 Mesh Generation

CFD pre-processing software Ansys-ICEM is used to generate mesh. The inlet, outlet, and wall surface are defined. 623923 unstructured tetrahedral 3-D meshes were constructed (Fig. 19.2). The fluid flowed into the simulated domain through the needle inlet and out of the domain through the root canal orifice where atmospheric pressure was imposed. No. of prism layer was 3 with initial height of 0.001.

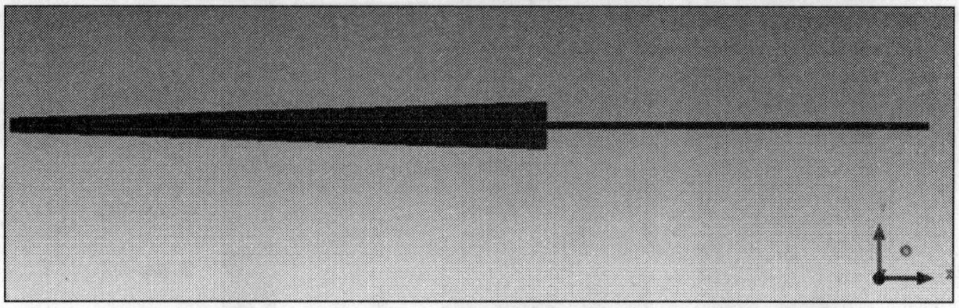

Fig. 19.2 Showing Mesh +Z View of cut section at Z = 0

19.2.3 Flow Parameters

No-slip boundary conditions were applied to the walls of the root canal and that of the needle, under the hypothesis of rigid, smooth and impermeable walls. Wall roughness is expected to have a limited effect on pressure drop as long as the flow remains laminar, but it may induce vortices or even turbulence in the flow. The fluid flowed into the simulated domain through the needle inlet and out of the domain through the root canal orifice, where atmospheric pressure was imposed. The canal and the needle were assumed to be completely filled with irrigant. Velocity inlet boundary condition was selected for the inlet of the needle. A flat velocity profile with a constant axial velocity of 8.6 m/s was imposed at the inlet, which is consistent with a clinically realistic irrigant flow rate of 0.26 mL/s through a 30G needle. The irrigant was distilled water, and it was modeled as an incompressible and homogenous Newtonian fluid with a density $\rho = 998.2\ kg/m^3$ and a constant viscosity $\mu = 1.0 \times 10^3$ kg/m-s. Gravity was included in the flow field in the direction of the negative Z-axis in first case and in positive Z-axis in second case.

The commercial CFD code Ansys-Fluent was used to set up and solve the problem and to analyses the results. The uncoupled Navier–Stokes equations which describe the time-dependent, three dimensional, incompressible flow were solved by an implicit iterative solver. Pressure based absolute velocity formulation was considered. An unsteady isothermal flow was assumed; steady-state solutions were obtained first, and then used as initial conditions for the unsteady simulations, to avoid transient effects. No turbulence model was used, as the flow under these conditions (Reynolds number Re = 1678 inside the needle lumen) was expected to be laminar (Boutsioukis et al. [10]). All transport equations were discretized to be at least second-order accurate. A time-step of 0.001 s was used throughout the calculations, which were carried out for a real flow time of 0.5 sec. The convergence criterion of the maximum scaled residuals was set at 10-4.

19.3 RESULT AND DISCUSSION

19.3.1 Validation of CFD Result

Calculation case 1 is selected for CFD validation. A good agreement between computational fluid dynamics and particle image velocimetry (PIV) results is evident in most of the domain studied. Figure 19.3 represents comparison between PIV result and CFD result for velocity. PIV results close to the wall of the root canal (dark blue stripes) are not considered accurate due to inherent limitations of the PIV method. The needle wall is coloured black in PIV result. The main characteristic bands occurred in regions where the velocity was high. The computational process adopted in this study was completely consistent with that proposed by Boutsioukis et al. [1].

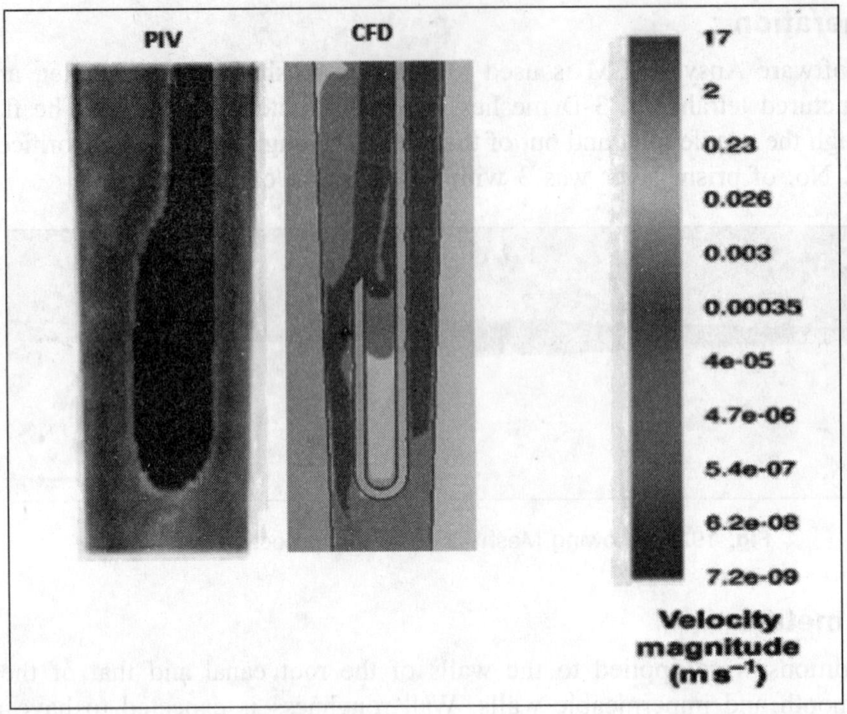

Fig. 19.3 Comparison of PIV and CFD result

19.3.2 Time Dependent Unsteady Result for Mandibular Teeth

For close resemble real condition the unsteady simulation was carried out and time dependent result was extracted. The flow parameters were averaged over a time period of 50 ms to distinguish the most stable characteristics of the flow, which are relevant for replacement of irrigant, thereby probably omitting transient flow characteristics.

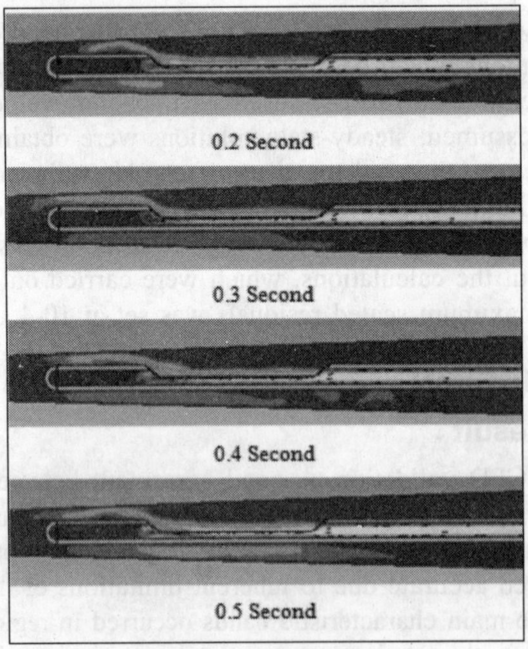

Fig. 19.4 Time dependent velocity contour

19.3.3 Comparison of Streamline for Velocity at the Needle Tip for Maxillary and mandibular teeth cases

Distance between the needle tip and end of apical root canal was taken as 3 mm. Direction of gravity was opposite for maxillary and mandibular teeth cases. Significant effect of gravity was captured during streamline representation. For upper teeth needle should be inserted at more depth for proper irrigation at end or velocity should be increased.

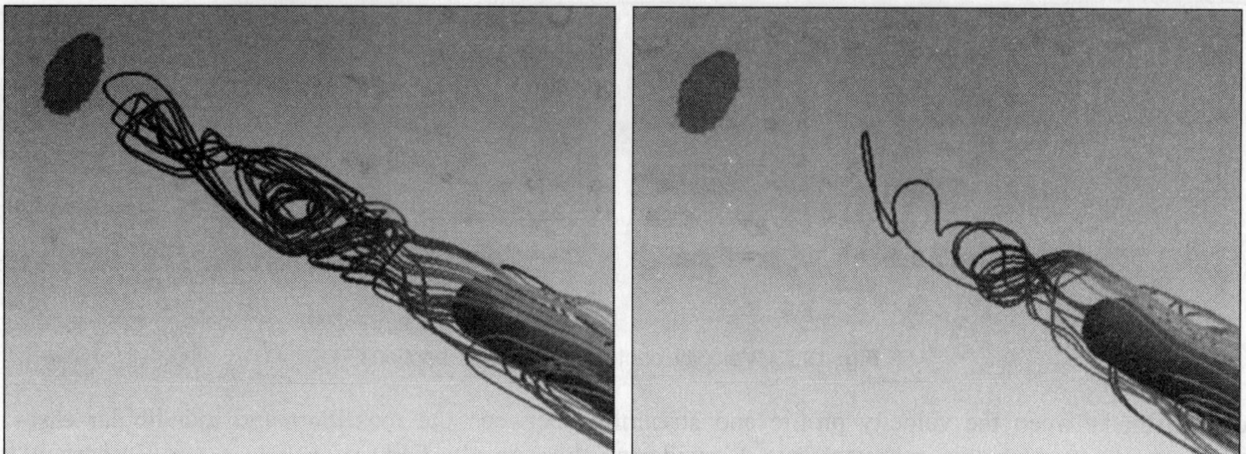

Fig. 19.5 Comparison of stream line for maxillary and mandibular case

19.3.4 Comparison of Wall Shear Stress

Wall Shear Stress (WSS) contour was presented at the canal wall. At present date no research work is there which can quantify value of wall shear stress suitable for endodontic irrigation. But it is basic assumption that wall shear stress has some role in irrigation. Comparison of WSS is presented in Fig. 19.6. Irrigant flowing out of the needle is producing high wall shear stress in maxillary teeth than mandibular teeth.

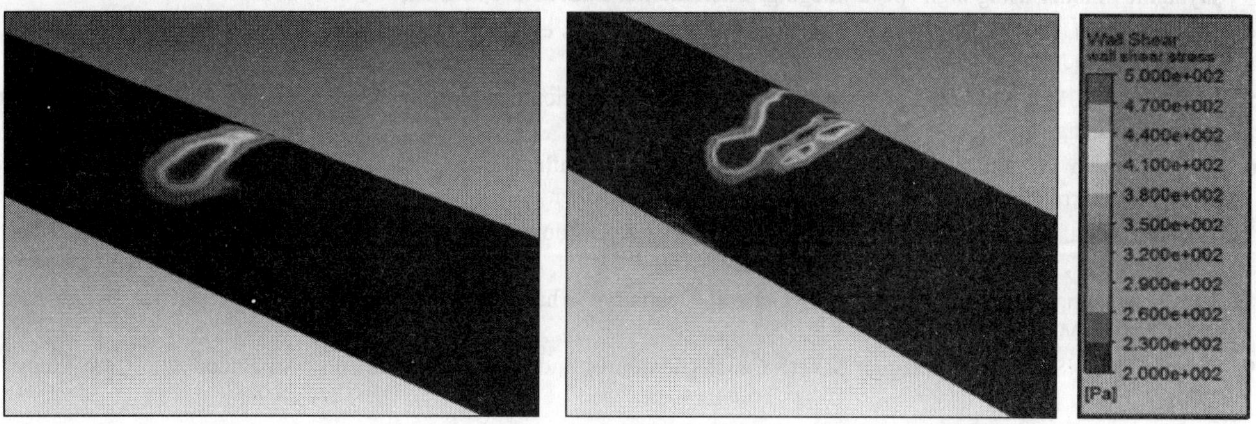

Fig. 19.6 Comparison of Wall shear stress contour at wall near needle outlet

19.3.5 Comparison of Velocity Contours at Different Sections Along Length

Figure 19.7 shows velocity contour at different sections along length for lower teeth. Different design of needle will have different effect. CFD can be a powerful tool to analyze the complex behavior of irrigant inside the root canal.

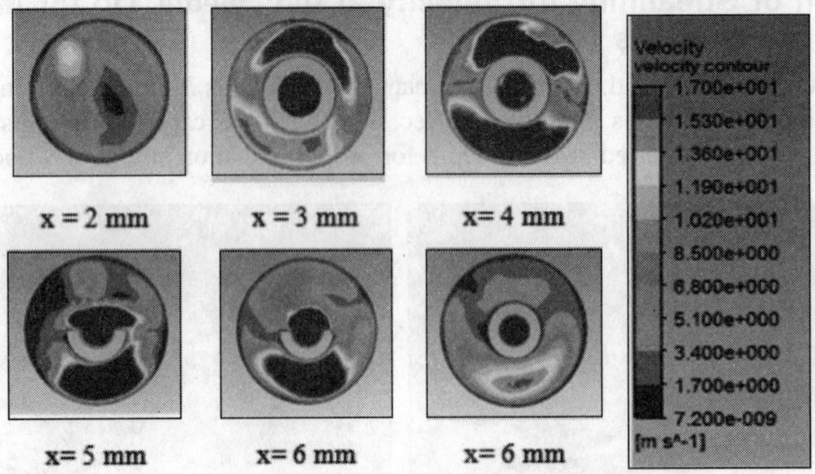

Fig. 19.7 Velocity contour at different sections

Comparison between the velocity profile and streamline between the maxillary and mandibular cases shows that effect of gravity plays an important role. Considering these results CFD proved to a very important tool for analysing flow behaviour inside the root canal and c\an be used as an important research tool in future for endodontic irrigation.

REFERENCES

1. Boutsioukis C. et al. (2007); Measurement of pressure and flow rates during irrigation of a root canal ex vivo with three endodontic needles, International Endodontic Journal, 40, 504–513.
2. Ertbruggen C.V. et al. (2010); validation of CFD predictions of flow in a 3d alveolated bend with experimental data, J Biomech.; 41(2): 399–405.
3. Boutsioukis C. et al. (2010); Irrigant flow in the root canal: experimental validation of an unsteady Computational Fluid Dynamics model using high-speed imaging, International Endodontic Journal, 43, 393–403.
4. Damir S. et al. (2012); Irrigation of human prepared root canal: ex vivo based computational fluid dynamics analysis, Croat Med J.; 53: 470-9.
5. Li P. et al. (2013); Numerical investigation of root canal irrigation adopting innovative needles with dimple and protrusion, Acta of Bioengineering and Biomechanics, Vol. 15, No. 1.
6. Falk, Kenneth W., and Sedgley, Christine M. (2005), "The Influence of Preparation Size on the Mechanical Efficacy of Root Canal Irrigation in Vitro" JOE—Vol. 31, No. 10, October.
7. Usman, Najia, Baumgartner J. Craig, and Marshall, J. Gordon (2004) "Influence of Instrument Size on Root Canal Debridement", JOE— Vol. 30, No. 2, February.
8. Factsheet compiled by the Academy of General Dentistry-"What is a Root Canal?" Featured in Oral Health Resources, Root Canal, March 30, 2007.
9. Buchanan, L. Stephen, "Managing Severe Canal Curvatures and Apical Impediments: An Endodontic Case Study", April 2005.
10. Boutsioukis, C., Lambrianidis, T., & Kastrinakis, E. (2009) "Irrigant flow within a prepared root canal using various flow rates-a CFD study", International Endodontic Journal, 42, 144–155.
11. Vinothkumar,Thilla Sekar, Kavitha, Sanjeev, Lakshminarayanan, Lakshmikanthan bharathi, Gomathi, Narayanan Shivaram, and Kumar, Vanaja (2007) "Influence of Irrigating Needle-Tip Designs in Removing Bacteria Inoculated Into Instrumented Root Canals Measured Using Single-Tube Luminometer", JOE—Vol. 33, No. 6, June.
12. Haapasalo, Markus, Shen, Ya, Qian, Wei, Gao, Yuan "Irrigation in Endodontics", Dent Clin N Am 54 (2010) 291–312.
13. Shen, Ya , Gao, Yuan, Qian, Wei , Ruse, N. Dorin, Zhou, Xuedong, Wu, Hongkun, and Haapasalo, Markus (2010) "Three-dimensional Numeric Simulation of Root Canal Irrigant Flow with Different Irrigation Needles", JOE Vol. 36, Number 5, May.
14. Khademi, Abbasali, Yazdizadeh, Mohammad, and Feizianfard, Mahboobe (2006), "Determination of the Minimum Instrumentation Size for Penetration of Irrigants to the Apical Third of Root Canal Systems", JOE — Vol. 32, N. 5, May.

20

Study of Shoulder Range of Motion of Tennis Players by Video Motion Analysis and Goniometer

C. S. Ram and Mitushi Dang

ITS College of Physiotherapy Murad Nagar, Ghaziabad, India

Email: csram@its.edu.in

ABSTRACT

Measurement of internal and external rotation range of motion (ROM) in dominant and non dominant shoulder of junior tennis players were studied by goniometry and motion analysis software. Sixteen male tennis players (15.69+1.65 yrs) participated in the study. Rotational ranges of motion of dominant and non dominant shoulder were measured using video analysis software and goniometry. Results showed significant differences in internal and external rotation (ROM) in the dominant arm in comparison to non dominant arm in junior tennis players. Results suggests that Eagle Eye Proviewer (EEPV) Software as an instrument for shoulder rotational ROM measurement can be used in clinical settings.

20.1 INTRODUCTION

Motion analysis of body parts is critical in the medical field. In postural and gait analysis, joint angles are used to track the location and orientation of body parts. Gait analysis is also used in sports to optimize athletic performance or to identify motions that may cause injury or strain. [1] In Two dimensional (2D) motion analyses recording of images take place through digital video cameras with digital video. Digital video camera produce 2D (i.e. flat) images at a rate of 50-60 Hz and are easily transferred in digital form to computers for processing. Digitization of visual images means conversion of parts of an image to numerical position data. In manual digitization the operator displays a single image on a video screen, moves a cursor in turn to the various points of interest (e.g. ankle, knee or hip joint). For automatic analysis, the body is fitted with small reflective markers before image capture at strategic points usually using an adhesive sticker of some sort. Digitizing software detects each reflective spot in the captured frame, calculates the centre of each marker, and transforms its relative position into a data point represented in either tabular or graphic form for analysis. In three-dimensional (3D) motion analyses a number of cameras need to be used so that "depth" data can be obtained as well as the planar information obtainable from a single camera. The aim of the study is (1) to compare internal and external rotation ROM in dominant and non-dominant shoulder of junior tennis players as measured by goniometer and motion analysis software. (2) Validate the accuracy of motion analysis software in measuring ROM in clinical settings.

Chandler et al found that alterations in the ROM of the shoulder joint and the strength of the rotator cuff musculature secondary to repetitive over-head activity occur [5]. Both have been considered potentially influencing the biomechanics of the joint and consequently leading to the development of shoulder pain in elite athletes. [6] A study by Darl W Vander Linden et al.[7] provides information regarding the reproducibility and accuracy of measurements of angles and distance obtained with the Motion Analysis system from reflective markers under static conditions. The findings are similar to those reported for other kinematic analysis systems and suggest that the system has acceptable error in calculating angles from reflective markers. R A States et al. [8] used a repeated measures design study to evaluate the precision and repeatability of the Optotrak 3020 by measuring distance and angle in three sessions, four distances and three conditions (motion, static vertical, and static tilted). Precision and repeatability were found to be excellent for both angle and distance although they decreased with increasing distance from the sensors and with tilt from the plane of the sensors.

20.2 METHODOLOGY

20.2.1 Sample Size

A total of 16 subjects were included in the study. They were recruited from Master's Tennis Academy, Punjabi Bagh Club. Written Informed consent was obtained from the subjects and a short performance and injury questionnaire was completed. Inclusion criteria: age group: 13-18 years, Sex: male only, Subjects playing tennis for two years or more. Exclusion criteria: (1) Subjects with a history of shoulder surgery or a previous shoulder injury that was not fully rehabilitated. (2) Any upper limb injury or pain in the preceding 6 months that held them out of training or competition. (3) Any other medical condition.

20.2.2 Instrumentation

Video camera with 7.2mpx, Body markers (2.5 × 2.5) cm., Full circle Goniometer, Eagle Eye Proviewer Software, Weighing machine, Measuring tape,

20.3 PROCEDURE OF MEASUREMENT

20.3.1 Goniometry

For goniometry the stationary arm of the goniometer was held perpendicular to the ground and the moving arm parallel was strapped to the lateral aspect of forearm.

30.3.2 Video Graphic Analysis

For videography the active rotational movements were recorded using a camera of 7.2 megapixels which was held at such a distance that both the markers were within the camera frame throughout the movement. The video was later analyzed frame by frame using the PROVIEWER SOFTWARE. The angle formed between the markers placed on the skin surface and the ground was measured.

20.4 DATA ANALYSIS

The data collected during the study was analyzed using GRAPHPAD statistical software package. Age, height, weight, years of play of the athletes was taken. Internal & External rotation ranges of motion were compared using paired t test between dominant and non-dominant arm, and between goniometry and video motion analysis method.

20.5 RESULTS & DISSCUSSION

Comparison of Shoulder Internal Rotation ROM: Mean of internal rotation ROM of dominant shoulder using goniometer and video analysis software are 64.63+8.59 and 62.88+16.43 respectively. The difference is not statistically significant ($p = 0.46$). Mean of internal rotation ROM of non-dominant shoulder using goniometer and video analysis software are 69.19+9.40 and 68.31+12.92 respectively. The difference is not statistically significant ($p = 0.62$). See Fig. 20.1.

Comparison of Shoulder External Rotation: Mean of external rotation ROM of dominant shoulder using goniometer and video analysis software are 91.56+9.26 and 88.31+6.61 respectively. The difference is not statistically significant ($p = 0.14$). Mean of external rotation ROM of non-dominant shoulder using goniometer and video analysis software are 85.50+6.39 and 82.13+7.74. The difference is not statistically significant. See Fig. 20.2.

Fig. 20.1 Digital camara used in the study

Fig. 20.2 Reflective markers (2.5 × 2.5) cm and full circle goniometer used in the study

While comparing the measurements taken with goniometer and video analysis, there were not significant differences in ROM in the dominant and non-dominant arm in either internal rotation or external rotation, thus indicating that Eagle Eye ProviewerR video analysis software is a valid instrument in measuring ROM in clinical settings. The findings are similar to those reported for other kinematic analysis systems by Darl W Vander Linden et al.[7] and suggest that the Eagel Eye Proviewer system is acceptable in calculating angles in clinical settings..

20.6 CONCLUSION

The study validated Eagle Eye Proviewer Software R as a valid instrument for shoulder rotation ROM measurement in clinical settings. The eagle eye previewer video analysis software could be viewed as an easy–to-use system for measurement of ROM, considering that the videos are re-accessible, repeatedly useable and the movement pattern and rhythm can also be analyzed through it. 2D video-based motion analysis is a simpler and less expensive analysis method, where all that is required is a digital video camera, and a digitizing software program that can be bought at a reasonable price. Further studies should be carried out in order to validate the remaining parts of the software. Further research is required to be done with large sample size.

REFERENCES

1. Aggarwal, J. K. and Q. Cai. (1999) Human Motion Analysis: A Review. Computer Vision and Image Understanding 73:428-440.
2. Ellenbecker T. S., Roetert E. P., Bailie D. S., Davies G. J., Brown S. W. (2002) Glenohumeral joint total rotation range of motion in elite tennis players & baseball pitchers. Med Sci Sports Exerc. 34: 2052-6.
3. Kibler W. B., Chandler T. J., Livingston B. P., Roetert E. P. (1996) Shoulder range of motion in elite tennis players. Effect of age and years of tournament play. Am J Sports Med.; 24: 279-85.
4. Vad V. B., Gebeh A., Dines D., Altchek D., Norris B. (2003) Hip and shoulder internal rotation range of motion deficits in professional tennis players. J Sci Med Sport; 6:, 71-5.
5. Chandler T. J., Kibler W. B., Uhl T. L., Wooten B., Kiser A., Stone E. (1990) Flexibility comparisons of junior elite tennis players to other athletes. Am J Sports Med. 18: 134-136.
6. Warner J. J., Micheli L. J., Arslanian L. E., Kennedy J., Kennedy R. (1990) Patterns of flexibility, laxity, and strength in normal shoulders and shoulders with instability and impingement. Am J Sports Med; 18: 366-375.
7. Darl W. Vander Linden, Shirley J. Carlson and Robin L. Hubbard. (1992) Reproducibility and Accuracy of Angle measurements Obtained Under Static Conditions with the Motion Analysis Video System Phys Ther.; 72: 300-305.
8. R.A. States, E Pappas (2005) Precision and repeatability of the Optotrak 3020 motion measurement system. Med Eng Phys. 27: 505-12.
9. Crockett H.C., Gross L. B., Wilk K. E., et al. (2002) Osseous adaptation and range of motion at the glenohumeral joint in professional baseball pitchers. Am J. Sports Med; 30: 20–26.
10. Ellenbecker T. S., Roetert E. P., Piorkowski P. A. (1993) Shoulder internal and external rotation range of motion of elite junior tennis players: a comparison of two protocols. J Orthop Sports Phys Ther; 17: 65.
11. Levangie P. K., Norkins Cythia C. (2005) Joint structure and function. A comprehensive analysis. Rev Ed F. A. Davis Company (United States).

12. Yoneda M., Nakagawa S., Mizuno N., et al. (2006) Arthroscopic capsular release for painful throwing shoulder with posterior capsular tightness. Arthroscopy; 22:801.
13. Kibler W.B., Chandler T.J. (2003) Range of motion in junior tennis players participating in an injury risk modification program. J. Sci Med Sport; 6(1):51-62.
14. Tuite M.J., Petersen B.D., Wise S.M., Fine J.P., Kaplan L.D., Orwin J.F. (2007) Shoulder MR arthrography of the posterior labrocapsular complex in overhead throwers with pathologic internal impingement and internal rotation deficit. Skeletal Radiol; 36:495-502.
15. Borich M.R., Bright J.M., Lorello D.J., Cieminski C.J., Buisman T., Ludewig P.M. (2006) Scapular angular positioning at end range internal rotation in cases of glenohumeral internal rotation deficit. J. Orthop Sports Phys Ther; 36:926-934.
16. Von Eisenhart-Rothe R.M., Jager A., Englmeier K.H., Vogl T., Graichen H. (2002) Relevance of arm position and muscle activity on three-dimensional glenohumeral translation in patients with traumatic and atraumatic shoulder instability. Am J Sports Med. 30: 514-519.

21

Gait Analysis of Ankle Joint Forces Using Force Platform

Protima Nomo Sudro and A. Thirugnanam

Department of Biotechnology and Medical Engineering
National Institute of Technology Rourkela, Rourkela–769008, Odisha, India

Email: protimanomosudro@gmail.com

ABSTRACT

The aim of this study is to analyze the ankle joint kinetics involve in daily activities. Each and every characteristic of gait (walking, running, jumping or standing) is performed with the help of the ankle-joint movement. This study investigates the ankle-joint kinetics between a healthy subject and a differently-abled subject using a multi-axial force platform. The data from the force platform are collected using a data acquisition system and processed using BiowareTM software. The profiles of velocity-time, acceleration-time, moment, friction, torque, displacement, power, work, impulse and force gradient obtained from ground reaction force-time curve demonstrates the dynamics of the ankle-joint during locomotion. The profiles obtained from the ground reaction force-time curve is compared to the two different subjects (normal, healthy and differently-abled) and investigate the role of ankle (dorsiflexion, plantarflexion, eversion and inversion) in gait analysis. Gait analysis of ankle-joint kinetics is accomplished by using the inverse dynamics method. Besides these performance parameters, the efficient motion of ankle joint also depends upon the subtalar joint-axis position. Investigation of the ankle-joint kinetics provides a wide range of applications including footwear design, ankle-foot orthoses for hemiplegia and guiding the clinicians by studying the biomechanics of the joint for different ankle related pathologies..

Keywords: Ankle-joint kinetics, dynamics, gait characteristics, orthoses.

21.1 INTRODUCTION

The ankle-joint is a bone joint specifically termed a hinge joint, located in the anterior part of the leg. The ankle-joint allows motion only in a single plane (up and down movement). The ankle-joint along with the foot form a complex of functional parts, enabling a person to perform a motor act like standing, walking, running and jumping. The ankle-joint and subtler-joint when combined together acts as a universal joint, the anterior and posterior ankle capsular ligaments surrounds the joint and protect the joint from injury by limiting the range of motion. The ankle-joint functions in all phases of motor activities, example, during standing the ankle-joint is neutral, during walking the ankle flexes with heel strike and extends with toe-off, thus depending upon the motor act it flexes and extends correspondingly. Since, motor act depends upon the proper functions of the ankle-joint, it is necessary to analyze the biomechanical aspects of the ankle-joint [6].

A body initiates the motor act with heel strike, during which the ankle is in a state of dorsiflexion. At this state the respective foot accepts the body weight and requires a certain amount of impact force to attain forward momentum [3]. After the heel strike, the ankle-joint becomes neutral, holding the whole body weight on a single foot known as midstance phase or single support phase. Later ankle joint starts to raise the heel by applying high impulse force in the forefoot to lift the leg, causing the ankle-joint to be plantar flexed. On the other hand, the opposite leg, which is in swinging phase starts to decelerate by landing on the ground, providing a double support phase (initiating a heel strike).

Thus, the functional performance of the ankle joint involves control and stabilization of leg on foot, by managing irregular surfaces, shock absorption during walking, running, jumping, elevating the body while standing on toes etc, subject to the entire body weight. During a heel strike phase (dorsiflexion) of bipedal gait, power absorption occurs in order to decelerate the center of mass of the body and maintain forward progression with vertical support.

The closely packed position of the ankle is the dorsiflexion when the anterior part of the body is in contact with the foot and the loosely packed position of the ankle is the plantarflexion when a narrow posterior part of the body is in contact with the foot. The ankle is found to be less stable during toe-off phase, i.e., plantarflexion and hence there is a high incidence of ankle sprain during plantar flexion rather than dorsiflexion.

The subtalar joint located in the hind foot, just behind the ankle joint also contributes in the stability of bipedal gait. The axis position of the subtalar joint determines the line of action for load transfer [2]. Inversion and eversion take place simultaneously during motor activities. However, inversion injuries are much common in the ankle-foot complex.

The purpose of the study is to analyze the kinetic parameters of gait events occurring in the lower extremity for a healthy subject and differently abled subject depending upon their gait patterns [1].

21.2 METHODS

Prior to the experiment, the study was approved by the Ethical Committee of National Institute of Technology, Rourkela. Informed consent is taken from the subjects before acquiring their gait patterns.

21.2.1 Subjects

For this study two subjects were consented. The patterns obtained from the subjects were used to analyze the kinetic parameters of the ankle-joint in motor act. The subjects (abled and differently abled) 23 and 20 years of age, 171 cm and 125 cm in height, and 52 and 65 kg in mass were chosen. The able subject is free from any neurological and orthopedic impairment, however differently abled subject has impairment in one of the leg facing difficulties in bipedal locomotion.

21.2.2 Equipment

The force data were collected using the multiaxial force platform (Kistler Instruments, Type 9260AA6). The geometric configuration of the force plate is shown in Fig. 21.1. The data are obtained at a sampling frequency of 1000 Hz. The subjects were properly instructed prior to the activities, and they were also asked to perform 2 to 3 trials before the acquisition of the data.

21.2.3 Data Collection

The force data initially was recorded for walking, running and jumping for a healthy able subject. Similarly, for differently abled subject the force data is recorded during standing and walking. Walking was recorded at their self-selected speed as a defined speed might change their natural gait pattern. After 2 to 3 trials the natural gait pattern of the subjects was recorded. The subjects were made to walk through a 330 cm long path for duration of 15 seconds. The static and dynamic data were collected for each and every conditions performed by the subjects.

21.2.4 Data Analysis

All the calculations were performed based on Newton's 2nd law of motion,

$$F_Z(t) = m \cdot a_z(t) \qquad \text{...(1)}$$

The profile parameters are computed by using the equations derived from Eq. 1. The acceleration –time record is obtained using body mass information and static acceleration as described in Eq. 2 below:

$$a_z(t) = \frac{F_Z(t)}{m} - a_{oZ} \qquad \text{...(2)}$$

where, a_{oZ} = static acceleration; a_z = acceleration along the vertical direction, m is body mass and F_Z = vertical force. Similarly, the velocity parameter is computed from initial velocity value and acceleration obtained above. The moment of force about Z-axis is calculated based on the direction represented as:

$$M_Z = b * (-f_x 12 + f_x 34) + a * (f_y 14 - f_y 23) \quad ...(3)$$

where M_Z = Moment of force along Z-axis, $f_x 12$ = force along X-axis measured by sensor 1 and 2, $f_x 34$ = force along X-axis measured by sensor 3 and 4, a and b are respective offsets. From Eq. 3 the torque is expressed as:

$$T_Z = M_Z - F_Y * ax + F_X ay \quad ...(4)$$

where T_Z = vertical torque, F_X and F_Y are the forces along X- and Y-axis.

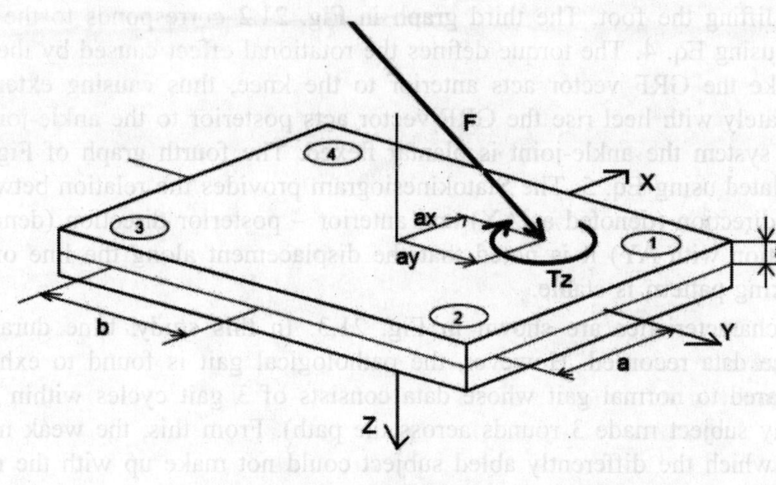

Fig. 21.1 Geometric configuration of a force plate

The center of pressure is expressed as:

$$ax = -\frac{M_y}{F_Z} \text{ and } a_y = \frac{M_x}{F_Z} \quad ...(5)$$

where a_x = centre of pressure along x direction, a_y = centre of pressure along y-direction, M_y = moment of force along Y-axis, M_x = moment of force along X-axis and F_Z = force exerted vertically. Accordingly the displacement, power, work done, impulse and force gradient equations are derived from Eqs. 1 and 2.

21.3 RESULTS AND DISCUSSION

At a different speed of walking, the characteristic ground reaction force curve shows two-humps for different phases of a motor act for a normal healthy subject which deviates in case of a differently able subject. The differently abled subject as reported is a patient of Spastic quadriplegia, which limits the range of motion resulting in different gait pattern. From Fig. 21.2 and 21.3, it is observed that the power absorption is comparatively higher during the ankle plantarflexion phase in both the subjects. It is also observed that the differently abled subject even though has lower mass compared to the normal healthy subject, uses more power than the healthy one which shows that the gait of a differently abled subject is very expensive. The power absorption results from the contraction of muscles, tendons and ligaments. The tendons of the knee and ankle are reported to be longer with shorter muscle fibers than the hip tendon. These muscles and tendons are responsible for providing energy for acceleration and deceleration as required. The leg muscles and tendons, especially the tendon of the knee and ankle located in front of the thigh and calf, lengthen and shorten to flex and extend the leg during any gait activities. The mechanism of the tendon is similar to that of a spring with magnificent elastic properties [8]. The Achilles tendon connecting the calf to the heel provides the elastic strain energy required for stretching the leg in gait. It is qualitatively observed in the differently abled subject that the muscles are weak due to which the subject walks by leaning to some degree in order to compensate the lifting of the other leg for forward propulsion. The muscles of the calf, hamstring and quadriceps are weak due to which the subject was unable to stretch the muscles as per required for gait accomplishment and thus utilizes the most amount of force to overcome the respective gait.

In Fig. 21.2, the first graph shows the variation of a ground reaction force with time for a moderate walking of a normal healthy subject. A two-hump force curve is seen during a moderate walk. The initial hump (peak) corresponds to the stance phase resulting into F = mg, again with the bending knees immediately causes the force to decrease (F < mg). Later the legs are stretch leading to a rise in the force (F > mg). Both the limbs follow the same pattern in all the cases. The second graph in Fig. 21.2 is the power variation with time for the normal healthy subject obtained with respect to the force exerted by the subject. From this graph it is noted that the power profile follows the pattern of force-time curve with a subsidiary hump at initial heel strike. The small subsidiary hump in the ankle dorsiflexion phase is due to the least amount of power required to decelerate the body. However a pronounced hump is seen in ankle plantarflexion phase as more power is required to overcome the frictional force and acceleration for lifting the foot. The third graph in Fig. 21.2 corresponds to the torque applied during normal walking calculated using Eq. 4. The torque defines the rotational effect caused by the ground reaction force (GRF) vector. At heel strike the GRF vector acts anterior to the knee, thus causing extension of the knee and ankle dorsiflexion. Immediately with heel rise the GRF vector acts posterior to the ankle-joint causing the knee to be flexed and like a lever system the ankle-joint is plantar flexed. The fourth graph of Fig. 21.2 is the graph of Statokinesiogram [7] calculated using Eq. 5. The Statokinesiogram provides the relation between center of pressure along the medial – lateral direction (denoted as AX) and anterior – posterior direction (denoted as AY). From the Statokinesigram (AX variation with AY) it is noted that the displacement along the line of action is less, which infers that the subject walking pattern is stable.

The pathological gait characteristics are shown in Fig. 21.3. In this study, time duration of 15 seconds is maintained for all the force data recorded. However, the pathological gait is found to exhibit only a single gait cycle in 15 seconds compared to normal gait whose data consists of 3 gait cycles within 15 seconds across the 330 m path (normal healthy subject made 3 rounds across the path). From this, the weak muscular activity effect is clearly observed due to which the differently abled subject could not make up with the normal healthy subject whose speed is 1/3 times ahead of the differently abled subject.

The first graph of Fig. 21.3 shows the force-time record of a differently abled subject. In this graph, an initial hump whose magnitude is lower than that of the hump exhibited by the front foot was observed. In normal gait, with initial contact the knee must be extended and ankle must be flexed and then slowly attains a stance phase where a sharp peak is acquired. But in this case the subject applies less force in stance phase because of limited elasticity of the muscles and tendons of the knee and ankle. Due to the limitation in stretching the tendons and muscle fibres, less energy is generated. Thus, the subject applied less force to decelerate during heel strike. From the observation, it is also seen that the Achilles tendon that surrounds the heel and calf has main limitations because the subject was facing problem in applying desired force during midstance phase.

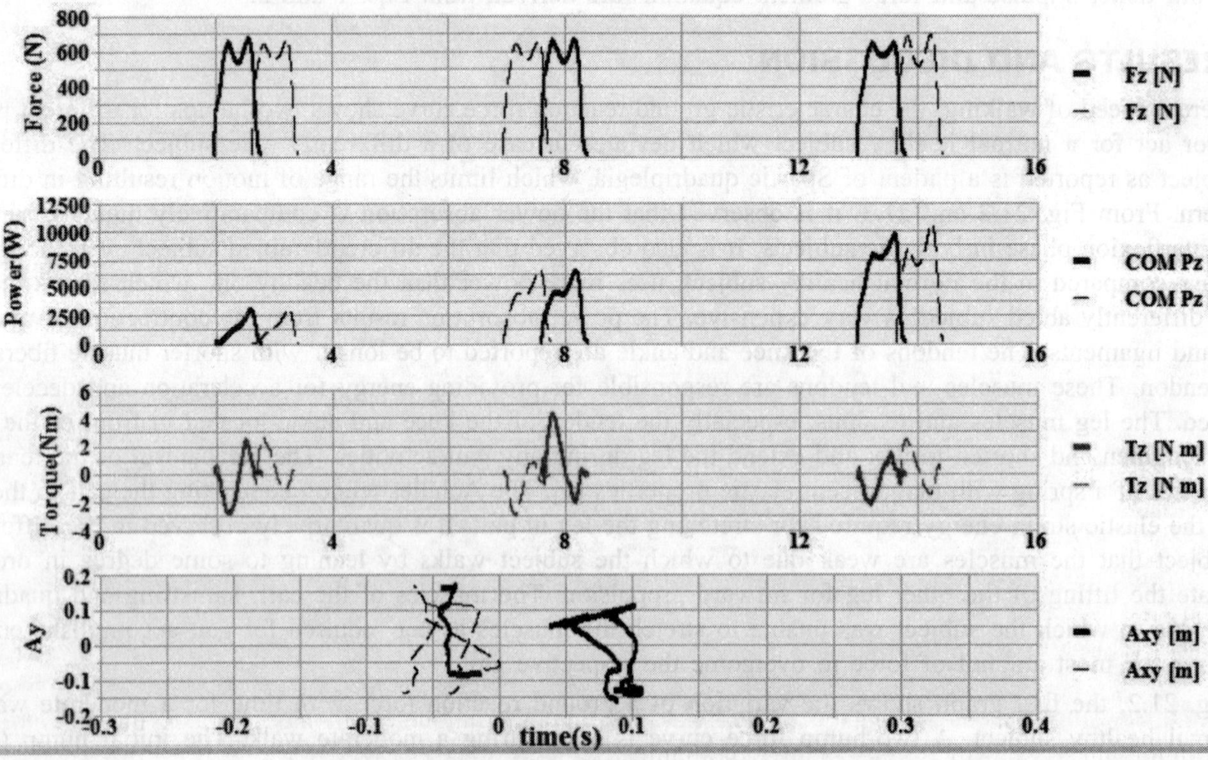

Fig. 21.2 Profile parameters of an abled subject

However, the second hump is respectively higher in magnitude but as observed it does not decreases rapidly, because the loading rate is very slow and the subject is spending more time to overcome the frictional force in order to apply a backward component for accelerating the body. Hence, a plateau level is seen during the push-off phase. Since both the limbs exert same characteristics, we discuss only the single limb behaviour.

The second graph of Fig. 21.3 shows the power variation with time. In this graph the power follows the profile of force exerted by the subject. Since, the subject impulse is higher in the push-off phase, power requirement is also high.

The third graph of Fig. 21.3 corresponds to the torque variation with time. It is observed that the rotational effect is higher in the ankle plantarflexion. This implies that the tendon of the ankle (Achilles tendon) shortens as required to attain the desired gait.

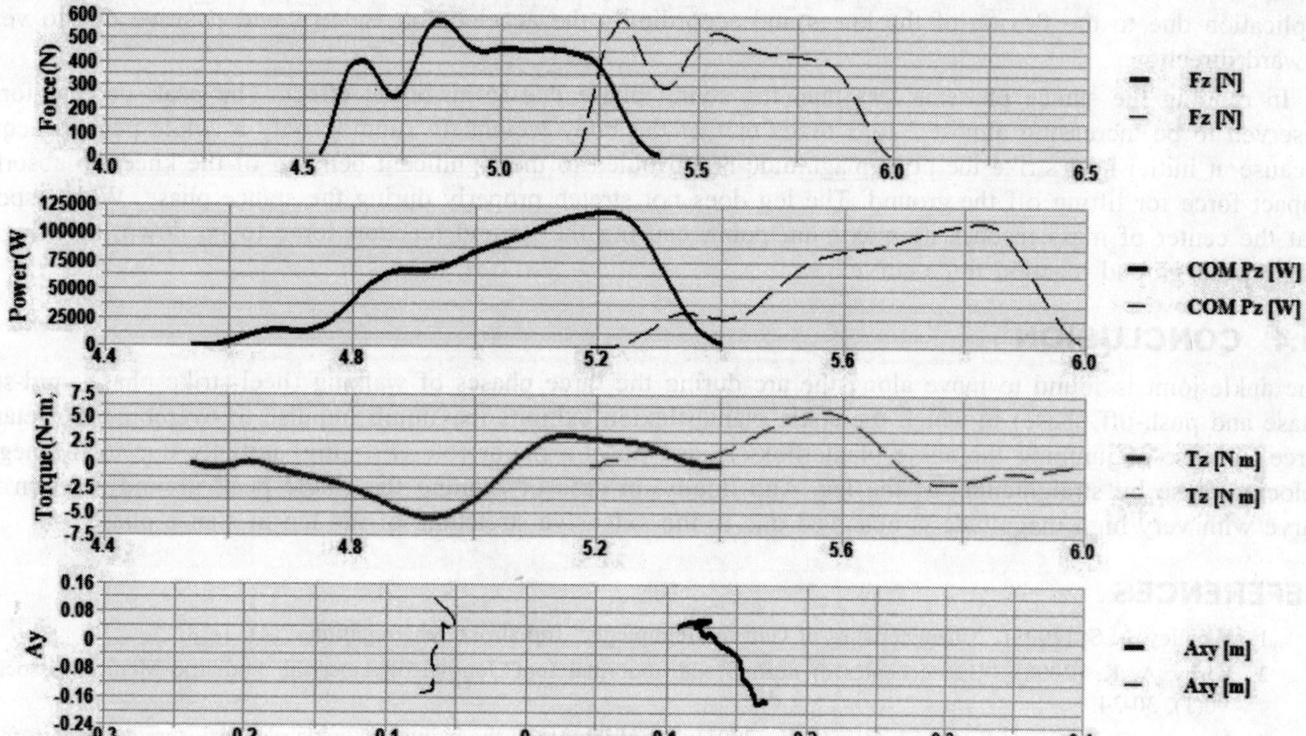

Fig. 21.3 Profile parameters of a differently abled subject

The fourth graph of Fig. 21.3 shows the Statokinesigram. The line of action implies that both the anterior-posterior and medial lateral direction the stability are maintained to a normal extent which is found similar to that in Fig. 21.2.

From the other parameters like, displacement, work done, impulse, force gradient, etc. it is observed that the gait of a normal healthy subject is cheaper compared to the differently abled subject who adjusts the gait in such a way that the energy costs is as low as possible but still expends more than the normal one. With limited knee and ankle flexion and reduced ankle plantarflexion, the subject finds problem in clearing the foot of the ground for the swing phase and hence result in longer impulse in push-off phase.

Beside the hip, knee and ankle joints, there is also one more joint termed "subtalar joint" whose position affects the load distribution in the foot along the line of action. The line of action is one on which the gravitational force is acting on the foot. The deviation of this joint deviates the direction of force, ultimately leading to the risk of fall causing some injury. The position-axis of subtalar-joint can affect normal healthy subjects as well, and it is easily observed in day to day life by looking at their shoes which used to get worn-out on a particular side. Depending on the lateral and medial axis subtalar joint deviation, the inversion and eversion movement of the foot will lead to ankle sprain and other pain related to lower extremity.

The gait parameters such as step length, single support phase also known as stance phase, swing phase and double support phase are defined. And these gait parameters distinguish pathological gait from normal gait. While walking a retarding force is required to attain a constant speed. Both in walking and running the retarding force arises from the impulse generated by the front foot while pushing forward on the ground. During push-off phase an impulse is observed which is equal and opposite to the impulse generated during heel-strike when the back foot pushes backward. So, ultimately the retarding force is the frictional force resulting from the contact of the foot with the ground. And it is due to the frictional force, a person prevents itself from sliding forward on the ground.

In countermovent jumping the impulse is found to be closely related to the height of the jump. The magnitude of the impact force can be reduced as well as increased by maintaining the bending and straightening of the knees [4, 5]. The motion of the centre of mass is assumed to fluctuate with the motion of the legs, initially the velocity of the centre of mass is found to be negative at contact but rapidly changes to zero with slight positive velocity application due to the flexion of the knees and accordingly the acceleration is large and positive due to vertical upward direction.

In running the stance phase is less than the swing phase due to air-borne effect. The peak vertical force is observed to be increasing almost 2 to 3 times that of the body weight. In running only a single peak is acquired because at initial heel strike the peak magnitude is attributed to the significant bending of the knees to absorb the impact force for lifting off the ground. The leg does not stretch properly during the stance phase. With respect to that the center of mass reaches its maximum point, causing the ground reaction force to go down, resulting in a single peak ground reaction force curve.

21.4 CONCLUSION

The ankle-joint is found to move along the arc during the three phases of walking (heel-strike phase, mid-stance phase and push-off phase) in which the ankle plantarflexion exhibits maximum impulse to overcome the retarding force. In case of jumping the ankle plantarflexion also plays a major role ($F > mg$) initially due to the negative velocity cause by straightening of the leg. And finally, in case of running the single peak ground reaction force curve with very high magnitude is observed due to the extensive stretching of the leg in stance phase.

REFERENCES

1. Woolley, M.S. (2001). "Characteristics of Gait in Hemiplegia".Top stroke rehabilitation, 7(4), 1-18.
2. Kirby, A. K. (2000). "Biomechanics of normal and abnormal foot".Journal of American Podiatric Medical Association, 90(1), 30-34.
3. Neptune, R. R. Kautz, S. A. and Zajac, F. E. (2001). "Contribution of the individual ankle plantarflexors to support, forward progression and swing initiation during walking". Journal of Biomechanics, 34, 1387-1398.
4. Cross, R. (1999). "Standing, walking, running and jumping on a force plate". American Journal of Physics, 67(4), 304-309.
5. Linthorne, P. N. (2001). "Analysis of Standing Vertical jump using a Force platform". American Journal of Physics, 69(11), 1198-1204.
6. Winter, D. A. (1995)."Human balance and posture control during standing and walking". Gait and Posture, 3, 193-214.
7. Duarte, M and Freitas, F. S. M. S. (2010). "Revision of Posturography based on force plate for balance evaluation". Revista Brasileira de Fisioterapia, 14(3), 183-192.
8. Alexander, N. M. R. (1984). "Walking and running", American Scientist, 72, 348-354.

22

A Study on Working Posture and Musculoskeletal Issues of Indian Medical Laboratory Technicians

Shreya Maulik, Rauf Iqbal, Amitabha De and H. V. Bhasin

National Institute of Industrial Engineering
Vihar Lake Road, Mumbai–400087
Maharashtra, India

Email: rauf.nitie@gmail.com

ABSTRACT

Dynamic and static activities in medical laboratory cause musculoskeletal disorders but laboratory technicians do not have much awareness about effects of postural risk factors on their health. This study was performed among 20 Indian medical laboratory technicians with the aim to determine Work Related Musculoskeletal Disorders (WRMSDs) prevalence; exposure assessment of WRMSDs risks especially through postural assessment. In this cross-sectional study 5 predominant tasks (pipetting, microscopy, biosafety cabinets, microtome and automated analyzers) carried out in the medical laboratory were selected for the evaluation of the risk factors and the postural assessment. The tools selected for the study were Modified Nordic Musculoskeletal Questionnaire (NMQ), Quick Exposure Checklist (QEC), Rapid Upper Limb Assessment (RULA) and Rapid Entire Body Assessment (REBA) technique to evaluate the laboratory tasks. The results revealed that the overall prevalence of WRMSDs was 95% in the last 12 months and the most affected body parts were low back and upper back (65%), followed by knees (30%), neck (25%) and shoulder wrists and hands (25%). Postural analysis using RULA and REBA indicates that the technicians are not working in the safe limit. The average RULA score of 7, i.e., action level 4 and REBA score ranging from 8-12 was found for technicians working with biosafety cabinets, automated analyzers and the microtome activity which indicates very high risk and urgent corrective measures are required. This is due to constrained posture of back, neck and shoulders, prolonged standing postures and also due to limited leg space while performing various activities in the laboratory. In conclusion these methods indicated that different body parts at specific postures are vulnerable to injury and musculoskeletal disorders and thereby need an immediate ergonomics intervention in terms of design of workstation and work.

Keywords: Posture, medical laboratory technicians, ergonomic risk factors, musculoskeletal problems, laboratory tasks.

22.1 INTRODUCTION

Musculoskeletal injuries associated with occupation are among the most important occupational problems worldwide. Unfavourable occupational body position is one of the most important factors in these injuries. Other risk factors include genetic predisposition, aging and emotional stress. Such injuries bear negative social, economical and political complications for human societies (Varmazyar et al., 2012). At present, WMSDs are one of the most important problems ergonomists encounter in the workplace around the world (Vanwontergem 1996). In many countries, preventing work-related musculoskeletal disorders (WMSDs) is considered a national priority (Spielholz et al., 2001). WMSDs are a worldwide concern present both in industrialized and industrially developing countries. Poor working conditions and the absence of an effective work injury prevention program in industrially developing countries has resulted in a very high rate of WMSDs (Jafry et al., 2000). Risk factors of WMSDs include workplace activities such as heavy load lifting, repetitive tasks and awkward working postures (Bernard, 2009; Haynes, 2008). Demographic characteristics and psychosocial factors are also known to be important predictive variables (Linton et al., 1989; Carter et al., 1994; Weiser, 1997). WMSDs occur more frequently in certain occupations characterized by job tasks involving exposure to physical risk factors, as found in some healthcare professions (Punnett et al., 2004). Among the various health-care professionals, medical laboratory technicians play a crucial role in the diagnosis of disease. In medical laboratories, technicians are involved in a fairly long hour of static work; predominant risk factors include awkward posture, force and repetitive movements. The prevalence of WMSD is high among medical laboratory technicians. Laboratory technicians – e.g., doing pipetting work–show high prevalence of neck pain, shoulder pain, elbow pain and hand pain (Bjorksten et al., 1994; David et al., 1997). Pain in shoulder and hand is associated with the duration of the pipetting work and neck pain is associated with the amount of fixed working postures when pipetting (Bjorksten et al., 1994; David et al., 1997). In addition sedentary work has been associated with an excess risk of low back pain (Thorbjornsson et al., 2000). Reduced workability and absence from work are common consequences of WMSD (Andersen et al., 2011). The daily routine of laboratory technicians includes several hours of microscope and computer work (George, 2010). Microscopes are often not ergonomically optimized and do not allow a neutral sitting position, requiring the technicians to bend forward [George, 2010, Kofler et al., 2002, Sillanpaa et al., 2003). A survey among 244 cytotechnologists, a typical example of microscope-using employees, demonstrated that more than 80% of participants suffered from musculoskeletal discomfort including headache, neck pain, stiffness, back pain and upper-extremity discomfort (Thompson et al., 2003). In this situation, high rate of WMSDs occurrence could be expected. Thus, initiatives to reduce the burden of WMSDs are needed. Despites these facts, there have not many studies done so far to find the prevalence of musculoskeletal disorders among medical laboratory technicians in India. Therefore, the present study was carried out with the objectives of (a) determining WMSD prevalence rate among medical laboratory technicians (b) assessing the level of technician's exposure to task specific WMSD risks and c) task specific postural analysis for laboratory technicians. The results of this study should be an appropriate basis for planning and implementing an interventional ergonomics program in the workplace and for improving technician's health in health care industry.

22.2 MATERIALS AND METHODS

22.2.1 Examined Population

This cross-sectional study was conducted among medical laboratory technicians in Mumbai. This study involves 20 laboratory technicians with at least one-year job tenure and was randomly selected for this study. Technicians with background diseases or accidents affecting their musculoskeletal system were excluded from this study. A written consent was obtained from all the subjects prior to the study. All the subjects were explained the detailed procedure of the study. The study was approved by the local ethics committee

22.2.2 Questionnaire

Data were collected via anonymous questionnaires, which covered (a) personal details (including age, weight, height, job tenure, education, health, and medical background); and (b) musculoskeletal problems in different body regions. The general Nordic questionnaire of musculoskeletal symptoms (Kuorinka et al., 1987) was used to examine reported cases of WMSDs among the study population. Reported WMSD symptoms were limited to the 12 months prior to the study. The questionnaires were completed on the basis of an interview with each laboratory technician.

22.2.3 Ergonomics Assessment

Physical exposure to musculoskeletal risks was assessed with a pen-and-paper observation method called a Quick Exposure Checklist (QEC) (Buckle and Li 1997, Li and Buckle 1997). QEC is a sensitive method for assessing physical exposure to musculoskeletal risks in the workplace with fair inter-intra observer reliability (Li and Buckle 1999). This technique includes an assessment of the back, shoulder/arm, wrist/hand and neck, regarding their position and repetitive movement. In QEC, task duration, maximum weight handled, hand force exertion, vibration, visual demand of the task, and subjective responses to work was also considered, whereas the required data were obtained from the subject. To find the most awkward postures with QEC, 5 predominant tasks (pipetting, microscopic activity, biosafety cabinets, microtome and automated analyzers) carried out in laboratory were selected and each worker was videotaped during their routine job activities for nearly 20 min. The tapes were then reviewed in a laboratory and the QEC score was calculated.

22.2.4 Postural Assessment

In order to evaluate angular measurements of the body parts and evaluation of the body status during work, direct observation, RULA (Rapid Upper Limb Assessment) and REBA (Rapid Entire Body Assessment) method was used. REBA method was specifically developed to be useful for assessing WMSD risks/working postures found in healthcare and other service industries (Hignet and McAtamney 2000). However, it can be used to assess a variety of tasks, in any setting, where the whole body is being used, the posture is static, dynamic, rapidly changing, or unstable, or animate or inanimate loads are being handled either frequently or infrequently (McAtamney and Corlett 1993). RULA is a quick survey method for use in ergonomic investigations of workplaces where WMSDs are reported. It is a screening tool that assesses biomechanical and postural loading on the body. It focuses on the neck, trunk and upper limbs, and is ideal for sedentary workers (McAtamney and Corlett 1993). It is a simple, quick and easy to complete. RULA scores indicate the level of intervention required to reduce WMSD risks.

22.2.5 Data Analysis

Data were anonymously coded and entered into a spreadsheet programme before being analyzed using the Statistical Package for the Social Sciences (SPSS 17). Appropriate statistical methods such as frequency, descriptive measures were used in the present study.

22.3 RESULTS AND DISCUSSION

The sample population consist of 20 subjects, of whom 15 (95%) were females and rest 5 (5%) were males. The mean age, height, weight, work experience and working hours in the studied population were 29.75±11.4 years, 155.70±11.9 cm, 52.9±9.7 kg, 9.2±11.2 years and 8.05±0.8 respectively (Table 22.1).

Table 22.1: An overview of demographic and occupational characteristics of the respondents

(n = 20)	*Mean*	*±SD**	*Range*
Age (years)	29.75	11.4	20-60
Height	155.70	11.9	121.92-170.18
Weight	52.9	9.7	35-69
Duration of employment (years)	9.2	11.2	1-36
Working hours	8.05	0.8	5-9

**SD* = Standard Deviation

According to NMQ, the overall prevalence of musculoskeletal symptoms in the last 12 months was 95% and in the last 7 days, it was 55%. Similar investigations also reported that the global reference to musculoskeletal symptoms in at least one anatomic segment was 86.7% in the last 12 months, 79.8% in the last 30 days and 70.2% in the last 7 days, regardless of the type of activity performed by the worker (Ramadan et al., 2006). The most frequent MSD was reported to occur in back, knees and neck region and the least frequent area involved was wrists/ hands and shoulder. The 12 month and 7 days prevalence is shown in Fig. 22.1.

Among the WMSDs reported by the study population, pains and disorders of the back region had the highest frequency both in last 12 months (65%) and last 7 days (40%), followed by knees in 12 months (30%) and 7 days (20%) and discomforts of the neck region (25%) in 12 months and (15%) in 7 days. A similar study conducted by Rasikeviciene (2005) among 112 lab technicians and reported that neck pain, back pain, waist pain and joint pain due to long lasting fixed position was rather frequent among laboratory technicians. Inconvenient work posture increased the risk for waist pain (OR = 1.65; $p > 0.05$), neck pain (OR = 2.32; $p > 0.05$) (Rasikeviciene 2005). The intensity of pain varied among the subjects. Majority of them reported mild as their pain intensity (70%), few of them reported severe (20%) and negligible percentage of respondents (5%) reported very severe. However, rest 5% of the respondents didn't report the prevelence of musculoskeletal symptoms. Undoubtedly, the reason for such similarity is the nature of occupational positions in laboratory technician's work. The possible reasons could be due to inadequate training, lack of continuous monitoring, lack of attention to the correct ergonomic guidelines, unfavourable designing of the work station and instruments, and work anxiety lead to unfavourable work positioning, giving rise to major discomforts. It is suggested that ergonomics should be taught in pre-clinical education as well as continuing educational programs.

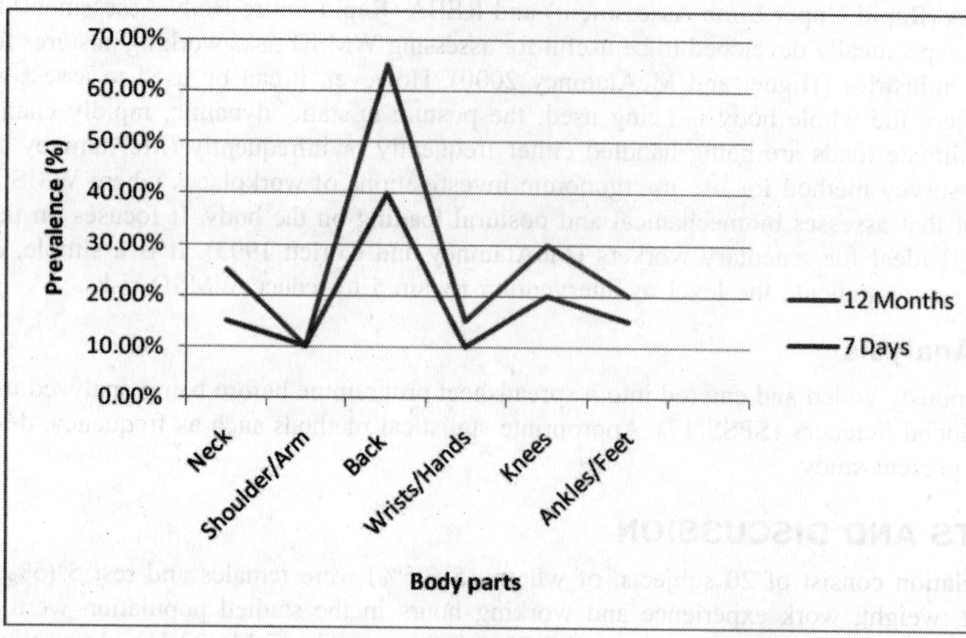

Fig. 22.1 Prevalence of musculoskeletal symtoms in the last 12 months and last 7 days

The prevalence of musculoskeletal symptoms among the technicians indicated the presence of risk factors in the laboratories. The ergonomic risk assessment was done with the help of Quick Exposure Checklist (QEC). QEC is a sensitive method for assessing physical exposure to musculoskeletal risks in the workplace with fair inter- intra-observer reliability (Li and Buckle, 1999). Figure 22.2 shows the results of assessment of physical exposure to work-related musculoskeletal risks in different tasks selected for the study.

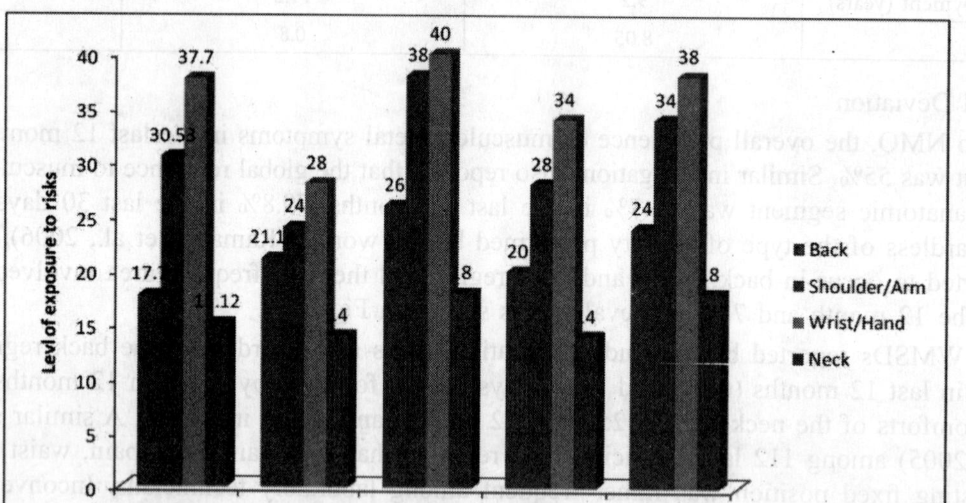

Fig. 22.2 Level of exposure to musculoskeletal risks

The results indicated that among all the tasks carried out in the medical laboratory, the overall level of exposure is highest at biosafety cabinets (132.75±7.6) followed by automated analyzers (130.22±5.02), pipetting (114.03±10.4), microtome (102±5.7), and microscopic activity (97.57±3.9). The exposure level at the back region was high while working with biosafety cabinets and automated analyzers however, the exposure becomes moderate while pipetting, microscopic activity and working with microtome. Moreover, it was found that the exposure level at shoulder or arm was high in almost all the activities except while working with microtome. However, the exposure level at wrist or hand was high in all the tasks except while working with microscopes. Further, it was found that the exposure level at neck was comparatively higher in all the activites. Literature has reported high prevalence of neck pain, shoulder pain, hand and elbow pain while pipetting (Bjorksten et al., 1994; David et al., 1997). Several large surveys have documented a high prevalence of musculoskeletal disorders (MSDs) among microscopists, 2 of which were based on surveys of cytotechnologists (Kalavar et al., 1996; Thompson et al., 2003) These workers reported a high prevalence of pain of the neck, upper back, lower back, shoulder, and upper extremities. Moreover, biosafety cabinets and laboratory workbenches present similar ergonomic hazards which are mostly due to lack of adjustability and leg room (CDC 2008). Further the scores for the assesment of the postural stress in laboratory technicians were analyzed with the help of using RULA and REBA technique.

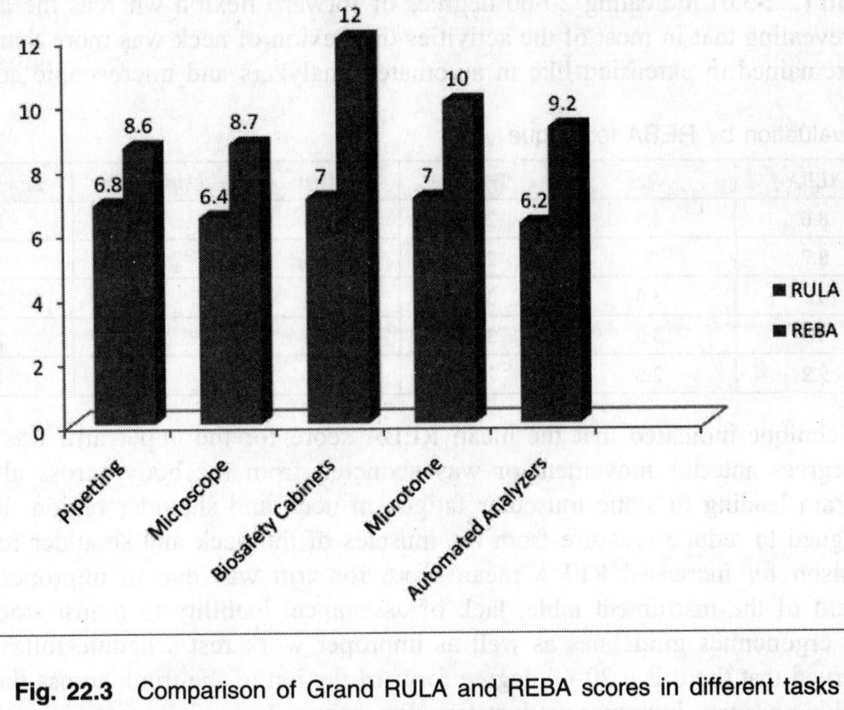

Fig. 22.3 Comparison of Grand RULA and REBA scores in different tasks

The above Fig. 22.3 shows the RULA and REBA scores for all the tasks which were being selected for the present study. The results indicated that postural scores of most technicians are well above the assumed safe scores hence immediate change or ergonomic intervention is needed in the current workplace. Rapid Entire Body Assessment (REBA) (Hignet and McAtamney, 2000) and Rapid Upper Limb Assessment (RULA) (McAtamney and Corlett, 1993) working posture analysis were used to analyze the postures while performing the tasks and the associated risks. RULA scores were high for the technicians who all were working with biosafety cabinets, microtome users, automated analyzers and pipetting activity which indicated that for these activities, the level of exposure to musculoskeletal risks need to be considered and change to be implemented. This result was further supported by the similar finding revealed from the REBA technique used to do the postural analysis while performing the above mentioned tasks in the laboratory. It indicated that the REBA scores also ranged very high for biosafety cabinets, microtome activity, automated analyzers and pipetting as well. This shows that the postures adopted during these tasks were at high risk and change to be implemented as early as possible. Table 22.2 shows the task wise postural load on each of the body regions carried out in medical laboratory.

Table 22.2: Postural evaluation by RULA technique

Tasks	*RULA*	*Upper arm*	*Lower arm*	*Wrist position*	*Wrist twist*	*Neck*	*Trunk*	*Legs*
Pipetting	6.8	2.8	3.0	4.0	1.5	3.5	2.3	1.6
Microscope	6.4	2.0	2.7	3.7	1.57	3.4	2.4	1.2
Biosafety cabinets	7	3.0	3.0	4.0	1.0	4.0	3.0	2.0
Microtome	7	3.0	3.0	4.0	1.0	4.0	3.0	2.0
Automated Analyzer	6.2	3.8	2.6	3.4	2.0	4.2	2.2	1.0

The results revealed that the grand RULA score for all the above mentioned tasks ranged between 6 and 7 (Table 22.2) which indicated the level of musculoskeletal risks is high and falls in the category of action level 4. However, if the REBA scores are concerned, they also give the similar finding indicating that for the above-mentioned activities the grand score ranged from 8-12 (Table 22.3) which shows that the level of musculoskeletal risks for these tasks are very high and immediate change needs to be implemented. The range of mean score of upper arm (2.0-3.8) shows the flexion between 45-90 degrees and above; or the arm was abducted, whereas for lower arm the score ranged from (2.6-3.0) indicating lower arm flexion of more than 100 degrees. The average score for trunk for all the tasks ranged from (2.2-3.0) indicating 20-60 degrees of forward flexion whereas the average score for neck ranged from (3.4-4.2), revealing that in most of the activities the flexion of neck was more than 20 degrees, however in few tasks, the neck remained in extension like in automated analyzers and microscopic activity (Table 22.2).

Table 22.3: Postural evaluation by REBA technique

Tasks	*REBA*	*Neck*	*Trunk*	*Legs*	*Upper arm*	*Lower Arm*	*Wrist*
Pipetting	8.6	2.5	2.2	1.5	2.9	2.0	2.8
Microscope	8.7	2	2.4	3.1	2.1	1.9	2.7
Biosafety cabinets	12	3.0	3.0	2.0	2.0	2.0	3.0
Microtome	10	3.0	3.0	2.0	3.0	2.0	3.0
Automated analyzer	9.2	2.6	2.2	2.2	3.8	1.8	2.4

Results of REBA technique indicated that the mean REBA score for the upper arm was 2.0-3.8 showing that it had more than 90 degrees anterior movement or was abducted from the body across all the activities which consequently created strain leading to static muscular fatigue in neck and shoulder region. It is suggested that an arm-rest should be designed to reduce pressure from the muscles of the neck and shoulder region (Haddad, 2007). The most important reason for increased REBA mean score for arm was due to improper work-station design, unfavourable arrangement of the instrument table, lack of assistance, inability to adjust stool height, and lack of working style based on ergonomics guidelines as well as improper work rest schedule followed by the laboratory technicians. It was observed that there is a 20-60 degree forward flexion of the trunk across the tasks. This might be due to the unnatural working posture, improper work station dimensions, lack of awareness about ergonomic guidelines among the technicians. More than 20 degrees of forward and sideward flexion of the neck is an interpretation of the REBA mean score of 2.0-3.0 for neck region, which resulted in unfavourable posture of the neck resulting in 25% of the disorders in the last 12 months. This might be because of improper height of lab bench which makes the technicians to work in neck flexion. More than a 15-degree angle for the hand or wrist toward radius and ulna is depicted by the mean REBA score of 2.4-3.0 (Table 22.3). Unfavourable working equipment, inadequate training, lack of positional monitoring and in some cases static activity of the body region leads to loss of strength, flexibility and endurance leading to musculoskeletal disorders.

22.4 CONCLUSION

The results suggested that the prevalence of musculoskeletal symptoms among medical laboratory technicians is high in India. The results indicated that the most affected body regions are back, knees and neck region. From the above discussions, we can conclude that unawareness about ergonomics is observed in medical laboratories in which work is undertaken. Musculoskeletal disorders are present in the various activities carried out in medical laboratories where a significant number of technicians are working in bad postures and it shows that there is a need to change the body postures. On the basis of analysis of results and scores obtained by the various tools applied in analysis of the working conditions and postures, it can be concluded that, there is a lack of ergonomics planning and methods in medical laboratories, employing thousands of technicians and which plays a very crucial role in Indian health-care industry. Thus, the technicians performing various types of tasks in the laboratory are under moderate to high risk of work related musculoskeletal disorders (WMSDs). Working with the biosafety cabinets, microtome machine, automated analyzers and the pipetting tasks make the technicians more prone to musculoskeletal problems as the risk exposure level for the laboratory tasks are high. The average scores of both the RULA and REBA indicated that none of the subjects are working in the safe limit, and this evaluation shows that there is a need of investigation and immediate changes are needed in the workplace arrangement. Adopting adequate postures in clinical practice and having a favourable work environment could reduce the muscular skeletal system problem. Therefore, it is of vital importance to promote the occupational health and prevention programs regarding ergonomic postures which must be acquired by the laboratory technicians during their clinical practices.

REFERENCES

1. Andersen, L.L.; Mortensen, O.S.; Hansen, J.V. and Burr, H. (2011). A prospective cohort study on severe pain as a risk factor for long-term sickness absence in blue and white-collar workers. Occup Environ Med, 68, 590–592.
2. Bernard, B.P. (1997). Musculoskeletal disorders and workplace factors (NIOSH Publication No. 97-141). Cincinnati, OH, USA: National Institute for Occupational Safety and Health (NIOSH). http://www cdc.gov/niosh/docs/97-141/.
3. Bjorksten, M.G.; Almby, B. and Jansson, E.S. (1994). Hand and shoulder ailments among laboratory technicians using modern plunger-operated pipettes. Appl Ergon, 25, 88–94.
4. Buckle, P. and Li, G. (1997). A practical approach musculoskeletal risk assessment in the real workplace. In: Seppälä P, Luopajärvi T, Nygård CH, Mattila M, editors. Proceedingsof the 13th Triennial Congress of the International Ergonomics Association, Tampere, Finland. Helsinki, Finland: Finnish Institute of Occupational Health, pp. 138–40.
5. Carter, J.B. and Banister, E.W. (1994). Musculoskeletal problems in VDT work: a review. Ergonomics, 37: 1623–48.
6. Centre for Disease Control and Prevention. (2008). Lab ergonomics.
7. David, G. and Buckle, P. (1997). A questionnaire survey of the ergonomic problems associated with pipettes and their usage with specific reference to work related upper limb disorders. Appl Ergon, 28, 257–262.
8. George, E. (2010). Occupational hazard for pathologists: microscope use and musculoskeletal disorders. Am J Clin Pathol, 133(4), 543–548.
9. Haddad, O. (2007). Evaluation of anchor under chest muscle trapeze electromyography dentist's activities. Iran first Inter Conf Ergonom. 165-169.
10. Haynes, S. and Williams, K. (2008). Impact of seating posture on user comfort and typing performance for people with chronic low back pain. Int J Ind Ergon. 38, 35–46.
11. Hignet and McAtamney.(2000) Technical Note: Rapid Entire Body Assessment (REBA), Applied Ergonomics, 31, 201-205.
12. Jafry, T. and O'Neill, D.H. (2000). The application of ergonomics in rural development: a review..Appl Ergon, 31, 263–8.
13. Kalavar, S. and Hunting, K.L. (1996). Musculoskeletal symptoms among cytotechnologists. Lab Med., 11,765-769.
14. Kofler, M., Kreczy, A. And Gschwendtner, A. (2002). Occupational backache" – surface electromyography demonstrates the advantage of an ergonomic versus a standard microscope workstation. Eur J Appl Physiol, 86(6), 492–497.
15. Kuorinka, I.; Jonsson, B.; Kilbom, A.; Vinterberg, H.; Biering.; Sorensen, F.; Andersson, G. and Jorgensen, K. (1987). Standardised Nordic questionnaires for the analysis of musculoskeletal symptoms. Appl Ergon, 18(3), 233-7.
16. Li, G. and Buckle, P. (1997). The development of practical tool for musculoskeletal risk assessment. In: Robertson SA, editor. Contemporary ergonomics. London, UK: Taylor & Francis, p. 442–7.
17. Li, G. and Buckle, P. (1999). Current techniques for assessing physical exposure to work-related musculoskeletal risks, with emphasis on posture-based methods. Ergonomics, 42(5), 674–95.
18. Linton, S.J. and Kamwendo, K. (1989). Risk factors in the psychosocial work environment for neck and shoulder pain in secretaries. J Occup Med., 31: 609–13.
19. McAtamney and Corlett, (1993). RULA. Applied ergonomics, 91-99.
20. Punnett, L. and Wegman, D.H. (2004). Work-related musculoskeletal disorders: the epidemiologic evidence and the debate. J Electromyogr Kinesiol, 1413-23.
21. Ramadan, P. and Mario Ferreira, J.R. (2006). Risk Factors Associated with the Reporting of Musculoskeletal Symptoms in Workers at a Laboratory of Clinical Pathology. Ann. Occup. Hyg., 50, (3), 297–303.

22. Raskeviciene, R. and Maroziene, S. (2005). Evaluation of occupational risk factors and laboratory workers' health in biochemical and clinical laboratories of hospitals of Kaunas city. Medicina (Kaunas) 41(6), 512-21.
23. Sillanpaa, J.; Nyberg, M. and Laippala, P. (2003). A new table for work with a microscope, a solution to ergonomic problems. Appl Ergon, 34(6), 621–628.
24. Spielholz, P.; Silverstein, B.; Morgan, M.; Checkoway, H. and Kaufman, J. (2001). Comparison of self-report, video observation and direct measurement methods for upper extremity musculoskeletal disorder physical risk factors. Ergonomics, 44(6), 588–613.
25. Thompson, S.K.; Mason, E.; Dukes, S. (2003). Ergonomics and cytotechnologists: reported musculoskeletal discomfort. Diagn Cytopathol, 29(6), 364–367.
26. Thorbjornsson, C.B.; Alfredsson, L.; Fredriksson, K.; Michelsen, H.; Punnett, L. and Vingard, E. (2000). Physical and psychosocial factors related to low back pain during a 24-year period. A nested case–control analysis. Spine, 25, 369–374.
27. Varmazyar, S.; Amini, M. and Kiafar, S. (2012). Ergonomic Evaluation of Work Conditions in Qazvin Dentists and its association with Musculoskeletal Disorders Using REBA Method. Journal of Islamic Dental Association of IRAN (JIDA), 24 (3), 182-187.
28. Vanwonterghem, K. (1996). Work-related musculoskeletal problems: some ergonomics considerations. J Hum Ergol (Tokyo), 25(1), 5–13.
29. Weiser, S. (1997). Psychosocial aspects of occupational musculoskeletal disorders. In: Nordin M, Andersson GBJ, Pope MH, editors. Musculoskeletal disorders in the workplace: principles and practice. St. Louis, MO, USA: Mosby-Year Book, p. 51–61.

23

Study of Pennes Bioheat Equation and Its Application to Cryosurgical Freezing of Tumors

B. N. Panigrahi, B. Mallik and M. K. Padhy

Deparment of Mechanical Engineering, Institute of Technical Education and Research
Shiksha O Anusandhan University, Bhubaneswar, India

Email: badripanigrahi@soaunivetrsity.ac.in

ABSTRACT

Bioheat transfer has been a field of study and research since long back. Different researchers have studied various bioheat models to study the relation of heat transfer in biological tissues. Here we have studied the first bioheat model that was developed by Henry Pennes which has been the base model in developing many other models. Here the Pennes model and all the parameters upon which the study depends are studied. The various plots were plotted varying these parameters to study the variation of local tissue temperature. The dependence of parameters like wind velocity, blood perfusion rate and varying ambient conditions on local tissue temperature is used to find the limiting radius of a lungs tumor that can be freezed using a single cryoprobe. Thus, an analytical solution is obtained that helps in curing lungs tumor in cryotherapy. This study also helps in determining the number of cryoprobes that needs to be used to freeze the entire tumor in optimal time. Since the results are found by taking real life data and mathematical simulation, the reliability of results is trustworthy. With these results the tumor can be healed without damaging the nearby healthy living tissues. The results obtained were compared with that given by other models, and the percentage of deviation was negligible. This explained the efficiency of Penne's model.

Keywords: Cryoablation, Penne's steady-state equation, tumor, cyoprobe.

23.1 INTRODUCTION

23.1.1 Bioheat Transfer

Bioheat transfer is the study of the transport of thermal energy in living systems. Life is a series of chemical reactions that consume energy, temperature is perhaps the most important parameter of these reactions. As the biochemical processes are temperature dependent, heat transfer plays a major role in living systems. Also, because the mass transport of blood through tissue causes a consequent thermal energy transfer, bioheat transfer methods are applicable for diagnostic and therapeutic applications involving either mass or heat transfer. Bioheat transfer today is a fast growing area of interest due to its widespread applications in areas of human body modeling, thermal comfort models, burn diagnosis and treatment and in tumor detection. Mathematical modeling of the human body thermal response is a valuable tool to understand the human body thermal behavior under different environmental conditions and activity levels. Thermal models of the human body are able to predict core and partitional body temperatures during time transients in any environment. [1] The single-segment and multi-segmental models for the human body and its thermoregulatory responses have been developed based on the theories of physiology, thermodynamics and transport processes for the prediction of thermal behavior of either the entire human body or a part of it. Thermal comfort is maintained when the heat generated by human metabolism is allowed to dissipate, thus maintaining thermal equilibrium with the surroundings. Thermoregulatory control mechanisms of the human body such as shivering, regulatory sweating, vasodilation, and vasoconstriction are to be considered in comfort analysis that significantly influence the thermal conditions of the human body. A serious problem in emergency medicine is the correct evaluation of skin burn depth in order to make the appropriate choice of treatment. Thermal injury is the irreversible change to living tissue caused by temperature. It occur when the tissue temperature exceeds the range between which normal life processes exist. Both high and low temperature can cause irreversible changes to biomolecules, resulting in injury. Hot or cold packs can be used to create therapeutic heating for injuries near the skin surface.

23.2 LITERATURE REVIEW

The bio-heat modeling is the development of models for the analysis of heat transfer in materials (e.g., tissues, fluids) and system derived from or related to currently or previously living organisms. The solution of the bio-heat equation as applied to particular model is most important, obviously, when those models are developed to explore potential technique for critical therapeutic applications(destroying cancer cells, killing tumors) [7].

In 1948, Penne's published – what most people consider the first of its kind – work on the mathematical coupling between mass transfer of blood perfusion and thermal heat transfer . The key to thermal modeling of blood perfused tissue is the formulation of an appropriate heat transfer equation. Such an equation must take into consideration three factors [7]:

23.2.1 Use of Penne's Equation in Various Fields of Bioheat Transfer

23.2.2.1 The Cardiovascular Cystem

The cardiovascular system is the key system by which heat is distributed throughout the body. The blood serves as the vehicle to transport heat from the areas of high heat to areas of lower heat. In general, this transfer takes place in the direction from body core to extremities such as the limbs and head. In the case where the extremities are hotter than the core body temperature, perfusion would serve to transport heat from the extremity back towards the body core. Whether or not the core temperature rises depends on many factors such as the duration of elevated local temperature at the extremity, temperature difference between extremity and core, volume of blood heated above core temperature, and the rate of blood perfusion. This same moderating phenomena of blood perfusion can be applied to local heat transfer problems, e.g., heating of a tumor [9].

23.2.2.2 Ablative Surgical Procedure

Ablative surgical techniques generally employ some method of energy deposition which destroys cells and tissue with a concomitant increase in temperature at the targeted site. Bioheat transport equations can be applied to the process of energy deposition into the tissue as well as the subsequent conduction/convection heat transport to cells neighboring the targeted site to predict a temperature history and distribution. Such a model could be used to select the most appropriate protocol for the surgery. [9]

23.2.2.3 Cryosurgery and Parameters

Blood flow through large vessel plays an important role in affecting the temperature profiles of the living tissues under cryosurgery. Besides, arresting of blood vessels due to freezing may possibly cause danger to the patient, which needs to be considered when operating the cryoprobe. However, such important issues received few attentions in the bioheat field even up to date. Pilot studies are performed to investigate the cryogenic heat transfer behaviors in biological tissues embedded with large blood vessels. Parametric studies are performed to test the influences of the blood vessel entrance temperature, the vessel diameter, the blood flow velocity and the vessel length, etc. to the whole region's temperature distribution. The critical tissue surface temperature to freeze the blood vessel is also theoretically predicted. To reveal the role of the countercurrent artery-vein blood flows to the transient phase change in living tissues subject to freezing, qualitative simulating and experiments are performed. The study raises an important issue to study the phase change heat transfer of blood vessels to the living tissues subject to cryosurgery, which may have significant clinical applications. The present method can also possibly be extended to wider fields such as heat transfer in buried pipes and collectors. [8]

23.2.2.4 Energy Balance of the Phase Change Process Across the Freezing Front of a Cryoprobe

The simple bioheat equation given by Penne's was used to calculate the limiting radius of the tumor which could be freezed [4].

Pennes' Equation

$$\rho_s C_S \frac{\partial T}{\partial t} = \frac{\partial}{\partial x_j}\left(k_s \frac{\partial T}{\partial x_j}\right) + \rho_f c_f \omega_{\text{pennes}} (T_a - T) + q_m \qquad \text{...(1)}$$

Here we have used the integral method to find an analytical expression for the limiting radius of the tumor, which one single cryoprobe can freeze at the maximum.

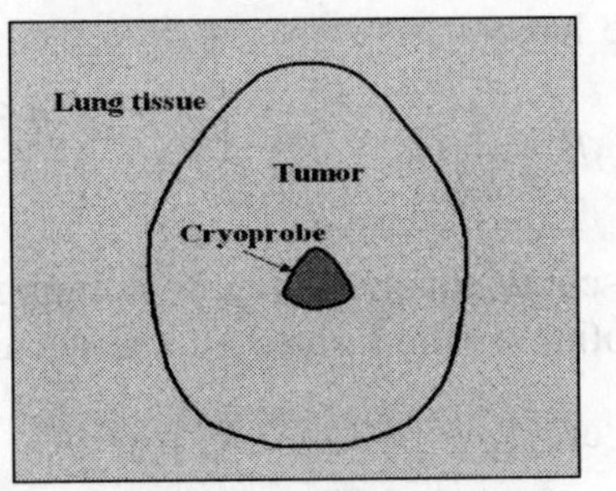

Numerical model for cryoablation

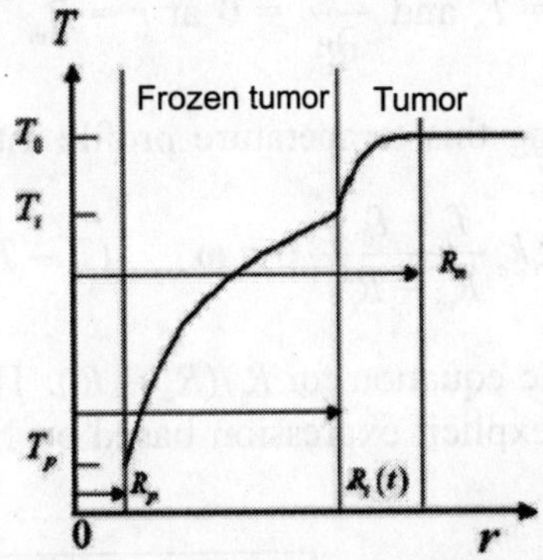

Temperature profile around a probe

Assumptions

- The probe is a right circular cylinder
- The tumor is very large and heat transfer to healthy lung tissue is negligible.
- Referring to the above figure, the energy balance relationship at $r = R_i$ the freezing front at is as follows:

$$\rho_i h_{sf} R_i \frac{dR_i}{dt} = R_i k_i \left(\frac{\partial T}{\partial r}\right) - R_i k_c \left(\frac{\partial T}{\partial r}\right) - \rho_C C_C (T_o - T_i) R_i \frac{dR_i}{dt} \qquad \text{...(2)}$$

In equation (2),

The first term on the right-hand side correspond to the conduction heat flux evaluated at the ice side.

The second term on the right-hand side corresponds to the conduction heat flux evaluated at the unfrozen side.

The third term on the right-hand side corresponds to sensible heat entering the interface at the freezing front.

The freezing front moves so slowly that a quasi-steady approximation may be valid. Assuming that the temperature profile within the frozen region follows the steady state

$$\frac{T - T_p}{T_i - T_p} = \frac{\ln (r/r_p)}{\ln (R_i/R_p)} : R_p \le r \le R_i \qquad \text{...(3)}$$

where we may estimate the first term on the right-hand side as

$$R_i k_i \left(\frac{\partial T}{\partial r} = k_i \frac{T_i - T_p}{\ln (R_i/R_p)}\right) \qquad \text{...(4)}$$

In order to estimate the second term on the right-hand side (representing the heat flux from the unfrozen tumor to the interface), we write the bioheat equation (3) for the unfrozen tumor region using the cylindrical coordinate system, which, under the quasi-steady approximation, may be integrated to give:

$$-R_i k_c\left(\frac{\partial T}{\partial r}\right) - \rho_c\, c_c\, \omega_{pennes}\int_{R_i}^{R_m} r(T - T_0)dr + q_m'''\int_{R_i}^{R_m} rdr = 0 \qquad \text{...(5)}$$

Let us assume that the temperature in this unfrozen region follows:

$$\frac{T - T_0}{T_i - T_0} = \left(1 - \frac{r - R_i}{R_m - R_i}\right)^2 : R_i \le r \le R_m \qquad \text{...(6)}$$

The equation satisfies:

$T = T_i$ at $r = R_i$ and

$T = T_0$ and $\dfrac{\partial T}{\partial r} = 0$ at $r = R_m$

Then substituting this temperature profile into Eq. (5), we have

$$2R_i k_c \frac{T_i - T_0}{R_m - R_i} - \rho_c c_c \omega_{pennes}(T_i - T_0)\left(\frac{1}{3}(R_m - R_i)R_i + \frac{1}{12}(R_m - R_i)^2\right) + q_m'''\left(\frac{R_m^{\,2} - R_i^{\,2}}{2}\right) = 0 \qquad \text{...(7)}$$

which forms a cubic equation for $R_i/(R_m - R_i)$. The root of the cubic equation is quite complex. However, it is found that the following explicit expression based on Newton's shooting method gives a quite accurate value for the root:

$$\frac{R_i}{R_m - R_i} = R_i\sqrt{\frac{1}{6}\left(\frac{\omega_{pennes}}{\alpha_c} + 3\frac{q_m}{k_c(T_o - T_i)}\right)} + \frac{1}{8}\left(\frac{\dfrac{\omega_{pennes}}{\alpha_c} + 6\dfrac{q_m'''}{k_c(T_o - T_i)}}{\dfrac{\omega_{pennes}}{\alpha_c} + 3\dfrac{q_m'''}{k_c(T_o - T_i)}}\right) \qquad \text{...(8)}$$

where,

Thus, the second term on the right-hand side of Eq. (3) may be estimated as:

$$R_i k_c\left(\frac{\partial T}{\partial r}\right) = 2k_c(T_o - T_i)\left(R_i\sqrt{\frac{1}{6}\left(\frac{\omega_{pennes}}{\alpha_c} + 3\frac{q_m}{k_c(T_o - T_i)}\right)} + \frac{1}{8}\left(\frac{\dfrac{\omega_{pennes}}{\alpha_c} + 6\dfrac{q_m'''}{k_c(T_o - T_i)}}{\dfrac{\omega_{pennes}}{\alpha_c} + 3\dfrac{q_m'''}{k_c(T_o - T_i)}}\right)\right) \qquad \text{...(9)}$$

Substituting equation (9) and (4) into Eq. (2), we have

$$(\rho_i h_{sf} + \rho_c c_c(T_o - T_i))R_i\frac{dR_i}{dt} = k_i\frac{T_i - T_p}{\ln(R_i/R_p)} - 2k_c(T_o - T_i)\left(R_i\sqrt{\frac{1}{6}\left(\frac{\omega_{pennes}}{\alpha_c} + 3\frac{q_m}{k_c(T_o - T_i)}\right)} + \frac{1}{8}\left(\frac{\dfrac{\omega_{pennes}}{\alpha_c} + 6\dfrac{q_m'''}{k_c(T_o - T_i)}}{\dfrac{\omega_{pennes}}{\alpha_c} + 3\dfrac{q_m'''}{k_c(T_o - T_i)}}\right)\right) \qquad \text{...(10)}$$

Which reduces to,

$$dt^* = \frac{1 + Sr}{Ste} \times \frac{R_i^* \ln R_i^*}{1 - \dfrac{\ln R_i}{Cr}}\left(\sqrt{\frac{2}{3}(\omega^* + 3Met)^{1/2} R_i^* + \frac{\omega^* + 6Met}{4(\omega^* + 3Met)}}\right) dR_i^* \qquad \text{...(11)}$$

where, $R_i^* = R_i/R_p$ ...(12)

and $t^* = \alpha_i t/R_p^{\,2}$ ...(13)

where, $a_i = k_i/\rho_i c_i$

Moreover, the following dimensionless parameters are introduced:

$$Ste = \frac{c_i(T_i - T_p)}{h_{sf}} \quad ...(14)$$

$$Sr = \frac{\rho_c c_c (T_o - T_i)}{\rho_i\, h_{sf}} \quad ...(15)$$

$$\omega^* = \frac{\omega_{pennes}\, R_p^{\,2}}{\alpha_c} \quad ...(16)$$

$$Cr = \frac{k_i(T_i - T_p)}{k_c\,(T_0 - T_i)} \quad ...(17)$$

$$Met = \frac{q_m'''\, R_p^{\,2}}{k_c(T_o - T_i)} \quad ...(18)$$

By integrating the above foregoing equation we can obtain the dimensionless time t^* required to freeze the given tumor of a limiting radius.

Here we see that the quasi-steady state assumption, which we made in the beginning holds good for most part of the freezing process, when $\frac{t^*\,Ste}{1+Sr} > 1$, which roughly gives: $t > 1$ s.

Now the fact is there exists a limiting radius of the tumor which can be freezed by a single cryoprobe.

where:

$$R_{lim}^* = R_{lim}/R_p \quad ...(19)$$

This value is obtained by setting: $dR_{lim}^*/dt^* = 0$

$$\frac{\ln R_{lim}^*}{Cr}\left(\sqrt{\frac{2}{3}}(\omega^* + 3Met)^{1/2} R_{lim}^* + \frac{\omega^* + 6Met}{4(\omega^* + 3Met)}\right) = 1 \quad ...(20)$$

Usually, is much larger than Met. For such cases, the following explicit expression based on Newton's shooting method may be used to give a reasonably accurate value for R_{lim}:

$$R_{lim}^* = \frac{\left(\sqrt{\frac{3}{2\omega^*}}Cr + R_0^*\right)R_0^* + \sqrt{\frac{3}{32\omega^*}}R_0^*(1 - \ln R_0^*)}{R_0^*(1 + \ln R_0^*) + \sqrt{\frac{3}{32\omega^*}}} \quad ...(21)$$

where,

$$R_0^* = \frac{\frac{3Cr^2}{\omega^*} + \frac{3Cr}{8\omega^*}\left(1 - \ln\sqrt{\frac{3}{2\omega^*}}Cr\right)}{\sqrt{\frac{3}{2\omega^*}}Cr\left(1 + \ln\sqrt{\frac{3}{2\omega^*}}Cr\right) + \sqrt{\frac{3}{32\omega^*}}} \quad ...(22)$$

23.3 RESULTS AND DISCUSSION

The analytical solution of the Pennes' bioheat equation gives the dimensionless temperature term, theta, that depends upon various parameters like wind velocity, blood perfusion, ambient temperature and alike. We have studied the effect of variation of these parameters on theta and the results have been plotted on different graphs using MATLAB simulation software. From the plots obtained the following observations were made:

1. We know that with change in wind velocity, heat transfer coefficient changes which affects the local tissue temperature. This phenomenon was observed in Plot-1. Here, at the core, with increase in wind velocity the local tissue temperature decreases till ambient temperature reaches 37°C. After that, there is an increase in Local tissue temperature with increase in wind velocity.
2. Plot-2 shows the same phenomenon as plot-1 at the body surface temperature, i.e, $R = r$.
3. Plot-3 shows that at the body surface, with a very small value of blood perfusion rate the local tissue temperature remains almost constant with varying ambient temperature till 37°C, after which T increases with increase in blood perfusion rate.
4. Plot-4 shows that at the core, with a small increase in blood perfusion rate the local tissue temperature slightly increases with the increase in ambient temperature till 37°C, after which it decreases.
5. Plot-5 shows that the local tissue temperature decreases with increase in ambient temperature at varying wind velocity.
6. Plot-6 shows that with increase in wind velocity the dimensionless temperature theta decreases with the dimensionless radius, zeta. At a low wind velocity value of theta almost remains constant with increase in zeta
7. Plot-7 shows that a low blood perfusion rate the value of theta remains almost constant with varying zeta. But as the blood perfusion rate increases the graph takes a shape of an increasing parabola.

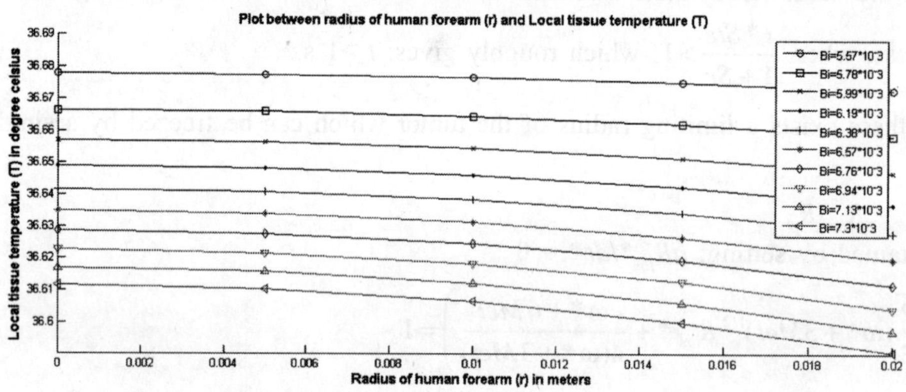

Fig. 23.1 (Tinf) & *T* at constant perfusion rate (*w*) at zeta = 0

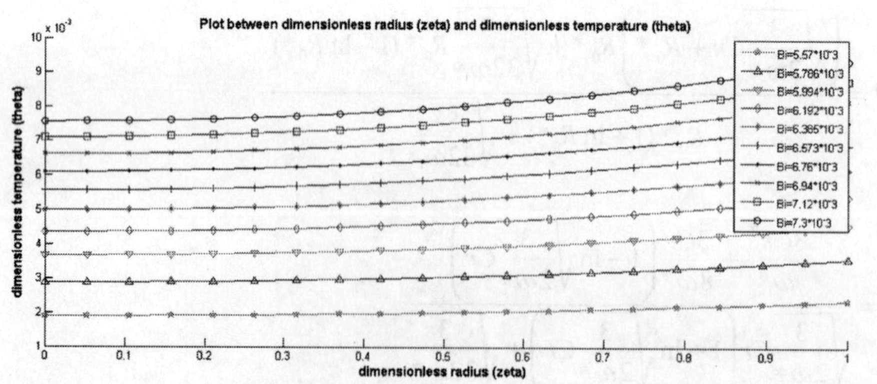

Fig. 23.2 (Tinf) and (*T*) at zeta = 1

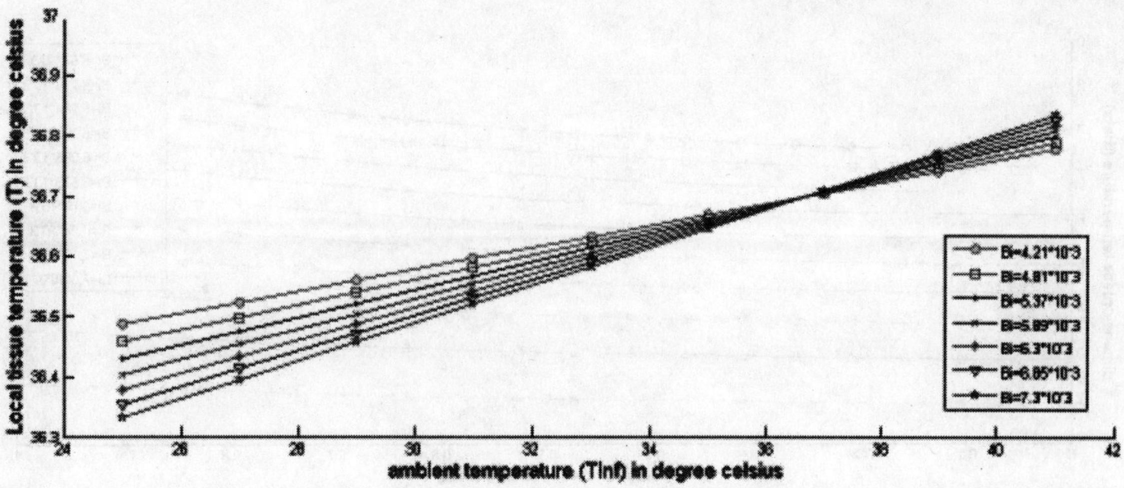

Fig. 23.3 Ambient temperature (Tinf) & tissue at constant at constant blood perfusion rate at zeta = 1

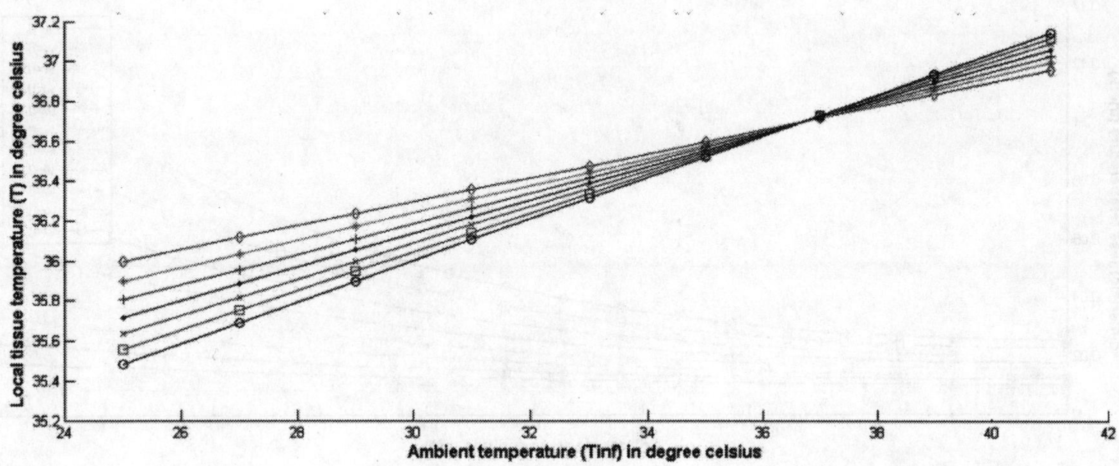

Fig. 23.4 Ambient temperature (Tinf) & tissue at constant at constant blood perfusion at zeta = 1

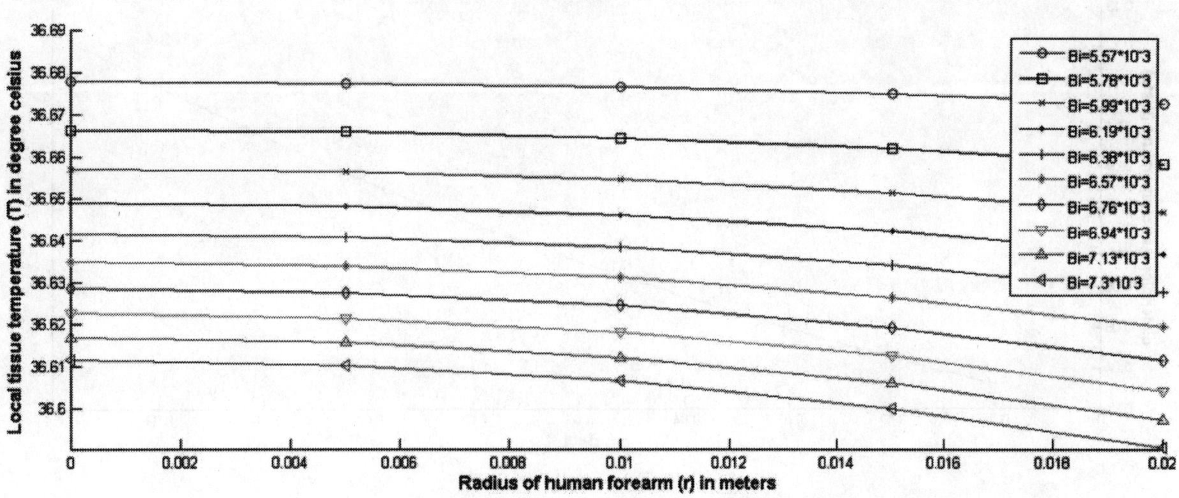

Fig. 23.5 Radius of forearm & local tissue temperature at different wind velocities

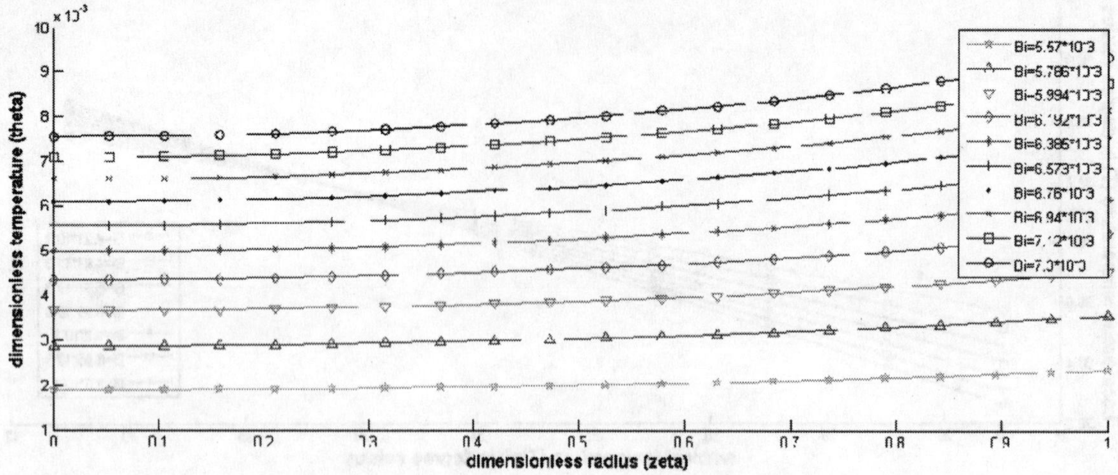

Fig. 23.6 Dimensionless radius (zeta) and dimensionless temperature (theta) at different wind velocities

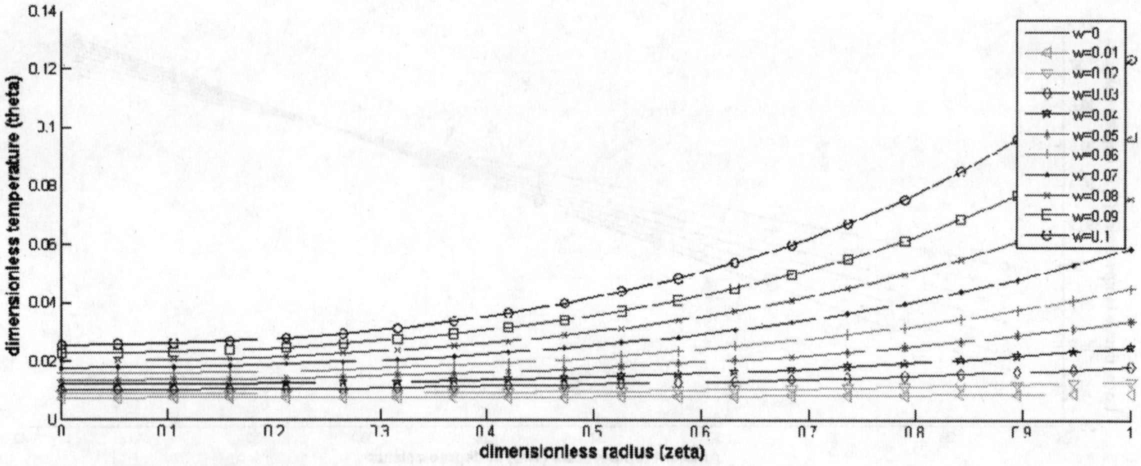

Fig 23.7 Dimensionless radius (zeta) and dimensionless (theta) at different perfusion rate (w)

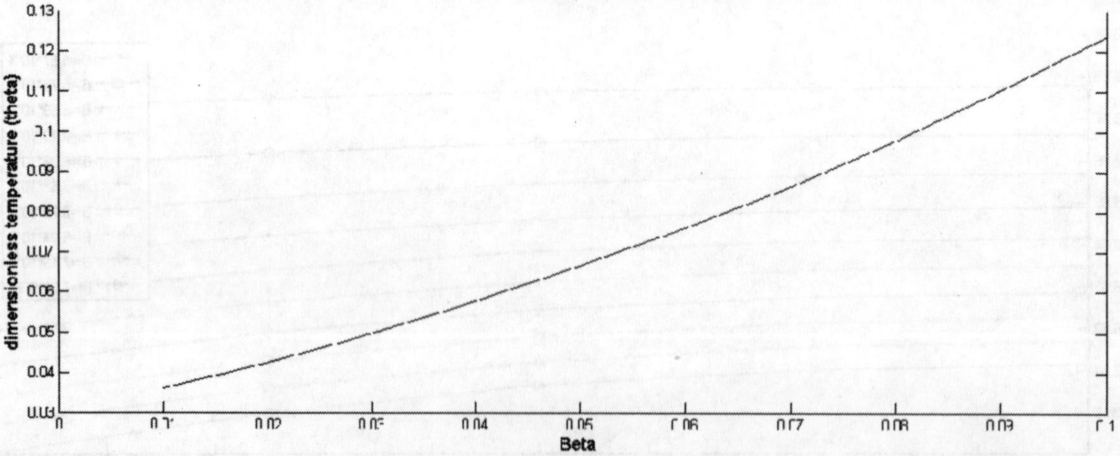

Fig. 23.8 Graph between beta (perfusion rate) and temperature dimensionless temperature (theta)

23.4 CALCULATIONS

From the equations (11) and (12), we can obtain the value of the limiting radius of the tumor by setting

$$\frac{dt^*}{dRi^*} = 0$$

The optimized value of R_{lim} obtained is:

$$R_{lim} = \frac{\left(\sqrt{\frac{3}{2\omega^*}Cr} + R_o^*\right)R_o^* + \sqrt{\frac{3}{32\omega^*}}R_o^*(1 - \ln R_o^*)}{R_o^*(1 + \ln R_o^*) + \sqrt{\frac{3}{32\omega^*}}}$$

where, $$R_o^* = \frac{\frac{3Cr^2}{\omega^*} + \frac{3Cr}{8\omega^*}\left(1 - \ln\sqrt{\frac{3}{2\omega^*}Cr}\right)}{R_i\sqrt{\frac{3}{2\omega^*}Cr}\left(1 + \ln\sqrt{\frac{3}{2\omega^*}Cr}\right) + \sqrt{\frac{3}{32\omega^*}}}$$

For Cr = 15.4: Values obtained for limiting radius corresponding to the following values used are shown in Table 23.1.

Table 23.1: Data of limiting radius

	ω_{pennes} (s-1)	ω^*	R_o^*	R_{lim}^*	R_{lim} (mm)
1	0.004	0.307	14.66	12.82	12.82
2	0.022	0.169	18.54	15.907	15.907
3	0.004	0.0307	36.68	30.06	30.06

23.5 ERROR CALCULATIONS

1. Limiting radius value as in the journal, referred, [4] R_{lim} = 12.9 mm,
   For $\omega_{pennes} = 0.04\ s^{-1}$
   Limiting radius value as obtained in the table, for $\omega_{pennes} = 0.04\ s^{-1}$
   R_{lim} = 12.82 mm
   Error = ((12.9-12.82)/12.92)*100 = 0.6%
2. Limiting radius value as in the journal referred: [4] R_{lim} = 29.92 mm,
   For $\omega_{pennes} = 0.004\ s^{-1}$
   Limiting radius value as obtained in the table, for $\omega_{pennes} = 0.004\ s^{-1}$
   R_{lim} = 30.06 mm
   Error = ((29.92–30.06)/29.92)100 = 0.5%

23.6 CONCLUSION

Although various models of bioheat transfer has been proposed by various authors, it is seen that the initial model proposed by Henry Pennes' still holds good and is widely accepted. Heat transfer affecting living organism under the influence of blood perfusion is given great and increasing attention in medicine today. One reason is the increasing use of thermal treatment methods in for example heart and neuro-surgery, treating cancerous cells, tumors and skin depth. Apart from its other advantages mentioned earlier in the report, we found it equally effective for the case of determining the limiting radius of the lung tumor that could be freezed using a single cryoprobe. When the results were compared with that of the other modified models the deviation was found out to be negligible. This verifies the efficiency of the model and justifies it. The work done in calculating the limiting radius gives a more reliable and logical approach to cryotherapy for curing lungs tumor. As we have found the relation between the limiting radius and the freezing time required, we can determine the number of cryoprobes required to freeze the entire tumor in an optimized time without harming the other surrounding tissues. As wind speed affects the rate at which the body loses heat by convection and in the results and discussion, we have seen that the wind velocity and the blood perfusion rate are two most important parameters which affects the local tissue temperature of the human body. These variation governs the study of human comfort making it more meaningful and acceptable. It shows that under particular conditions of blood perfusion rate and wind velocity how the local tissue temperature varies with the change in ambient temperature. Thus, by setting proper wind velocity we can control the temperature as per our will. Since the report gives results calculated analytically taking real-world data, we obtain more feasible and reliable effects which are unbiased by assumptions.

The difference in limiting radii obtained by the two methods is only 0.6%. This would suggest that using a more complicated model does not give any appreciable difference from using Pennes model. Also, our estimate is the smaller of the two values and would thus be a more conservative estimate while planning a cryosurgical procedure.

REFERENCES

1. Jiji. Latif M., (2009) Heat Conduction, Newyork, Springer.
2. Rubinsky. B.,"Numerical bio-heat transfer", 2006, ISBN: 978-0-471-34878-8, Willey.
3. Jonathan W. Valvano , (2005) "Bioheat Transfer", University of Texas Austin, Texas.
4. Yusuke Kuwahara, Kansei Iwata, Masafumi Kawamura, (2008). "The Limiting Radius for Freezing a Tumor During Percutaneous Cryoablation." J. Heat Transfer, Vol. 130 (11), 111101.
5. M. Sallouma, N. Gaddara, K. Ghalib (2007). "A new transient bio-heat model of the human body and its integration to clothing models" 46(4), 371–384
6. Roger W. Pryor, Pub: Jones & Bartlett learning, (2009), Multiphysics Modeling using COSMOS.
7. Charlotte S. Daniels*, Boris Rubinsky , Cryosurgery with Pulsed Electric Fields,
8. Department of Mechanical Engineering, University of California, Berkeley, Berkeley, California, United States of America.
9. Tzu-Ching Shih , Ping Yuan, Win-Li Lin, Hong-Sen Kou, (2006). "Analytical analysis of the Pennesbioheat transfer equation with sinusoidal heat flux condition on skin surface", Vol. 29(9), pp. 946-953
10. Tzu-Ching Shih, Hong-Sen Kou, Chihng-TsungLiauh, Win-Li Lina, 2005. "The impact of thermal wave characteristics on thermal dose distribution during thermal therapy: A numerical study", Vol. 32(9), pp. 3029-3036.

24

A Correlation Study on Fractals Extracted During Fatigue and Non-fatigue Condition with Anthropometric Data

J. Jibin, M. Navaneethakrishna and S. Ramakrishnan

Non-invasive Imaging and Diagnostics Laboratory
Dept. of Applied Mechanics, Indian Institute of Technology, Madras, Chennai, India
Email: jibinj@gmail.com

ABSTRACT

In this work, an attempt is made to correlate the fractal dimension of surface electromyography signals under fatigue and non-fatigue conditions with the anthropometric data. Surface electromyography signals are recorded from the biceps brachii muscles of 20 healthy subjects with no history of neuromuscular disease. Anthropometric data such as standing height, weight, biceps circumference, and forearm circumference are measured for the same subject. Based on the measured anthropometric parameters arm muscle area and total muscle mass is calculated. The acquired signals are pre-processed and Higuchi fractal based feature is extracted. The result shows no significant difference in the acquired signals under fatigue and non-fatigue conditions. The extracted fractal features show poor separability. An increase in correlation between the fractal dimension and anthropometric data is observed for fatigue conditions. It appears that this analysis can be used to study the relationship between the anthropometry and rate of fatigue.

Keywords: Fractal, surface electromyography, biceps brachii, anthropometry.

24.1 INTRODUCTION

Surface Electromyography (sEMG) is a technique, used to analyze the electrical activity of skeletal muscle. It is non-invasive, safe, simple and can be used in static and dynamic conditions. The sEMG is one of most widely used technique to assess and characterize muscle fatigue. Muscle fatigue is a neuromuscular condition in which the muscle is unable to maintain the required or expected force [4].

In fatigue condition the motor unit recruitment increases and motor unit firing decreases, which leads to an increase in amplitude and reduction in frequency. Therefore, time domain and frequency domain based features have been reported in the literature [4]; [8].

Fractal dimension (FD) is a powerful mathematical tool for modelling many physical and biological time series signal with high complexity and irregularity. It has been reported that FD are very sensitive to small changes in Time-frequency distribution. Several studies with sEMG and FD are reported in the literature [8].

Anthropometric data involves measurement of the body dimension and other physical characteristic such as standing height, total body muscle mass and mass of the fore arm. Phinyomark et al. (2013) have showed correlation between sEMG and anthropometry [9]. EMG force relationship has been established by Marras et al. (1991). This relates muscle size, cross-sectional area and length, to their contribution in muscle force. Anthropometric variables are considered for studying force variations during muscle fatigue conditions.

In this study, sEMG signals under non-fatigue (NF) and fatigue (F) conditions are recorded and analyzed. Anthropometric data is measured. Features are extracted from recorded signal. A correlation study is carried out between extracted features and various anthropometric parameters .

24.2 MATERIALS AND METHODS

20 healthy volunteers with no history of neuromuscular disorders participated in this study (avg. age 21, avg height 176 cm, average weight 65 kg).

24.2.1 sEMG Data Acquisition and Exercise Protocol

Ag-Ag Cl Electrodes are placed on the bulk of the BB muscle with an inter electrode distance of 3 cm after skin preparation. Biopac MP36 is used for this study. The signals are recorded at a 24 bit resolution and a sampling rate of 10 kHz. The subjects are made to stand upright on an insulated wooden platform with a weight of 6 kg. They are asked to perform biceps curl exercise. Encouragement is given to continue the experiment until they experienced unacceptable BB discomfort [4]. The signals are acquired during isometric contraction at an angle of 90° with vertical axis. The angle measurement is ensured using mechanical goniometers. The data acquired is mean removed, filtered (5 – 400 Hz butterworth band pass IIR, 50 Hz notch) offline [6].

24.2.2 Anthropometric Measurement

In this study four static anthropometric variables such as biceps circumference (cm), forearm circumference (cm), arm muscle area (cm^2) and total body muscle mass (kg) considered. Standing height, biceps circumference, forearm circumference (cm) and triceps skin fold thickness are measured using stature meter, measurement tape and a calliper, from dominant arm of the subject following a standard protocol defined by Phinyomark et al. (2013). Arm muscle area and total body muscle mass were estimated.

24.2.2.1 Arm Muscle Cross-sectional Area (AMA)

AMA is defined as an estimation of the cross-sectional area of the muscle portions of the upper arm [12].

where MAC = Mid Arm Circumference or Biceps Circumference in cm

TSF = Triceps Skin Fold Thickness in cm

24.2.2.2 Total Body Muscle Mass (TMM)

TMM is defined as the combined weight of all the muscles present in the whole body [12].

where *ht* = Standing Height in cm

24.2.3 Feature Extraction

The signal is processed with hamming windows of length 200 ms with 30% overlap. Higuchi fractal dimension is extracted in each epoch and mean is computed.

24.2.3.1 Higuchi Fractal Dimension

Higuchi fractal dimension (HFD) gives the complexity of the signal [2].

Given a one-dimensional signal the HFD is calculated in the following algorithm

1. Form k new time series defined by k is the discrete time interval between points and $m = 1, 2, 3, \ldots, k$ is the initial time value
2. Length of each new time series is given as where N is length of the original time series X and is a normalization factor.

The length of the curve for each time interval k is defined as the average of the k values $L(m, k)$, for $m = 1, 2, \ldots, k$:

$L(k)$ is plotted against 1/k on a double logarithmic scale with $k = 1, 2, \ldots k$ max and the data should fall on the straight line with slope FD of X.

HFD = slope of the line that fits the pairs $\{\ln [L(k)], \ln (1/k)\}$ in a least square sense.

In this study k max is taken as 128.

The mean of the all the HFD is computed.

24.3 RESULTS AND DISCUSSION

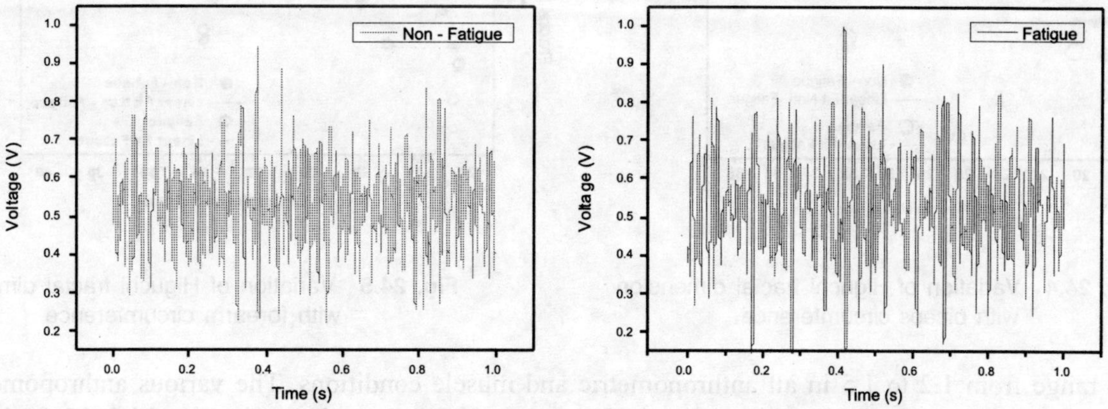

Fig. 24.1 Representative EMG signal after filtering

Figure 24.1 shows the typical representation of recorded sEMG signal before and after fatigue. Subjective evaluation does not provide any clear distinction between signals under NF and F condition.

Table 24.1: Anthropometric parameters variations

Anthropometric Parameters	*Mean ±Standard deviation*
BC (cm)	29.0578 ± 4.6614
AMA (cm²)	40.2105 ± 14.0022
TMM (Kg)	24.4421 ± 2.0605
FC (cm)	24.3509 ± 6.5696

The mean and standard deviation of the measured anthropometric data are calculated and is shown in Table 24.1.

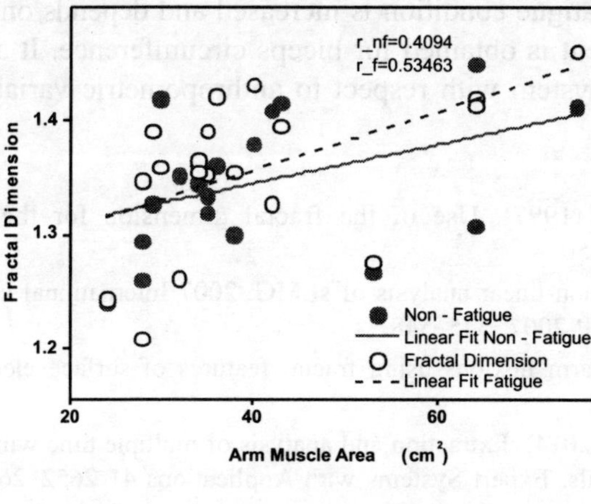

Fig. 24.2 Variation of Higuchi fractal dimension with dimension with arm muscle area

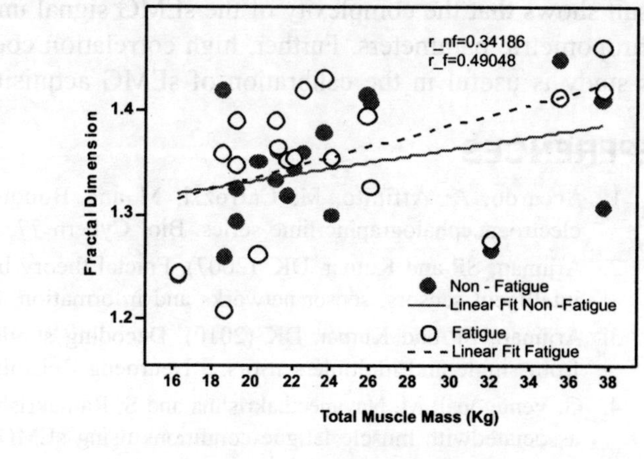

Fig. 24.3 Variation of Higuchi fractal total body muscle mass

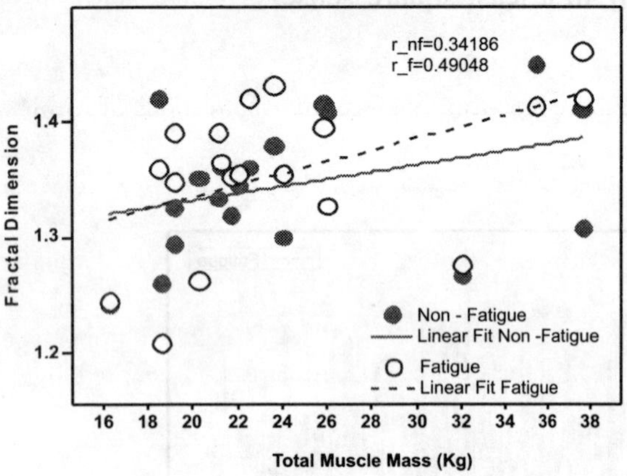

Fig. 24.4 Variation of Higuchi fractal dimension with biceps circumference

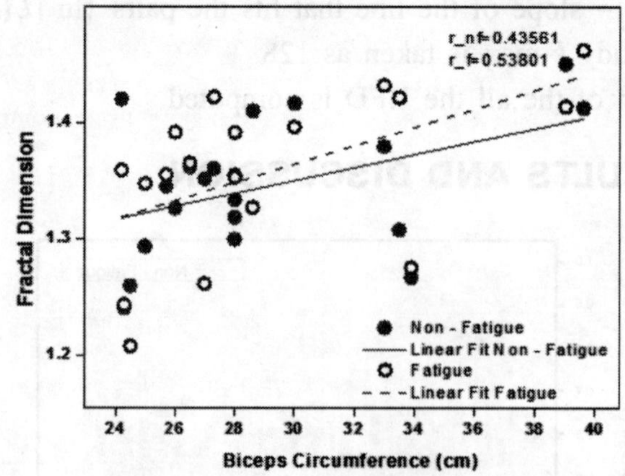

Fig. 24.5 Variation of Higuchi fractal dimension with forearm circumference

The FD range from 1.2 to 1.5 in all anthropometric and muscle conditions. The various anthropometric data is plotted against the FD and Pearson correlation coefficient is calculated as shown in Fig. 24.5. Multiple-regression analysis and variance are used to determine the equality of slopes and intercepts for the regression lines.

A positive slope is obtained for all linear regression fits. It is found from the Fig. 24.5 that with the increase in the anthropometry parameters such as biceps circumference and arm muscle area the FD increases. Several studies conducted previously using techniques such as biopsy, MRI, CT have shown that there is some correlation between the anthropometry and the number of muscle fibers [11]; [7]. It can be inferred that when number of muscle fibers increases the FD also increases. The increased value in FD shows that the complexity of the signal increases in fatigue conditions. This may be due to the increase in the number of motor units recruited during fatigue [4].

24.4 CONCLUSION

In this work, a correlation study is carried out using extracted features of sEMG signal and anthropometric parameters. For this experiment sEMG signals are recorded from biceps brachii under fatigue and non fatigue conditions. Anthropometric data are also measured. sEMG signals are pre-processed for noise removal and FD is extracted. The anthropometric data and extracted features are correlated and the corresponding correlation coefficients are obtained. Result shows that the complexity of the sEMG signal under fatigue condition is increased and depends on considered anthropometric parameters. Further, high correlation coefficient is obtained for biceps circumference. It appears that this study is useful in the calibration of sEMG acquisition system with respect to anthropometric variables.

REFERENCES

1. Accardo, A; Affinito, M; Carrozzi, M and Bouquet F. (1997). Use of the fractal dimension for the analysis of electroencephalographic time series. Biol Cybern 77, 339–350.
2. Arjunan, SP and Kumar, DK. (2007). Fractal theory based non-linear analysis of sEMG. 2007 International conference on intelligent sensors, sensor networks and information, ISSNIP 2007, 545–548.
3. Arjunan, SP and Kumar, DK (2010). Decoding subtle forearm flexions using fractal features of surface electromyogram from single and multiple sensors. J Neuroeng Rehabil 7, 53.
4. G, Venugopal; M, Navaneethakrishna and S, Ramakrishnan. (2014). Extraction and analysis of multiple time window features associatedwith muscle fatigue conditions using sEMG signals. Expert Systems with Applications 41 2652–2659.
5. Housh, D.J.; Housh, T.J; Weir, J.P; Weir, L.L; Johnson, G.O and Stout, J.R. (1995). Anthropometric estimation of thigh muscle cross-sectional area. Med. Sci. Sports Exerc., 27, 784–791.
6. J.A., Burne; V L, Carleton and N J, O'Dwyer. (2005). The spasiticity paradox: movement disorder or disorder of resting limbs?. Journal of Neural Neurosurgery Psychiatry, 76, 47-54.
7. Klein, Cliff S, Greg D Marsh, Robert J Petrella, and Charles L Rice. (2013). Muscle Fiber Number in the Biceps Brachii Muscle of Young and Old Men. Muscle & Nerve 28(1), 62–68.

8. Phinyomark, A; P, Phukpattaranont and C, Limsakul, (2012). Fractal analysis features for weak and single-channel upper-limb EMG signals. Expert Systems with Applications: An International Journal, 39(12), 11156-11163.
9. Phinyomark, A; Quaine, F; Charbonnier, S; Serviere, C; Tarpin-Bernard, F and Laurillau Y. (2013). A feasibility study on the use of anthropometric variables to make muscle–computer interface more practical, Engineering Applications of Artificial Intelligence, 26, 1681–1688.
10. Marras, W.S. and Sommerich, C.M. (1991). A three-dimensional motion model of loads on the lumbar spine: I Model structure, Hum. Factors, 33, 123–137.
11. MacDougall, J.D.; D.G., Sale; S.E., Alway, and J. R., Sutton. (1984). Muscle Fiber Number in Biceps Brachii in Bodybuilders and Control Subjects. Journal of Applied Physiology, 57:1399-1403.
12. Steven B Heymsfield, Clifford Mc Manus, Janet Smith, Victoria Stevens and Daniel W Nixon. (1982) Anthropometric measurement of muscle mass: revised equations for calculating bone-free arm muscle area. The American Journal of Clinical Nutrition, 36, 680-690.
13. Zhengquan Xu and Shaojun Xiao. (1997). Fractal Dimension of Surface EMG and its Determinants, Proceedings - 19th International Conference - IEEE/EMBS Chicago, IL. USA, 1570-1573.

25

Finite Element Analysis of Segmental Cervical Spine Facet Strain and Disc Stress After Total Disc Replacement

R. Pradhan, A. Roychowdhury and S. Majumder

Department of Aerospace Engineering & Applied Mechanics
Bengal Engineering and Science University Shibpur, Howrah–711103, West Bengal, India

Email: rururaj.mech@gmail.com

ABSTRACT

Degeneration of intervertebral disc results in severe back pain and limited mobility. Current surgical treatments focus on anterior cervical disectomy and fusion (ACDF) at the concerned levels. However, spinal fusion may be responsible for progressive degeneration at the adjacent motion segments over time. ACDF also ceases the range of motion (ROM) at the cervical disc level resulting decrease in mobility. It is postulated through both clinical and bio mechanical studies that ACDF produces increase in stress on disc above and below fusion level leading to disc herniation and other degenerative diseases. Several alternatives to fusion have been studied previously, such as allograft transplantation and total disc replacement (TDR). The purpose of this document is to compare the disc stress and cervical body strain between the spine models containing natural and artificial intervertebral disc.

Keywords: Ankle joint, ground reaction, achilles tendons, stance phase.

25.1 INTRODUCTION

The human cervical spine is a multi-component mechanical feature with widely varying structural anatomy, material properties and functional requirements. The principal function of this part of vertebral column during physiologic loading is to maintain the structural integrity and interrelationship between the various components, and protect the spinal cord. The applied external load (or moment) to the cervical spine, as might be expected, is resisted unequally by its components. Cervical spinal instability due to ligamentous injury, degenerated disc and facetectomy is a subject of great interest. There are few analytical investigation reported on the biomechanical response of cervical spine in these respects. Clinically, the spinal motion segment is defined to be unstable if it exhibits abnormally large increase in rotational or translational displacements under physiological load. The sole objective of the work is to design patient specific human cervical spine implant on the basis of evaluation of morphology and computed tomography scan. The work is to be carried out in two phases. Primarily an 3-dimensional geometrically and mechanically accurate finite element model of human cervical spine (C3-C7) is to be developed as well as parametric study on the roles of ligaments, facets, and disc nucleus of human cervical spine (C3–C6) has to be conducted. In the subsequent phase implant design has to be carried out based on the morphometric and structural analysis of the cervical spine of Indian population. The intervertebral discs, undergo dramatic changes in structure, composition, and mechanical function with age. The intervertebral discs are also susceptible to degenerative disc disease. The effects of normal aging and of degenerative disc disease are very similar and difficult to differentiate. Cervical disc degeneration is a common pathology which may require surgical intervention as a final treatment. The intervertebral discs play a very important role in mobility and load transfer through the spinal column. Any load through the spinal column is transmitted to the intervertebral disc from the vertebral body. The normal intervertebral disc is anisotropic in structure, the jelly-like nucleus pulposus acts like a fluid filled bag and swells under pressure. The interaction between the intervertebral disc components is similar to a thick-walled pressure vessel, and allows the intervertebral discs to act as shock absorbers, absorbing and transmitting the loads experienced by the spine. With degeneration however, the biomechanical properties of the disc are altered. Once degeneration sets in, the intervertebral disc goes through a cascade of degenerative changes resulting in a loss of demarcation between the nucleus pulposus and annulus fibrosus, loss of disc height, a decrease in height as the intervertebral disc loses its ability to rehydrate after loading, and altered loading on the intervertebral disc and surrounding tissues . Though, the exact pathogenesis of the degenerative process is still unknown, several factors that might cause degeneration are: aging, mechanical factors due to occupational exposure, abnormal loading conditions, and the loss of nutrition to the disc.

25.2 MATERIALS AND METHODS

A 3D model of cervical spine was generated based on the scans of specimen maintaining the accuracy as much as possible. Computed tomography (CT) scan data of 50 years old male subject in the DICOM (0.208 × 0208 × 2.0) format were collected. Relevant geometric parameters were examined to ensure that the specific specimen represented the population mean with respect to disc height, vertebral height and orientation, cartilaginous endplate thickness, facet shape, size, and orientation, foraminal areas, and sagittal alignment. Contours were generated using a threshold value of 600 HU.11 These CT slices were threshold within the gray value range of the cervical vertebrae bone. CT slices were used with MIMICS image processing software (Materializes Interactive Medical Image control System) to create the cervical vertebrae (C0-T1). These zones were then stacked to create the 3D model of the vertebrae bone, and exported to a finite elements package ANSYS. Cortical bone was represented as an isotropic, elastic material with an elastic modulus (E) of 10,000 MPa and a Poisson ratio (v) of 0.3. Trabecular bone was also represented as an isotropic elastic material with an elastic modulus based linearly on Hounsfield attenuation per Teo et al. (2001). The spatially varying trabecular bone properties were based on the computed tomography data, which was registered automatically to the finite element model by the help of MIMICS.

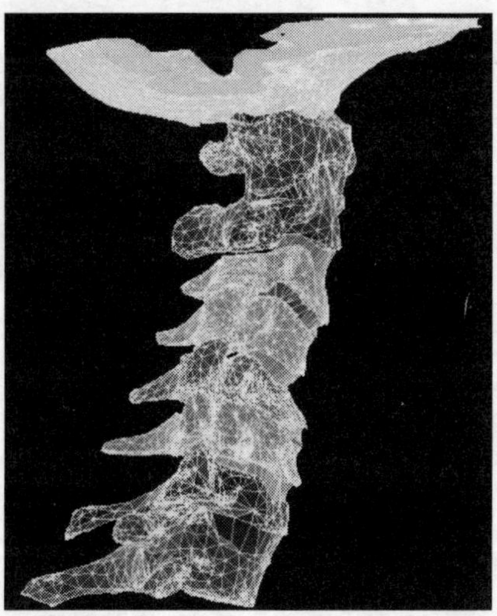

Fig. 25.1 FE model of cervical spine

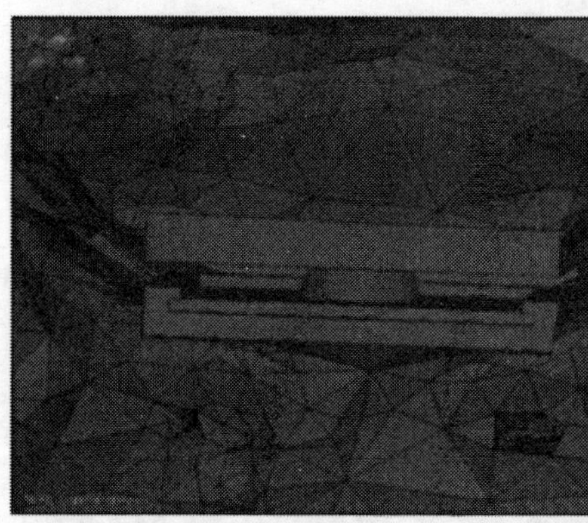

Fig. 25.2 FE model of artificial disc

The 3D model of intact vertebrae was discretised with the help of commercial package ANSYS. Tetrahedral solid element (SOLID 92) was selected for discretization process. Mesh was generated keeping the element size fixed as 5 resulting 208322 numbers of nodes and 131877 numbers of elements out of the total volume.

The intact bone model was modified to simulate a TDR at C5-C6. First, a proper size disc was chosen by analyzing the CT scan data. Then the ALL at the C5-C6 level was removed. Next, the nucleus and anterior and posterior parts of the annulus were removed in order to create space for the disc. The maximum possible portions of the lateral annulus and uncinate processes were preserved. The endplates were removed, and a platform fitting the ARC1 disc was created in the vertebrae simulating the milling operation performed during the surgical procedure. The C5 vertebra was distracted by 1.5 mm and the ARC1 model was then inserted in the disc space. A 16 mm disc was used based on the CT image analysis. Perfect implant-bone fixation was assumed. This was achieved by using the "merge" command in ANSYS which ensured that there was no relative motion between the implant and the vertebral endplates. Solid tetrahedral elements used for meshing ARC1 and a set of contact pairs created between the top endplate and middle part simulating a ball and socket joint. Young's modulus of endplate and UHMPE layer were taken as 110000 Mpa and 900 Mpa respectively. Poisson's ratio for both the material were taken as 0.3. Coefficient of friction between the contact pair was 0.2.

25.3 RESULTS AND DISCUSSION

The two FE model were simulated for flexion movement by fixing all degrees of freedom of the bottom nodes of T1 vertebrae. Both models are subjected to moments of 0.5 Nm, 1 Nm, 1.5 Nm, 2 Nm separately at the C2 level along with vertical load of head of 50 N.

Results from the analysis indicated good agreement between the FEM and previously reported experimental data. The FEM was generally able to predict cortical and endplate strains within the ranges reported in the literature. Each of the following subsections discusses range of motion, moments, disc stresses and facet stain in four different moment condition (0.5 Nm, 1 Nm, 1.5 Nm, 2 Nm) simulating the flexion movement of the vertebrae.

The disc stresses section details the different level annular stresses before and after the insertion of ARC1 disc. Stresses were recorded as the average stresses at each level of the vertebrae. Different vertebrae level interface strains are analyzed in the .the facet strain section. Strains were recorded in the superior and inferior regions of each disc level. This section compares the FE analysis of the natural bone and bone implanted with artificial disc at C5-C6 vertebrae level. Results were shown and discussed in the subsections below.

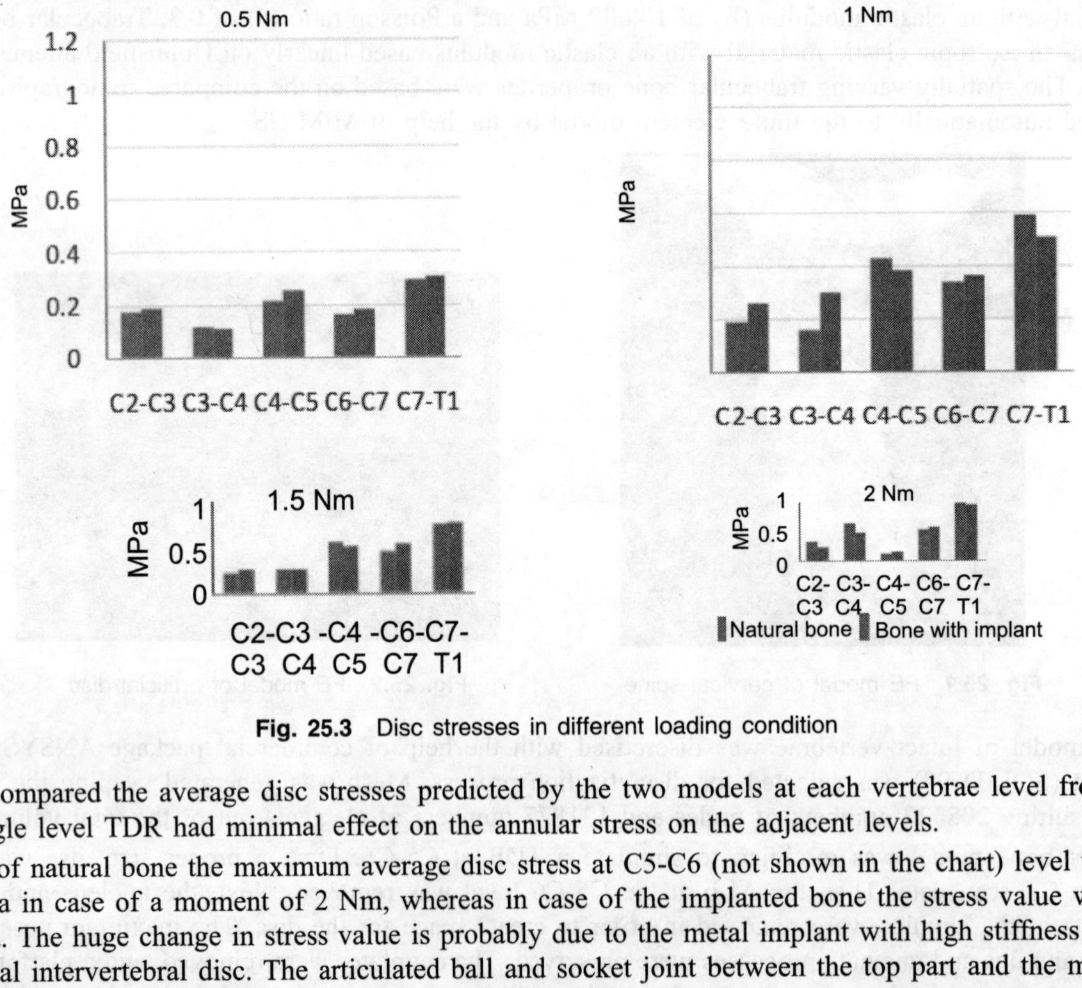

Fig. 25.3 Disc stresses in different loading condition

Figure compared the average disc stresses predicted by the two models at each vertebrae level from C2-C3 to C7-T1. Single level TDR had minimal effect on the annular stress on the adjacent levels.

In case of natural bone the maximum average disc stress at C5-C6 (not shown in the chart) level was recorded as 0.24 MPa in case of a moment of 2 Nm, whereas in case of the implanted bone the stress value was increased upto 2 MPa. The huge change in stress value is probably due to the metal implant with high stiffness as compared to the natural intervertebral disc. The articulated ball and socket joint between the top part and the middle part of ARC1 disc has less area of contact resulting in increase in stress value. At the adjacent levels (C3- C4 and C6-C7) there was merely 1% change.

Facet contact strain for the altered level (C5-C6) and adjacent levels for intact and implanted bone are shown in the figure. The implanted model showed a decrease in contact strain at the TDR level and the magnitude decreases slightly in both of the cases as the load increased. In different loading conditions the maximum value of the strain obtained at the level below C5-C6. The value of strain obtained were of the order 10-5

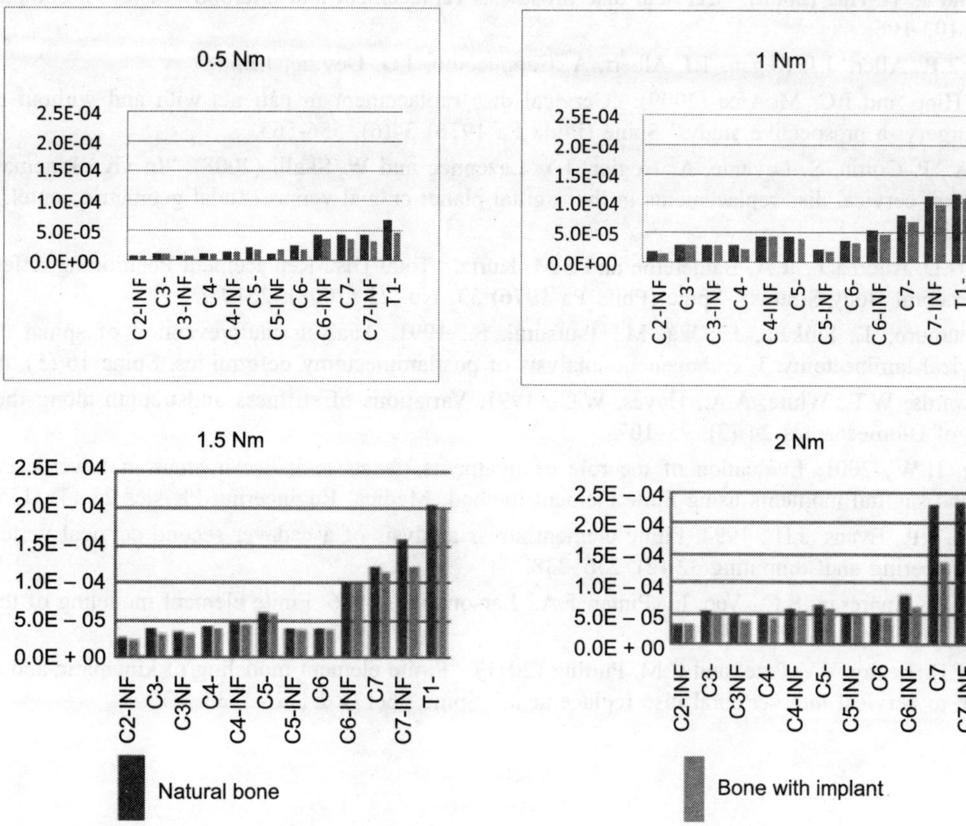

Fig. 25.4 Facet strain in different loading condition

REFERENCES

1. Auerbach, J.D., B.P. Wills, T.C. McIntosh and R.A. Balderston (2007). Evaluation of spinal kinematics following lumbar total disc replacement and circumferential fusion using in vivo fluoroscopy. Spine 32: 527-536.
2. Bauman, J.A., N.V. Jaumard, B.B. Guarino, C.L. Weisshaar, D.E. Lipschutz, W.C. Welch and B.A. Winkelstein (2012). "Facet joint contact pressure is not significantly affected by ProDisc cervical disc arthroplasty in sagittal bending: a single-level cadaveric study." Spine: 949-959.
3. Bozic kj., Keyak J.H., Skinner H.B., Beuf H.U., Bradford D.S., Three dimensional finite element modeling of a cervical vertebrae; an investigation of burst fracture mechanism. J Spinal Disorders 1994; 7: 102-110.
4. Jaumard, N.V., J.A. Bauman, B.B. GuarinoWeisshaar et al. (2013). "ProDisc cervical arthroplasty does not alter facet joint contact pressure during lateral bending or axial torsion." Spine 38 E84-93.
5. Kleinberger M. 1993. Application of finite element techniques to the study of cervical spine mechanics, In Proc 37th Stapp Car Crash Conf. Society of Automotive Engineers, Inc., San Antanico, TX.
6. Kumaresan, S., Yoganandan, N., Pintar, F.A., 1999. Finite element analysis of the cervical spine: a material property sensitivity study. Clinical Biomechanics 14 (1), 41–53 (Bristol, Avon).
7. Martin, S., A.J. Ghanayem, M.N. Tzermiadianos, L.I. Voronov, R.M. Havey, S.M.
8. Renner, G. Carandang, C. Abjornson and A.G. Patwardhan. "Kinematics of Cervical Total Disc Replacement Adjacent to a Two-Level, Straight Versus Lordotic Fusion." Spine (Phila Pa 1976) 36, No. 17 (2011): 1359-66.
9. Park, J.J., M. Quirno, M.R. Cunningham, R. Schwarzkopf, J.A. Bendo, J.M. Spivak and J.A. Goldstein. "Analysis of Segmental Cervical Spine Vertebral Motion after Prodisc- C Cervical Disc Replacement." Spine (Phila Pa 1976) 35, No. 8 (2010): E285-9.

10. Banerjee P, Roychoudhury A, Karmakar S.K. (2012). Morphometric analysis of the cervical spine of Indian population by using computerized tomography. J Me d A l l i e d S c I, 2: 66-76.
11. Banerjee P, Roychoudhury A, Karmakar SK. 2012. Biomechanical Remedies for Degeneration of Cervical Spine: A Review of Literature. J Med Imaging Health Informatics 2: 343–351.
12. Peng-Fei, S. and J. Yu-Hua (2008). "Cervical disc prosthesis replacement and interbody fusion: a comparative study." Int Orthop 32(1): 103-106.
13. Phillips, F.M., T.R. Allen, J.J. Regan, T.J. Albert, A. Cappuccino, J.G. Devine, J.E.
14. Ahrens, J.A. Hipp and P.C. McAfee (2009). "Cervical disc replacement in patients with and without previous adjacent level fusion surgery: a prospective study." Spine (Phila Pa 1976) 34(6): 556-565.
15. Rousseau, M.A., P. Cottin, S. Levante, A. Nogier, J.Y. Lazennec and W. Skalli (2008). "In vivo kinematics of two types of ball-and-socket cervical disc replacements in the sagittal plane: cranial versus caudal geometric center." Spine (Phila Pa 1976) 33(1): E6-9.
16. Rundell, S.A., J.D. Auerbach, R.A. Balderston and S.M. Kurtz. "Total Disc Replacement Positioning Affects Facet Contact Forces and Vertebral Body Strains." Spine (Phila Pa 1976) 33, No. 23 (2008): 2510-7.
17. Saito, T., Yamamuro, T., Shikata, J., Oka, M., Tsutsumi, S., 1991. Analysis andprevention of spinal column deformity following cervical laminectomy. I. Pathogenetic analysis of postlaminectomy deformities. Spine 16 (5), 494–502.
18. Shea, M., Edwards, W.T., White, A.A., Hayes, W.C., 1991. Variations of stiffness andstrength along the human cervical spine. Journal of Biomechanics 24 (2), 95–107.
19. Teo, E.C., Ng, H.W., 2001. Evaluation of the role of ligaments, facets and disc nucleus in lower cervical spine under compression and sagittal moments using finite element method. Medical Engineering Physics 23 (3), 155–164.
20. Teo, E.C., Paul, J.P., Evans, J.H., 1994. Finite element stress analysis of a cadaver second cervical vertebra. Medical and Biological Engineering andComputing 32 (2), 236–238.
21. Yoganandan, N., Kumaresan, S.C., Voo, L., Pintar, F.A., Larson, S.J., 1996. Finite element modeling of the C4-C6 cervical spine unit. MedEng Phys 18 (7), 569–574.
22. Womack, W., P.D. Leahy, V.V. Patel and C.M. Puttlitz (2011). "Finite element modeling of kinematic and load transmission alterations due to cervical intervertebral disc replacement." Spine 36(17): E1126-1133. 23.

26

Forces Analysis of Foot and Ankle During Stance Phase

R. Patra, A. K. Dash, H. C. Das, J. Sahoo and D. K. Agarwalla

Department of Mechanical Engineering & Department of Orthopedics, & SUM Hospital
Siksha O Anusandhan University, Bhubaneswar, Odisha, India

Email: omkar_patra@yahoo.com

ABSTRACT

Ankle and foot of human body experiences maximum stress during motion and functional stability during locaomotion, hence it is necessary to do detailed study of of different forces acting on ankle and foot during motion. It is the ankle which works more or less like a bevel gear of a gear train in human body. Muscles and tendons make the foot to have different desired motion in different planes. The present study has given insight to the mechanical stability of foot during stance phase considering the different forces acting on the foot and ankle. Equation has been developed based on engineering mechanics for the stability.

Keywords: Ankle joint, ground reaction, Achilles tendons, stance phase.

26.1 INTRODUCTION

Although biomechanics of foot is very complicated which gives the foot motion in sagital, medial and transverse planes but stability of upper limbs and body is done basically by two lower limbs and the lower limbs is supported by foot. Foot is balanced by tendons, ground reaction and body weight. Erdemir et al. [1] found the relation between plantar aponeurosis tension and Achilles tendon force. Mann et al. [2] studied the differect technique in the treatment of arthritic conditions of the ankle and hind foot. Moment generated by Achilles tendon and ground reaction are the two important forces which balances each other at ankle joint and brings stability to foot. Achilles tendon [3] which connects calf muscles to the calcaneus bone of foot. Cynthia et al. [4] have found the effect of vibration for maintance of balace of ankle and foot. Procter P. et al. [5] have studied the ankle biomechanics and have made multidimensional analysis of ankle joint. Hicks JH et al. [6] had measured maximum angle of movement during plantarflexion and dorsiflexion and load transmission.Todd M Hoagland et al. [7] have studied ankle joint anatomy. Donatelli RA et al. [8] had studied normal foot biomechanics. Zimi Sawacha et al. [9] have studied diabetic foot ulser. Howard J. Hillstrom et al. [10] had taken different types of foot of different age and gender and concluded that all foots have same gait pattern and different foots exhibit different foot pressure, force respectively.

26.2 BIOMECHANICAL ANALYSIS OF FOOT

Different parameters shown in Fig. 26.1 is listed below.

F_1 = Force due to Achilles tendons.

F_2 = Force due to ground reaction

R_1 = Perpendicular distance from ankle joint to direction of force of achiller tendon.

R_2 = Perpendicular distance from ankle joint to direction of force of ground reaction.

W = Force vector due to weight.

θ = Angle which force vector of achiller tendon makes with line joining ankle joint and talus.

Φ = Angle which ground reaction makes with the line joining ankle joint to point of ground reaction.

$F_1 \text{COS } \theta$ = Horizontal component of force vector of achiller tendon which acts along the ankle joint.

$F_2 \text{COS } \Phi$ = Horizontal component of force vector of ground reaction which acts along the ankle joint.

α = Angle between $F_2 \text{COS } \Phi$ and weight vector W at ankle joint, as shown in Fig. 26.2.

γ = Angle between $F_2 \text{COS } \Phi$ and $F_1 \text{COS } \theta$ at ankle joint, as shown in Fig. 26.2.

β = Angle between $F_1 \text{COS } \theta$ and weight vector W at ankle joint, as shown in Fig. 26.2.

26.3 CONDITIONS FOR FOOT STABILITY

Condition-I $F_1 \times R_1 = F_2 \times R_2$

Condition-II $F_2 \text{COS } \Phi$, $F_1 \text{COS } \theta$ and Weight Vector W must meet at one point.

Condtion-III Appling Lami's theorem for the forces $F_1 \text{COS } \Theta$, $F_2 \text{COS } \Phi$ and W which acts at common point is: $F_1 \text{COS } \theta / \text{COS } \alpha = F_2 \text{COS } \Phi / \text{COS } \beta = W / \text{COS } \gamma$

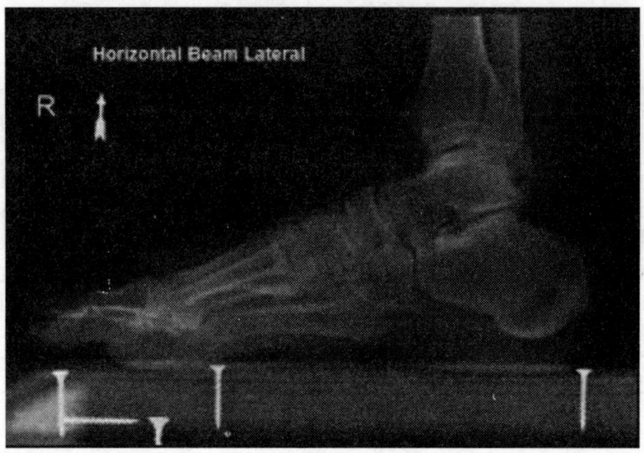

Fig. 26.1

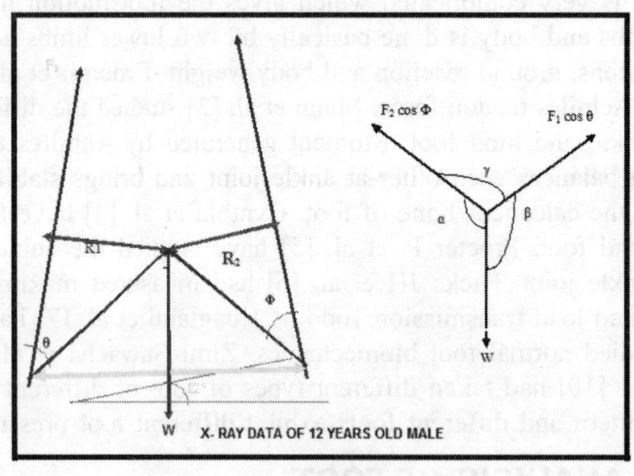

Fig. 26.2

26.4 RESULTS AND DISCUSSION

Higher the "perpendicular distance from ankle joint to direction of force of ground reaction", lesser ground reaction will be required for the mobility of the foot. Achilles tendon being most powerful one and the perpendicular distance from ankle joint to direction of force of Achilles tendon is a fixed value. Only two factors which could change are ground reaction and perpendicular distance from ankle joint to direction of force of ground reaction. In the current research, the equation that has been derived for stability of the body is in concerned to only the stance condition of human body. For other phases of gait equation may differ.

REFERENCES

1. Erdemir A., Hamel A.J., Fauth AR, Piazza SJ, Sharkey NA. (2004) Dynamic loading of the plantar aponeurosis in walking. J. Bone Joint Surg (A). 86:546-52.
2. Mann R.A., Haskell A. Biomechanics of the Foot and Ankle. In: Coughlin M.J., Mann R.A., Saltzman C.L., Eds. Surgery of the Foot and Ankle. Philadelphia, PA: 2007, p. 3-44.
3. Giddings, V.L., Beaupré, G.S; Whalen, R.T; Carter, D.R. (2000). Calcaneal loading during walking and 4. Medicine and science in sports and exercise 32: 627–634.
4. Cynthia Thompson, Marc Belangwer, Joyce Fung. (2007) Effects of bilateral Achilles tendon vibration on postural orientation and balance during standing. Intern. Fed. of Clinical Neurophysiology. Elsevier Inc.
5. Procter P., Paul JP.PMD: 7174695; PubMed-Indexed for MEDLINE.
6. Hicks J.H. The mechanics of the foot. (1953) I. the joints. J. Anat 3:345-357.
7. Ankle joint Anatomy. Todd M. Hoagland, PhD; Chief Editor: Thomas R Gest.
8. Donatelli R.A. (1985) Normal biomechanics of the foot and ankle. J. Ortho Sports, 7:91-5.
9. Integrated kinematics-kinetics-planter pressure data analysis: A useful tool for characterizing diabetic foot biomechanics Original Research Article; Gait & Posture, Vol. 36. Issuse 1, May 2012, pp. 20-26.
10. Zimi Sawacha, Gabriella Guarneri, Giuseppe Cristoferi, Annamaria Guiotto, Angelo Avogaro, Claudio Cobelli. (2013) Foot type biomechanics Part 1: Structure and function of the asymptomatic foot . Gait & Posture, 37:445-451.

27

Biomechanical Analysis of Crutch Gait

Ranjan Das and M. D. Burman Swami

Swami Vivekanand National Institute of Rehabilitation Training and Research
Olatpur, P.O. Bairoi, Dist. Cuttack, Odisha-754010, India
Email: dirnirtar@ori.nic.in

ABSTRACT

Crutches provide mobility aids for transferring the subject from one position to other, when the lower limb cannot provide sufficient support to body. Crutches are meant for compensating loss of muscle power, increasing stability and balance, reduce or completely eliminate stress on weight bearing joints. There are different types of crutch gait. The crutch walking the subject have to be taken by the shoulder girdle and the upper extremities and five muscle groups needed are needed for manipulation of crutches to forward, followed by strengthening exercises of flexors and extensor of arms, gripping the both arms rising the head and shoulder from bed and reaching forward with the hand as much as possible. Elderly patient using crutches or canes reduced walking velocity to maintain tolerable energy expenditure. Energy demand is particularly high with the swing through gait because of muscular demands of the upper limbs to lift the trunk and legs to initiates body swing phase and then provide stability while the patient swings through the line of crutches. The development of new material along with better understanding of a particular subject need, goals, future potentials, relative of his or her particular disease have enabled the health care team to prescribe more effectively and efficiently the appropriate assistive device for the subject.

Keywords: Crutches, gait pattern, assistive devices, cane and metal materials.

27.1 INTRODUCTION

Crutches are typically used when there is an injury to the lower extremities such as a fracture, pulled muscle (strain), ankle sprain or for post-operative recovery following orthopedic surgeries to the hip, knee or ankle increasing the size of an individual's base of support. Depending on the severity of the injury, a temporary period of partial- or non-weight bearing of a limb may be required to allow time to heal. This should be determined by the client's surgeon, physician or physical therapist. Crutches help by broadening a client's base of support, thereby increasing their lateral stability. Crutches may not be the most appropriate device for persons who have visual impairment, upper extremity weakness, nerve damage, severe balance disturbances, and impaired cognition affecting safety judgment. The earliest known evidence of the use of a supportive device to ancient Egypt about 2830 BC remains a simple *T* design. Around 1800 made it possible to split the staff, spread the two halves and insert a hand peace. There are three types of crutches: A-axillary; B-forearm; C. platform. Selection of gait pattern depends on the patient's ability to move the feet reciprocally, tolerate full load on each leg, push the body off the floor by pressing on the hands and maintaining balance. Force transducer in the device can register the applied load in the feet. Bone and joint disabilities applied significantly more force to the cane than do those with neuropathies. A single cane support bears approximately a quarter of body weight where as a forearm crutch can bear almost half of one's weight. When greater weight relief is required the patients needs axillary crutches. The selection of appropriate assistive device needs of a person with disabilities must carefully assess. Exercise and gait training may be required along with training in use the gait aid. The patient's age, mental status physical endurance is also factors. As the patient gain strength they may progress from parallel bar to a walker, crutches.

1. Crutches provide more stability and supports other than any other walking gaits. The ability to use crutches efficiently requires a systematic program with proper instruction.
2. The proper selection of the crutches, accessories and correct measurement.
3. A muscle test to judge the patient joint movement and strength.
4. Exercise to develop the muscles groups needed to crutch measurement.
5. Determination of the crutch gait best suited to meet the patients need.
6. Determining the gait aids best suited to an individual patient depends on:
   - The patients general functional ability
   - The amount of weight bearing permitted on the lower limb.
   - The patients balance.
   - Upper limb function.

27.2 GAIT PATTERN

There are several different gait or walking patterns that an individual using crutches may use as follows:

1. Point Gait
2. Swing Gait.

27.2.1 Four Point Gait

Sequence: right crutch, left foot, left crutch, right foot. This is slow and difficult to learn but is very stable and often used by ataxic patient.

27.2.2 Two Point Alternating Gait

Sequence Right crutch and left foot, left crutch and right foot. The stable gait pattern can also used by ataxic patient and significantly reduce weight bearing.

27.2.3 Three Point Gait

Sequence Both crutches and disabled limb, then sound lower extremity. This gait is used when weight bearing is permitted fractures of lower limb.

27.2.4 Tri-pod Gait

The crutches can be advance either individually or together and the patient then drags the body forward. Paraplegics often starts ambulating with this slow labourious gait pattern. The patient fitted with lower limb orthosis may ambulate with this gait.

27.3 SWING GAIT

Swing to–Both crutches are advance and the body is then lifted and swing to the crutches.

Swing through–The body is lifted and swing beyond the level of advance crutches, when patient is having more balance and stable itself.

27.4 NORMAL MAINTENANCE OF MOBILITY

Motion in joint and soft tissues is maintained by the normal movement of the parts of the body including joint capsules, muscles, subcutaneous tissues and ligaments though full range of motion many times in each day. In the course of the day this movements transfer the full range of motion. If for any reason the range of daily motion is restricted, tightness developed and restricted area of the motion.

It is easier to prevent tightness by frequently repeated activity than to correct it after it has developed. Limitation of the range of motion is of the greatest importance when it interfers with habitual posture or activities.

Tightness that prevents normal standing without muscular support is disabling. In the normal standing posture, extention of back, the hip and the knee is maintain by positioning the centre of gravity of the body above this joint, so that the weight of the body holds the joints extended against restricting ligaments and extensor muscle are relaxed. Muscular activity is used for balance or motion rather than to support the weight of the body.

Maintenance of the balance is provided mainly the activity of soleus muscle and intermittent counter balancing activity of the anterior tibial muscles. During relax standing the centre of gravity of the head, arm and trunk falls slightly posterior (0.5 to 1.0 cm) to the centre of motion of the acetabulum for anterior (3 to 5 cm.) to the centre of motion of the knee, and through the centre of the torsal arch approximately between the point of contact of the heel and heads of the metatarsals.

If tightness prevents complete extension of these weight bearing joints, relax standing cannot occur. When the joints are not fully extended muscles must support a part of the weight of the body during quite standing. The patient fatigues more rapidly and has less endurance for standing or ambulatory activity than those a person with normal mobility. In addition the contraction of muscles to hold the body irrect places more compressive stress on the joint surface, resulting in more rapid wear and more likely hood of joint pain.

27.5 CRUTCHES

Material Wood, Metal (Indigenously available)

It provides better stability and they have two point of contact with the body. In crutch walking the weight of the subject must be taken shoulder girdle and upper extremities. The five muscle group needed particularly for manipulation of crutches forward.

Felexors of the arm: To move the crutch forward

Extensors of the forearms: To hold the elbow at the desired angle when the weight of the body is placed on the hands and when the body weight is raised from the floor.

Fingers and thumb flexors: To permit grasping the crutches.

Dorsiflexor of the wrist: To keep the hand in the correct position of the hand piece.

Shoulder girdle depressors and downward rotators: To support the body by means of the crutches when it lives the floor.

Two basic types of crutches are:

1. Axillary
2. Non-axillary

Axillary crutches provide better trunk support and can transmit as much as 80% of body weight.

Non-axillary crutches can transfer 40 to 50% of body weight.

To achieve maximum function axilary crutches must be adjusted to the proper length and head level. The patient is measured supine from the heel to the anterior axillary fold and 5 cm. is added to accommodate shoe.

Measurement can also be made in an upright position from the anterior axilla fold to a point 15 cm. lateral to the 5th toe. The hand piece should be adjusted so the elbow is in 30 degree of flexion.

Forearm Crutches: It designed so the usual body weight can be tranfered to the forearm. The metal tubular shaft is adjusted and the cuffs for forearm can be fixed or hinged. Hinged cuffs make it easier for users to open a door or avoid injury in case of falls.

Children with spina-bifida who have absent hip abductor and extensors and who walk with ankle foot orthosis often used forearm crutches. They usually have good upper limb strength and the height of the forearm crutches provide sufficient stability for greater walking speed.

Axillary crutches The upper ends of axillary crutches rest in the axilla. They should be separated from the apex of the axilla by two or three finger breadths. A patient who 'hangs' from the axilary crutches is in danger of developing a crutch palsy. The weight should be borne primarily on the hands. The upper ends for stability, not weight bearing. An advantage of axilary crutches that user can free their hand for such activities as opening doors by learing against the shoulder piece.

REFERENCES

1. Joyce, B.M., & Kirby, R.L. (1991). Canes, crutches and walkers. American Family Physician, 43: 535-542.
2. Handbook of Physical Medicine and Rehabilitation Basics: (Krusen) (2nd Edition) Susan J. Garrison (editor), Lippincott Williams & Wilkins, NewYork, 2008.
3. Atlas of Orthotics: Biomechanical Principles and Application: (1976), American Academy of Orthopaedic Surgeons (AAOS) (Author), USA.
4. Foot Orthoses. Kent K.Wu. Lippincott Williams and Wilkins, New York, London (1990).
5. Iezzoni L.I., E.P. McCarthy, R.B. Davis, and H. Siebens. (2000) Mobility Impairments and Use of Screening and Preventive Services. American Journal of Public Health 90:955-961.
6. Larsen McNeal, M.A., L. Carrothers, and B. Premo. (2002) Fall. Providing Primary Health Care for People with Physical Disabilities: A Survey of California. Physicians. http://www.cdihp.org /pdf /ProvPrimeCare.
7. Longmore, P.K., and L. Umansky. 2001. The New Disability History: American Perspectives. New York: New York University Press.
8. Rimmer, J.H. (1999) Health Promotion for People with Disabilities: The Emerging Paradigm Shift from Disability Prevention to Prevention of Secondary Conditions. Physical Therapy 79:495-502.
9. Dejong, G. 1997. Primary Care for Persons with Disabilities: An Overview of the Problem. American Journal of Physical Medicine and Rehabilitation76: Supplement: S2-S8.
10. Cocchiarella, L., and G.B.J. Andersson, Eds. (2001) Guides to the Evaluation of Permanent Impairment. 5th ed. Chicago: American Medical Association.

28

Ergonomic Evaluation of Welders and Crane Operators in Steel Industry

Trinath Pachal

Steel Authority of India Limited

Bokaro Steel Plant, Bokaro–827009, Jharkhand

Email: dr_tpachal@yahoo.com

ABSTRACT

The harsh operating conditions prevail more or less in every unit of an integrated steel industry. Mechanical or manual carrying of difficult loads, complex handling materials, awkward static working postures, unmatched man-machine systems in high or extremely fluctuating ambient environments demand high levels of safety and functional reliability in every operation. Any failure or malfunction in any such operations due to man-machine adjustment therefore, result in serious accidents causing loss of life and big breakdown as well as huge medical expenses and unfortunate disability. As a matter of fact, in any steel production industry, safety and health has always been of paramount importance when it comes to people and processes. In Bokaro Steel Plant since 1990, Ergonomics unit has used 'Biomechanics' as a tool for reducing human errors, accidents, chronic musculoskeletal disorders (MSD), absenteeism as well as for increasing efficiency, well being and overall productivity. Present paper will highlight some of the successful applications of bio-mechanics in changing the work or the working posture, benefits of which is being measured in term of safety, health and well beings of the employees. As a matter of fact, HRD of Bokaro steel has incorporated Biomechanics in employee's training curriculum. The paper will focus on easy application of bio-mechanics through low cost evaluation procedures in steel industry.

Keywords: Steel industry, biomechanics, safety, musculoskeletal disorders.

28.1 INTRODUCTION

The field of applied Biomechanics in industry is devoted to the safe body movements at work and to the improvement of the person's performances on the job through application of conceptual and mathematical tools. This goal has to be shared by both management of the industry as well as by workers on the line. To achieve this goal, changes are to be made in the job that people do, in their work places and work spaces, in their working methodology, in the equipment and tools they use. These changes reduce injuries, errors, defects and certainly production cost. It also helps to reduce absenteeism. These changes thereby improve efficiency, quality and overall productivity. In the beginning of this century, International Labour Organization (ILO) calculated that 'the biggest single reason for economic losses' will be the Musculoskeletal Disorders (MSD) due to ignorance and /or neglect of biomechanical principles both in organized and unorganized industrial sectors. In 2003, ILO cautioned that new strategies and solutions need to be developed and applied both for well-known hazards and risks, as well as for emerging issues such as psychosocial hazards and musculoskeletal disorders." The musculo-skeletal disorders include a wide range of inflammatory and degenerative conditions often diagnosed by Occupational Physicians as repetitive strain injuries, chronic Back-Pain, neck strain, slip disk, etc. which resulted in a significant man-hour loss with huge medical expenses including partial disability of productive manpower. The inherent cause of all those injuries or illness must have an association with biomechanics and ergonomics.

Till the last decade, it was a common belief both for the scientists and the workers that the physical aspects of work like lifting, carrying, loading and unloading of heavy objects were the main reasons for MSD as it violated the Laws of Work Science. As a matter of fact, majority of applied biomechanical research studies on occupational health and safety were dealt with the load aspects of industrial work. But now the scenario is changing by the day. Today, MDS due to non-physical aspects exceed the physical aspects, due to which now a computer operator is much more vulnerable than an unskilled labour, where violation of basic biomechanical rules is concerned. As a matter of fact, the new challenges for higher productivity with lesser manpower, forces us to rethink about the finest mutual adjustment of workers (valued human resources) and their work to get a safe, qualitative productivity in Indian industrial scenario. Now at this juncture of globalization & modernization, time has come for all of us to understand the magnitude of biomechanical stress at work and also to develop an effective risk management system, so that we can create a healthy productive work atmosphere in our industry. In many industries, sincere efforts are being made either through department of Safety, Occupational Health and Industrial Engineering or through Quality Control. But success has not been achieved as desired. One school of thought believes that the main hindrance may be the cost of both equipment as well as human resources in the field of biomechanics. Other school believes that use of biomechanical principles is a costly affair in industry. Being an industrial ergonomics I have different sets of explanation regarding application of biomechanics in Indian industries. Let us observe few cases of a large integrated steel plant.

28.2 REPETITIVE STRESS INJURY & BIOMECHANICS

28.2.1 The Incidence

Nearly seven years back, Sri X Turi of 40 years of age, working as Operator at Blast Furnace, used to take pain killer tablets/ injectables regularly before starting his job. His complaint of Chronic 'Back Pain' had already been investigated by physicians and nothing abnormal was detected in his back. Gradually he became a habitual absentee.

28.2.2 Problem Identified

Upon investigation it was noticed that due to short Stature (155 cm), Mr. X had to operate a Joystick control with upward extension of his body and balancing it on toe. After each operation he returned to original position with a jerk. Repeated upward extension of trunk and jerk on spine were causing RSI (repeated stress injuries) on Vertebral column which led to severe spinal strain followed by physical abnormalities. It was a classical example of violating basic principles of biomechanics for extension of body parts at work. Only advanced medical diagnostics (e.g., MRI) can confirm the RSI on spine.

28.2.3 The Biomechanical Solution

A simple 25 inches high wooden platform has changed the whole scenario including the chronic back pain of Mr. X. One can easily estimate the value of such simple workplace modifications which can directly help in production. Such modifications need only basic knowledge on biomechanics & work science.

28.3 MUSCULOSKELETAL DISODERS IN WELDING & BIOMECHANICAL SOLUTION

28.3.1 The Incidence

In late nineties, Structural Shop of BSL experienced higher rate of MSD amongst the welders. Occupational Health Services (OHS) statistical report said that out of 48 welders 42.64% suffered with Cr. Back Pain and 52.63% suffered with other form of Musculoskeletal disorder. It was a challenge to OHS personnel for combating the diseases which definitely hindered productivity including huge medical expenses.

28.3.2 Problem Identified

Upon investigation, three major findings were identified:

a. Prolonged isometric welding operations were causing MSD
b. Age (mean 48.38±0.13) was a contributing factor for such strain
c. BMI <19 and >25 was other contributing factor for MSD in structural shop.

After analyzing free body diagram for standing welding operation, the biomechanical stress at spine was calculated as 12.55 kgm^2. It was also seen that minor change in spine elevation could reduce nearly half the spine stress. Important information was received from some welders that use of a small stool relieved back pain during welding operation. Biomechanical analysis of sitting welding operation with a small stool opened the new vistas in the area of welding ergonomics. A small stool can make welding operation stress free, specifically at low back.

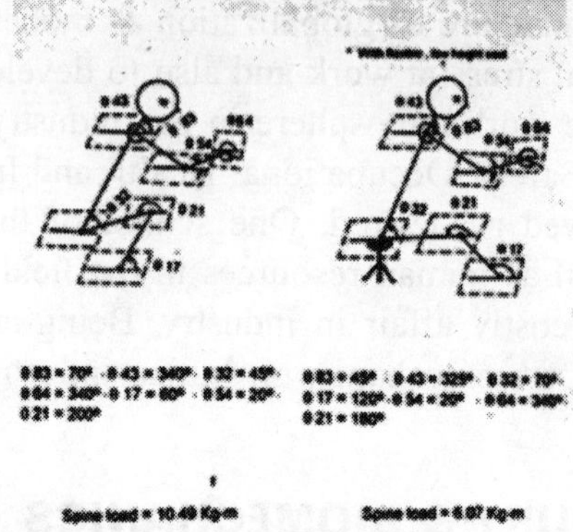

Fig. 28.1 Biomechanical stress during sitting welding operation

28.3.3 The Biomechanical Solution

As it was biomechanically proved that change of posture and use of a small stool can be the first line of treatment for all welders with MSD, some focused shop floor ergonomic trainings and demonstrations have changed the scenario. Till date, welders of structural shop of BSL are using self made small stools during sitting welding operations. For further prevention of such disorders, the following recommendations were given:

Special emphasis should be laid on anthropometric measurements during appointment of welders

Training for correct welding postures

Performing back-extension exercise after job

Regular medical check-up for aged workers

Low height and flexible sitting arrangements for welders

28.4 CRANE OPERATION & SPINE STRESS

28.4.1 The Incidence

If an industry experienced a huge loss in last quarter due to its defective production which has a close relation with mishandling of steel slab by some over head cranes, what actions management would take? As it has a direct relation with company's tangible economy, a high level inspection is obviously followed by administrative changes in the very shop including change of crane operator also. If same situations occur in next quarter then changes in existing crane may be a second thought. Can it be a rare imagination that at this juncture some one may think of biomechanical interventions to solve the problem?

28.4.2 Problem Identified

Occupational health survey indicated that severe physical and mental stress amongst the crane operators were the primary cause for such deteriorating work scenario in soaking pit of Slabbing Mill. Report showed that amongst 45 Soaker Crane operators, 48.9% suffered from Cr. Back / Neck Pain, 42.2% - Eye Strain, 31.1% - Mental Exhaustion and 15.6% suffered from Physical fatigue. Reasons for such high prevalence of physical and mental stress were Adverse head postures, Excessive head rotations, Frequent neck movements, Acute vibration exposure, Improper Illumination, Inappropriate seat design, Dirt & Scratches on the Glass Panes, Wrong & Unsafe Work Practices and Poor communication at soaking pit area, most of which were biomechanical stresses.

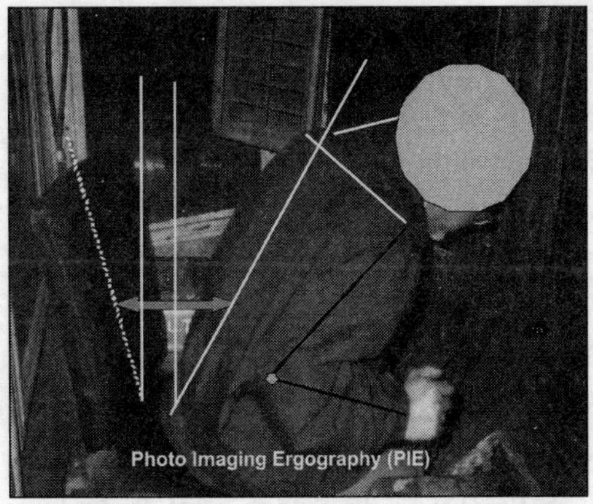

Fig. 28.2 Adverse head and neck movement at soaker crane

28.4.3 The Solution

After detailed analysis with Photo Imaging Ergography (PIE), a multidimensional ergonomic and biomechanical solution was recommended including use of Flexible Crane Chair, correction of operating postures, advanced Communication System, Regular Cleaning or use of Heat Resistant Wiper System for all glasspanes, Redesign of Pit Operators' Cabin and provision of Reflecting Mirrors in and outside of the crane to see the position of red hot steel slab. It is interesting to note that within six month of recommendation, five out of seven tasks were implemented. Success of the very project has initiated several new biomechanical studies on EOT cranes in the steel plant.

28.5 SAND SHOVELLING IN FOUNDRY AND BIOMECHANICS

28.5.1 The Incidence

Epidemiological findings confirmed that sand shovellers of Steel Foundry experienced more lumbo-sacral sprain than any other shovellers including coke sorting at Coke Ovens which seems to be one of the strenuous manual operation in Steel plant.

28.5.2 Problem Identified

A biomechanical model was developed to identify the net reaction forces and the moment load at lumbo-sacral joint which was the main contributing factor for chronic back-pain of sand shovellers. The model was tested on two major task:

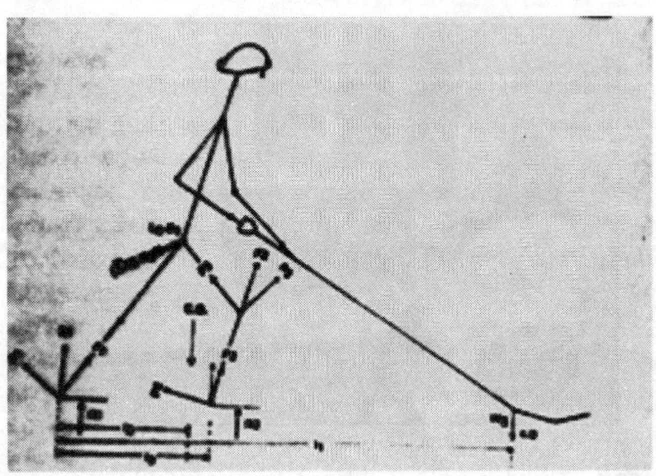

Fig. 28.3 Biomechanical model for shovelling showing reaction force and moment load

Sand Pouring in Foundry – ΣM = 17.722 kg M and R = 16.81 kg

Coke Sorting in Coke Oven – ΣM = 9.106 kg M and R = 10.177 kg

(where ΣM is Net moment and R is Net reaction force)

Three noticeable facts were identified through the model:

1. Sand shovelling is much more strenuous than coke sorting as both the moment and reaction force at lumbo-sacral joint is more in sand shovellers.
2. Nature of material to be shovelled is the major contributing factor for spine stress.
3. Postural adjustment is necessary during sand shoveling as maximum reaction force is experienced at the front foot.

28.5.3 The Biomechanical Solution

Postural training for sand shovellers at Steel foundry along with spinal exercise caused mild improvement of the situation. Recommendation for redesign of shovel in Foundry was given to Design department for further improvements.

28.6. COMPUTERISATION & BIOMECHANICAL STRESS

28.6.1 The Incidence

The introduction of computers in all Pharmacy counters of Bokaro General Hospital has made a fundamental change in the nature of pharmacy work, both cognitively and physically. Due to these changes in the nature of work, a need for detailed man-machine study was felt for better coordination in pharmacy activities because VDT work was causing mismatch between the task and the user, resulting in increased biomechanical stress leading to several discomforts, fatigue and job dissatisfaction to the users.

28.6.2 Problem Identified

Biomechanical stresses were noticed both in human-computer interaction as well as in manual task. Dissatisfaction was obvious due to sudden change in conventional methodologies. The major physical discomforts were noticed in the area of neck pain (30.76%), eye strain (23.00%) and hand strain (15.38%). The height of the key board was not acceptable to 38.46% and their angulations were not acceptable to 46.15% pharmacists. 84.61% of them felt that height of VDT was high and that was the main source of discomfort. Moreover only 15.38% pharmacists accepted the height of the working chairs. No foot rest and document stands were provided. 69.23% felt discomfort due to reflection of light at VDT.

Biomechanical observations indicated that excessive neck movements during numbering and data feedings of dispensary slips were the main causes for neck strain. Ergonomic error was identified in design of dispensary slip.

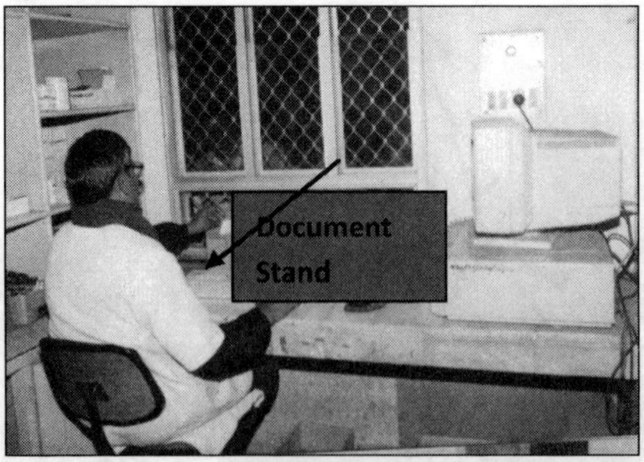

Fig. 28.4 The document stand and its user in pharmacy counter

28.6.3 The Biomechanical Solution

As the neck movement was the key ergonomic problem, a low height dispensary slip stand was recommended and some common designs were shown to them. A surprising change was noticed amongst the mindset of the pharmacists regarding acceptance of computerization of the counters. On the very next day of the discussion, the pharmacist of counter no.4 prepared his own document stand, using simple packaging material. He counseled other pharmacists to prepare their own stands as it reduced neck pain by half in a single day. That was a classic example of participatory ergonomics where users need to know only the biomechanical principles of related work and workplaces.

28.7 CONCLUSION

As an industrial researcher, it is felt by the author that there exists a fundamental difference between managing biomechanical problems in industry and doing academic research. Solution of each biomechanical problem is unique and varies from shop to shop, industry to industry. Scientists of such areas must have basic understanding of various industrial processes as well as have substantial depth in Safety & Occupational Health. In present scenario, majority of Indian industries do not have any idea about the domain of applied biomechanics as well as its capabilities. Value and prospects of biomechanics in Indian industries cannot be denied. Managements often consider it as costly affair. But the above discussion and extracts of few real case studies help us to conclude the following:

Every biomechanical solution is not costly. Simple change of work pattern or simple modification of work place can improve the whole scenario. Several musculoskeletal disorders can be cured/ alleviated only through biomechanical modification/adjustment.

Biomechanics can be utilized in industry for improving its Quality, Safety, Occupational Health and Well being of its employees and its overall productivity. Practicing applied biomechanics in Indian Steel Industries is not difficult, but it must be molded according to the industries demand.

REFERENCES

1. Alexander, D.C. (1986) The practice and management of industrial ergonomics', Prentice-Hall.
2. Garg, A and Chaffin, D.B. (1975) A biomechanical computerized simulation of human strength, AIIE Transactions, 7:1-15.
3. International Organization for standardization, 1996, Ergonomics, 13,180.
4. Kragt, H. (1995) Enhancing Human Performance, Ergonomics, 38:1674-1685.
5. Kumar, S. (1992) Advances in Industrial Ergonomics & Safety. Taylor & Francis, USA.
6. Pachal, T. (2000) Managing back pain through computer aided Ergography in metal works. Advances in Ergonomics, Occupational health and safety, Ed. Majumdar,D & Selvamurthy. W. New Age International Pvt. Ltd., pp. 51-56.
7. Pachal, T., R.K. Pathak, B.S. Sastri. (2005) Ergonomics intervention in Soaking pit during OHSAS: 18001 certification at Slabbing Mill of Bokaro Steel Plant. Proceedings HWWE, 2005, Allied Publishers Pvt. Ltd, New Delhi.
8. Pachal, T., and Sastri, B.S. (1999) Evaluating ergonomic stress amongst welders in steel industry, Occupational and Environmental Ergonomics, ISE, pp. 153-158.
9. Smith, J.L and Mc Laughlin. (1982) A biomechanical analysis of industrial manual material handling. Ergonomics 25:299-308
10. Tichauer, E.R. (1978) The biomechanical basis of ergonomics. John Wiley & Sons.
11. Wisner, A. (1989) Variety of physical characteristics in industrially developing countries- ergonomic consequences. Int, J Ind . Ergonomics 4:7-138.

29

Radiographic Image Based Analysis of Trabecular Architecture in Human Femur Bones Using Lacunarity

S. Sangeetha, C. M. Sujatha and D. Manamalli

Department of Instrumentation Engineering & Department of ECE, CEG

MIT Campus, Anna University, Chennai–600 044, India

Email: s.sangeethame@gmail.com

ABSTRACT

Estimation of mechanical strength of bones is an essential component in assessment of onset and progress of osteoporosis and fracture. Analyses of bone mineral density and soft bone architecture of femur are considered clinically important in explaining the mechanical variability of bone and in determining bone stiffness and fracture risk in osteoporosis. In this work, an attempt has been made to characterize trabecular femur bone architecture using lacunarity analysis. The compressive and tensile regions in the images are delineated and are subjected to this multiscale spatial analysis using two different methods of lacunarity analysis. Further the derived lacunarity values are correlated with apparent porosity of normal and abnormal samples of the strength regions. Result shows that lacunarity values are able to differentiate normal and abnormal images. These values show higher variations in both the strength regions at smaller box size. Also the values are found to be uniform for higher box sizes. The distinct variation between normal and abnormal samples is observed in tensile region. This may be due to the presence of heterogeneous patterns in this region. As lacunarity analysis could extract distinct variation between normal and abnormal samples of human femur trabecular bone, this study seems to be clinically relevant for mass screening of bone disorders using radiographs.

Keywords: Radiographic images, femur, trabeculae, apparent porosity, lacunarity.

29.1 INTRODUCTION

Trabecular femur bone strength is not only a function of femoral bone mineral density, but also a function of the spatial distribution of bone mass intrinsic in structural properties. Trabecular bone is a hierarchically organized structure, formed by an interconnecting network of plates and struts [1]. It is a highly complex material, being anisotropic with different strengths in compression and tension and is considered as the primary site for early bone loss. Perturbation in bone mineral homeostatic by hormone loss of menopause or by extended exposure to microgravity may cause a remodeling imbalance resulting in structural and mechanical impairment of the bone due to architectural deterioration [2]. It is apparent that bone disorder affects both bone density and bone quality. The measurement of bone mineral density reflect bone density at the organ or whole skeleton level, but are not informative on the micro-architecture of bone [3]. Therefore, there is a need for evaluation of micro-architectural characteristics of bone.

Conventional radiography plays a central role in the assessment of the fracture status of the osteoporotic patient. The availability of plain radiographs makes them the most commonly used tools in the evaluation of osteoporosis like disorders despite known limitations in detecting early disease and subtle changes over time [4]. X-ray images are advantageous in contrast to computed tomography or fluoroscopy because they exhibit less radiation, higher resolution and lower costs.

Image-based analysis of human trabecular bone to characterize its complex micro-structure is found to be useful in the diagnosis of bone disorders. Textural patterns are often complex, exhibit scale-dependent changes in structure and are difficult to identify [5]. Lacunarity has been applied recently to complex textures, since it quantifies spatial heterogeneity and anisotropy of textural patterns [6]. It uses multiscale windowing to measure the degree of translational invariance of the analyzed patterns. Lacunarity estimation methods for gray-scale image surfaces are proposed by various authors based on the gliding-box algorithm and differential box counting method [7- 9].

Lacunarity is used effectively in many medical applications. It has been applied in mammographic images to quantify the degree of heterogeneity of the parenchymal pattern for effective prediction of cancer risk [10]. Similarly, this analysis is useful in differentiating the benign lesion and malignant lesion from digital mammograms and in automated assessment of melanocytic naevi and melanoma from microscopic images [11, 12]. It is also used in the analysis of the temporal distribution of earthquakes, rock unit mapping, simulation and quantification of forest canopy structure, texture classification of urban and in analysis of porosity in naturally fractured media [13, 14].

In this work, micro-architectural variations on compressive and tensile strength regions of femur radiographic trabeculae images are analyzed using two methods of lacunarity.

29.2 METHODOLOGY

29.2.1 Image Acquisition

Digitized pelvis images ($N = 44$) recorded using a clinical X-ray unit (Siemens 500 mA Polyscope) is considered for this study. Two strength regions namely, compressive and tensile are identified as proposed by Singh index [15, 16]. The qualitative analysis is also performed on the delineated images to derive apparent porosity which is ratio of void area to bone area.

29.2.2 Lacunarity Analysis

The strength regions are subjected to two different methods of lacunarity analysis which include Lacunarity Gray (LG) and Lacunarity Range (LR) methods.

In lacunarity gray method, a box of size $L \times L$ is placed over the upper left corner of the image. The box mass M_a is calculated as the summation of pixel intensities inside the placed box. Then the box is moved one column to the right, and the box mass M_a is counted again for all the boxes and process is repeated for the entire image. Average of all box masses is computed as expected mass $E\ (M_a)$.

The values of lacunarity (*LG*) are calculated using

$$LG = E\left\{\left(\frac{M_{(a)}}{E(M_{(a)})} - 1\right)^2\right\} \qquad \text{...(1)}$$

where M_a is the mass and $E\ \{\cdot\}$ denotes the expected value of the entire box mass M_a.

The gradient-based method lacunarity range is implemented using the following procedure. For a given image of size $m \times m$ pixels, a window referred to as 'box' of size $r \times r$ is placed upon one corner of the image. The range of the values, also known as box mass, is obtained by finding the difference between maximum and minimum pixel intensity values within the box of size $r \times r$. The box is shifted by one pixel, so that its new position overlaps with the previous one. The contained range of values in a box is recorded in a cell of resulting matrix A_r. The entire region of interest is processed until all possible $r \times r$-neighborhoods are accounted for and A_r reaches an extent of $(m - r + 1) \times (m - r + 1)$ pixels.

The lacunarity of the matrix A_r for box size r is finally calculated according to the following formula:

$$LR = \frac{S^2(A_r)}{\bar{S}^2(A_r)} + 1 \qquad \text{...(2)}$$

where S^2 is the variance and S is arithmetic mean. Thus, the lacunarity value is obtained for each box size r. The mean lacunarity is calculated by taking average of lacunarity values for different box sizes.

29.3 RESULTS AND DISCUSSION

The typical binarized planar radiographic images of normal and abnormal femur trabecular bone are shown in Fig. 29.1 (a) and (b) respectively. In normal images, it is observed that all the trabecular patterns are distinct and are closely arranged and in abnormal, spacing is more with high discontinuities.

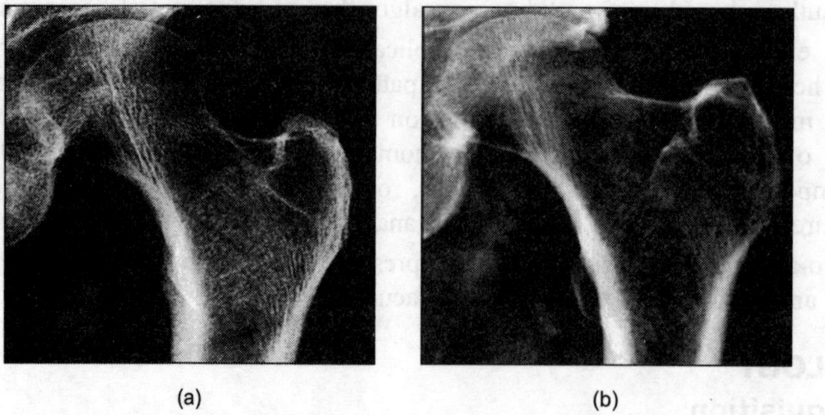

Fig. 29.1 Representative images (a) normal and (b) abnormal femur bone

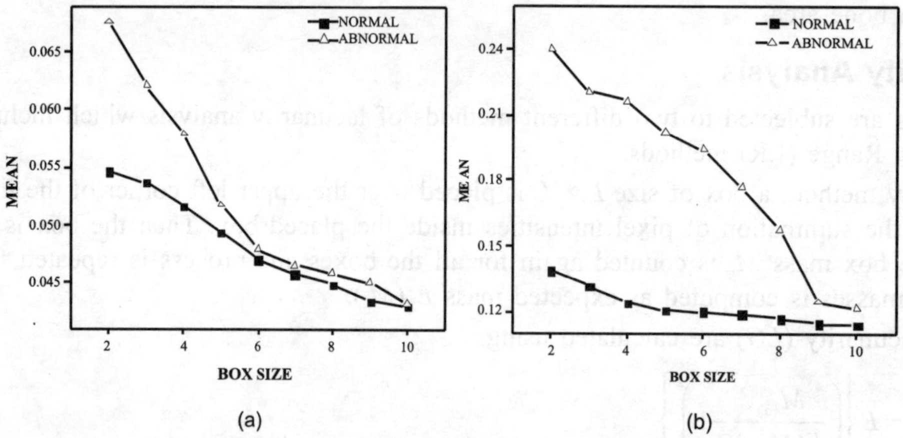

Fig. 29.2 Variation of mean lacunarity gray with varying box sizes of normal and abnormal samples (a) compressive and (b) tensile strength regions

Figure 29.2 (a) and (b) show the variations of mean lacunarity gray values of various box sizes for compressive and strength regions respectively. The mean values of lacunarity are found to be high for abnormal subjects for both the regions. It is also observed that there is rapid decline in the values of lacunarity for higher box sizes. This could be attributed to the sparse distribution of the trabeculae patterns in abnormal images. It is also found that mean lacunarity values exhibit a distinct separation between normal and abnormal subjects in both the regions at smaller box sizes. This may be due to the presence of micro patterns in the trabecular structure. The values of lacunarity are found to be high for both the subjects in the tensile region. Also, it is observed that the variation in lacunarity values between normal and abnormal subjects is more for this region. This could be due to the presence of more architectural variations in tensile region.

The scattergrams in Fig. 29.3 (a) and (b) show the variations of lacunarity gray values of normal and abnormal images at box size $r = 2$ for compressive and tensile strength regions respectively. The values of lacunarity are found to be high in abnormal subjects and low in case of normals for both the regions. This may be due to the structure alterations present in the abnormal trabecular bone. It is observed that the values of lacunarity derived from tensile region show distinct variation between normal and abnormal subjects and are found to be highly significant with P value of 0.001. This could be due to the trabecular structure alterations exhibited by tensile region.

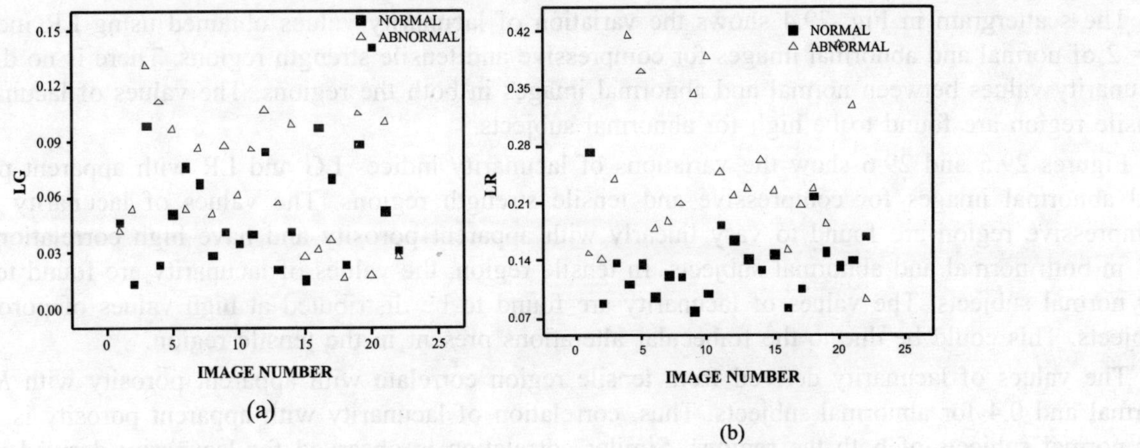

Fig. 29.3 Variation of LG of normal and abnormal images for (a) compressive and (b) tensile strength regions of box size $r = 2$

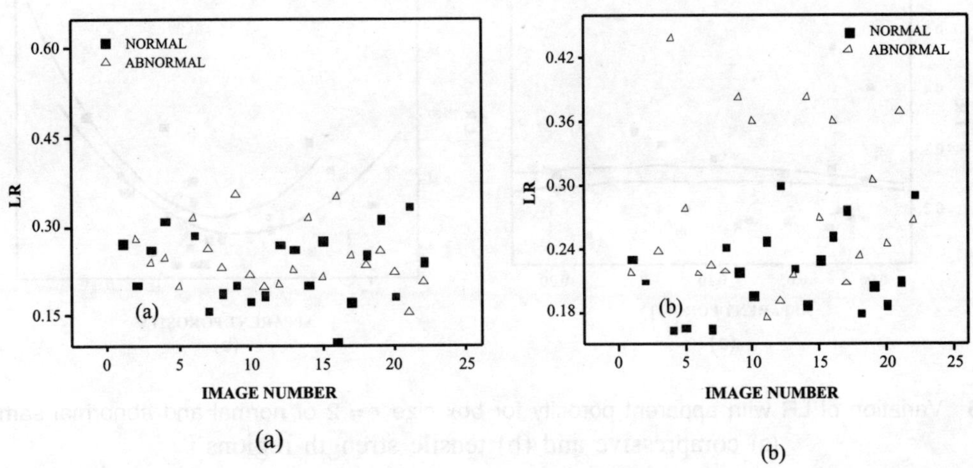

Fig. 29.4 Variation of LR of normal and abnormal images for (a) compressive and (b) tensile strength regions of box size $r = 2$

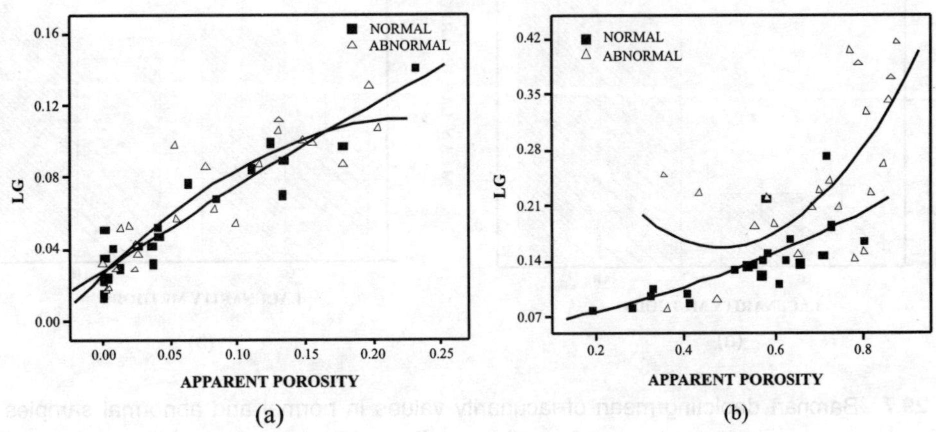

Fig. 29.5 Variation of LG with apparent porosity for box size $r = 2$ of normal and abnormal samples for (a) compressive and (b) tensile strength regions

The scattergram in Fig. 29.4 shows the variation of lacunarity values obtained using LR method for box size $L = 2$ of normal and abnormal images for compressive and tensile strength regions. There is no distinct variation in lacunarity values between normal and abnormal images in both the regions. The values of lacunarity derived from tensile region are found to be high for abnormal subjects.

Figures 29.5 and 29.6 show the variations of lacunarity indices LG and LR with apparent porosity of normal and abnormal images for compressive and tensile strength regions. The values of lacunarity LG derived from compressive region are found to vary linearly with apparent porosity and have high correlation with R value of 0.8 in both normal and abnormal subjects. In tensile region, the values of lacunarity are found to be less scattered for normal subjects. The values of lacunarity are found to be distributed at high values of porosity for abnormal subjects. This could be due to the trabecular alterations present in the tensile region.

The values of lacunarity derived from tensile region correlate with apparent porosity with R value of 0.5 for normal and 0.4 for abnormal subjects. Thus, correlation of lacunarity with apparent porosity is found to be more for normal subjects of both the regions. Similar correlation is observed for lacunarity derived using LR method. Hence, these lacunarity values could be used as indices to differentiate normal and abnormal subjects.

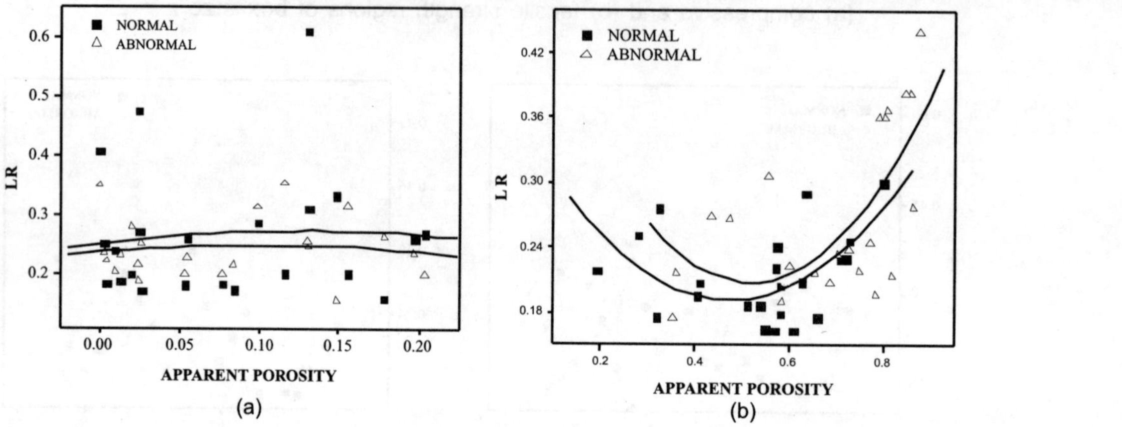

Fig. 29.6 Variation of LR with apparent porosity for box size $r = 2$ of normal and abnormal samples for (a) compressive and (b) tensile strength regions

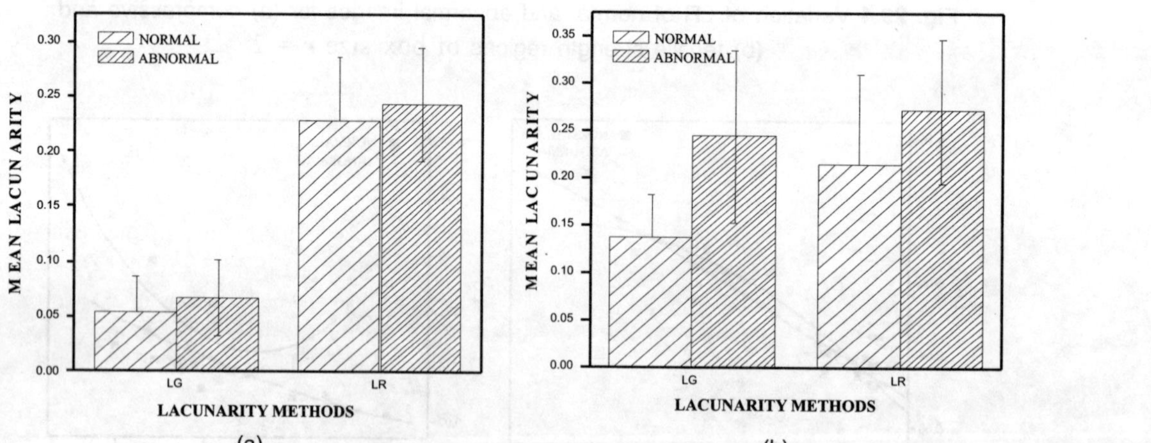

Fig. 29.7 Barchart depicting mean of lacunarity values in normal and abnormal samples for (a) compressive and (b) tensile strength regions

Figure 29.7 shows the bar graph depicting mean values of lacunarity for both compressive and tensile regions of normal and abnormal samples. The mean values of LG and LR are found to be high for abnormal samples than normals in both tensile and compressive regions.

This demonstrates the ability of lacunarity analysis to extract the heterogeneity exhibited by abnormal trabecular structure. The mean values of LG are found to be high for tensile region whereas mean LR values have high magnitude for compressive region in both the normal and abnormal subjects. The standard deviation of lacunarity derived using LG and LR methods are found to be low in compressive region of both normal and abnormal images. This could be due to homogeneous structure depicted in compressive region. It is also observed that the standard deviation of LG and LR are more for abnormal subjects in both the strength regions. This could be due to large variations exhibited in trabecular structure due to the pathology. In both the compressive and tensile regions, the mean lacunarity values derived using LG method show distinct variation between normal and abnormal samples. These values are also found to be highly statistical significant. Hence mean lacunarity value derived using LG method could be as an index to differentiate normal and abnormal subjects.

29.4 CONCLUSION

Evaluation of trabecular bone architecture is one of the most important factors contributing to estimation of bone strength and is useful in early diagnosis of osteoporosis like disorders [17]. It has been shown that non-invasive assessment of structural parameters can provide a considerable predictive ability of bone disorders.

In this study, trabecular structure and its mechanical strength distribution on human femur bone are analyzed using planar radiographs. Acquired digital images of proximal femur trabecular bone are subjected to auto threshold binarization to minimize the irregularities in images due to uneven exposure conditions. From the binarized image, two strength regions, namely, compressive and tensile are delineated and are comprehensively analyzed using two different methods of lacunarity analysis. Quantitative analysis is also done to estimate apparent porosity.

Results show that the mean values of lacunarity derived using LG and LR methods are able to differentiate normal and abnormal subjects. It is found that the separation is more distinct for lower box size. Both the LG and LR values exhibit distinct variation between normals and abnormals in tensile region. This could be attributed to trabecular alterations occurring in this region due to loading forces. It is also observed that LG method could achieve better differentiation in both the strength regions. Hence, heterogeneity in patterns can be reliably measured using LG lacunarity method as it could capture the gray level variations in an image. Also, the lacunarity values of LG and LR methods are found to correlate well with the apparent porosity. These lacunarity values show high correlation for normals than abnormals. It is also found that the LG values have high correlation in both the strength regions.

It appears that the region specific architectural analysis is useful to identify strength of femur bone from planar radiographic images. This automated analysis can be used to enhance the diagnosis without much human intervention, which would be useful for mass screening of osteoporosis and bone mass disorders.

REFERENCES

1. Magland, J. F.; Zhang, N.; Rajapakse, C.S. and Wehrli, F.W. (2012). Computationally optimized bone mechanical modeling from high-resolution structural images. Plos one, 7(4), e35525.
2. Rachid, J.; Ahmad, A.; Gabriel, A. and Eric L. (2012). 3D shape-dependent thinning method for trabecular bone characterization. Medical Physics, 39(1), 168-178.
3. Delmas, P. D. and Seeman, E. (2004). Changes in bone mineral density explain little of the reduction in vertebral or nonvertebral fracture risk with anti-resorptive therapy. Bone, 34(4), 599-604.
4. Pothuaud, L.; Carceller, P. and Hans, D. (2008). A Correlations between grey-level variations in 2D projection images (TBS) and 3D microarchitecture: Applications in the study of human trabecular bone microarchitecture. Bone, 42, 775-787, 2008.
5. Nektarios, D.; Valousa, W.S.; Paul, A. and Fernando, M. (2010). The use of lacunarity for visual texture characterization of pre-sliced cooked pork ham surface intensities. Food Res Int., 43, 387–395.
6. Rodrigues, P.; Barbosa, M. S. and Costa, L.F. (2005). Self-referred approach to lacunarity. Phys Revi E., 72, 016707.
7. Allain, C. and Cloitre, M. (1991) Characterizing the lacunarity of random and determined fractal set. Phys Revi E., 44, 3552–3558.
8. Sarkar N. and Chaudhuri, B. B. (1992). An efficient approach to estimate fractal dimension of textural images. Pattern recognition, 25,1035–1041.
9. Kemal, K and Rahib, H. A. (2011) Exploiting the synergy between fractal dimension and lacunarity for improved texture recognition. Signal. Process., 91, 2332–2344.

10. Torres-Mejia, G.; De Stavola, B.; Allen, D. S.; Perez-Gavilan, J. J.; Ferreira, J. M.; Fentiman, I.S. and Dos, S. I. (2005). Mammographic features and subsequent risk of breast cancer: a comparison of qualitative and quantitative evaluations in the Guernsey prospective studies. Cancer Epidemiol Biomarkers Prev., 14(5), 1052–1059.
11. Qi, G.; Jiaqing, S. and Virginie, F. R. (2009). Characterization and classification of tumor lesions using computerized fractal-based texture analysis and support vector machines in digital mammograms. Int J CARS., 4, 11–25.
12. Manousaki, G.; Manios, A. G.; Tsompanaki, E.I. and Tosca, A.D. (2005) Use of color texture in determining the nature of melanocytic skin lesions – a qualitative and quantitative approach," Comp Biol Med., 36, 419–427.
13. Dong, P. (2000) Test of a new lacunarity estimation method for image texture analysis. International Journal of Remote Sensing, 21, 3369–3373.
14. Miranda-Martinez, E.; Oleschko, K.; Parrot, J. F.; Castrejon-Vacio, F.; Hind, T. and Brambila-Pazl, F. (2006). Porosity in naturally fractured media: a fractal classification. Rev Mex Cienc Geol., 23, 199–214.
15. Singh, A. R. N. and Maini, P.S. (1970). Changes in trabecular pattern of the upper end of the femur as an index of osteoporosis. Journal of Bone Joint Surgery, 52, 457– 467.
16. Jesu Christopher, J. and Ramakrishnan, S. (2007). Assessment and classification of mechanical strength of human femur trabeculae bone using texture analysis and neural networks. Journal of Medical Systems, 35, 117–122.
17. Kreider, M. and Goldstein, S. A. (2009). Trabecular bone mechanical properties in patients with fragility fractures. Clinical Orthopedic Related Resarch, 467, 1955-1963.

30

Correlation of Muscle Work with Hand Dexterity in Dental Surgeons

Tejashree A. Dabholkar, Pradnya Sawant and Sujata Yardi

Dept. of Physiotherapy Padmashri D.Y. Patil University, Nerul, Navi Mumbai

Email: kamble.teju@yahoo.co.in

ABSTRACT

Hand dexterity is a pre-requisite to dental department. There is paucity of literature available to know whether there exist any correlation between type of muscle work in dental procedure and dexterity .hence this study aims at doing so using surface electromyogram. In this Co-relational study Design, Conservative dentistry and Endodontic department, Dental department of Pad Dr D.Y. Patil University, Navi Mumbai. Postgraduate students and dental faculty of Conservative dentistry and Endodontics department were evaluated for muscle work of flexor Carpi ulnaris, extensor carpi radialis and brevis, hand dexterity, grip strength and pinch strength. Study shows, ECRLB has maximum work in all three dental procedures, no significant correlation of muscle work with dexterity in any dental procedure ($p < 0.005$).

Keywords: Muscle work, dexterity, flexor carpi ulnaris, extensor carpi radialis and brevis.

30.1 INTRODUCTION

Hand dexterity is important for humans to interact with the external environment. Hand dexterity is a job pre-requisite to many professions amongst which is dentistry. Dentists require good visual acuity, hearing, depth perception, psychomotor skills, manual dexterity, and ability to maintain occupational postures over long periods[1]. It is postulated that upper limb muscles and skeleton are more implicated during dental operation rather than other sites of the body[2,3,4,5,6,7]. Amongst the many branches of dentistry this study mainly focuses on Conservative Dentistry and Endodontic as it included fillings, root canal treatment, endodontic implants, bleaching & management of traumatized teeth which are most of the commonest procedures performed. Occupational reasons leading to MSDs include repetitive bending movements of wrist, to catch devices forcefully with ulnar deviation of hand and repetitive pressure on palm which is used during exfoliation and canal cleaning that can compress median nerve[8]. Since dentists need good manual dexterity as well as hand skills, studying the muscle activity in the forearm during dexterity tasks would give us an idea about type of muscle work involved. Muscle work is studied using integrated surface electromyographic [EMG] studies. Also there is paucity of literature available to know whether or not the hand dexterity of the dentists has a correlation with the amount of motor activity happening in the forearm muscles. Thus the objective of this study is to determine the amount of work happening in the forearm musculature during the various dental procedure activities like drilling, filling and pulp excavation (P.E.) performed by the dentists, using electromyographic studies and also to find out if the hand dexterity of the dentists is affected owing to the same reason.

30.2 MATERIALS & METHODOLOGY

The sample size was 17 between the age group of 22 to 30. The mean age of the study sample was 25.4705 ± 1.772 and the mean of the work experience, i.e. number of practicing years was 3.911±1.822. There were 5(29.41%)

males and 12 (76.47%) females who participated in the study Dental postgraduate students and practitioners from Conservative Dentistry and Endodontic Dept. were included in study whereas those with musculoskeletal problems of upper quadrant; neurological problems of upper quadrant; cervical pathology leading to symptoms in upper quadrant; Subjects involved in any upper extremity training programme were excluded from study. The study was approved by institutional ethics committee and review board of Pad. Dr. D.Y. Patil University, Navi Mumbai, and informed consent was obtained from all participants before the study. Hand grip strength and pinch grip strength of the dominant hand was calculated using a Jamar hand held dynamometer and pinch grip dynamometer according to the recommendations of the American Hand Society of Hand Therapists.. Hand dexterity was assessed using Jebsen Hand Function Test (JHFT).

The muscle w ork during the activities of drilling, filling and Pulp excavation. was carried out using 2 channel integrated EMG. During dental procedures, most commonly adapted position of wrist was extension along with radial deviation. Also it has been studied that Flexor Carpi Ulnaris is one of the dynamic stabilizers at the wrist.[9, 10]. The electrodes were fixed on these muscles as per the SURFACE EMG MADE EASY: A Beginner's Guide for Rehabilitation while the subjects performed the activity.

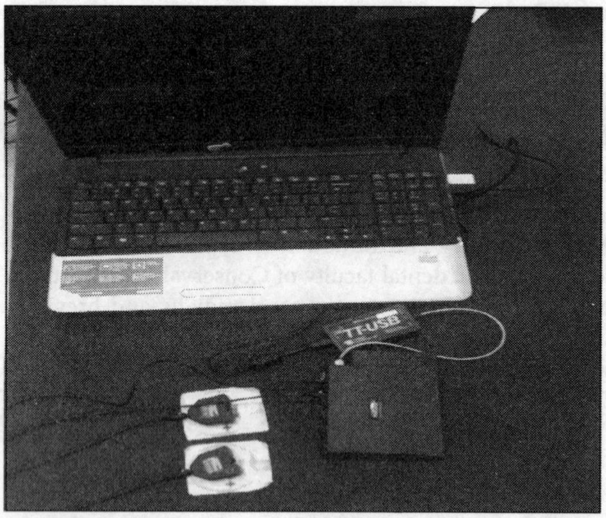

Fig. 30.1 Surface electromyogram with two triode electrode

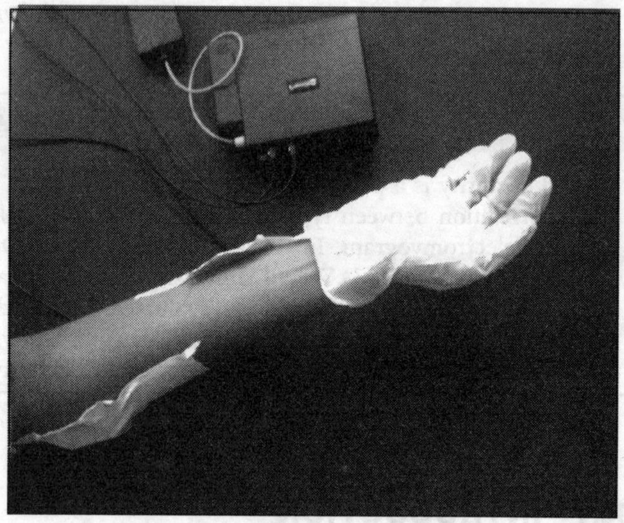

Fig. 30.2 Electrode placement

Each activity was noted for 10 seconds with gloves on and the reading was repeated three times for each activity and the best possible outcome was noted.

Statistical analysis was done using Graph pad Instat 3 to find out if there existed a correlation between the muscle work and hand dexterity in dentists.

Table 30.1: Muscle work correlation with dexterity in three dental procedures

Parameters FCU				*ECRL/B*		
P test	'r'value	95%CI	'p'value	'r'value0.419	95%CI	'p'value 0.12#
Drilling	0.149	−0.11 to 0.76	0.12#	0.031	−0.48 to 0.53	0.91#
Pulp excavation	0.34	−0.19 to −0.73	0.20#	0.26	−0.28 to 0.68	0.34#
# not significant	0.22	0.32 to 0.66	0.41#	−0.02	0.51 to 0.51	0.99#
# not significance						

30.3 DATA ANALYSIS & RESULTS

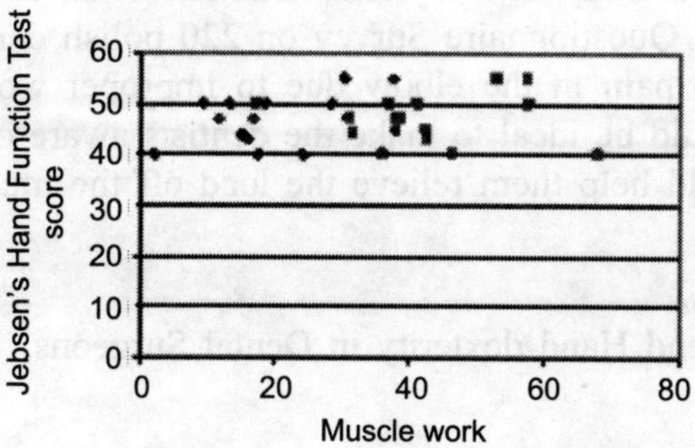

Muscle Work and Hand Dexterity (Drilling)

Graph 1

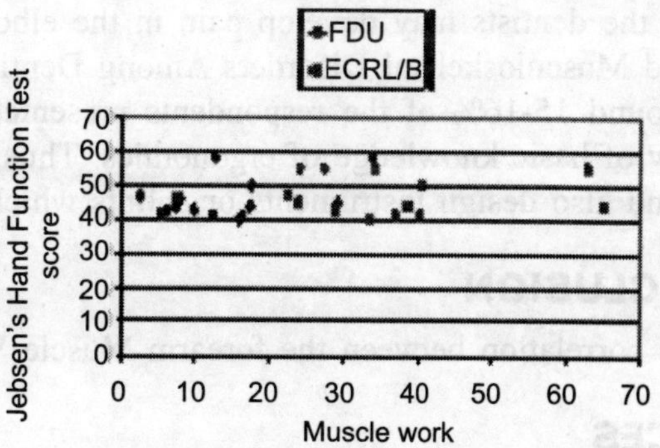

Muscle Work and Hand Dexterity (Filling)

Graph 2

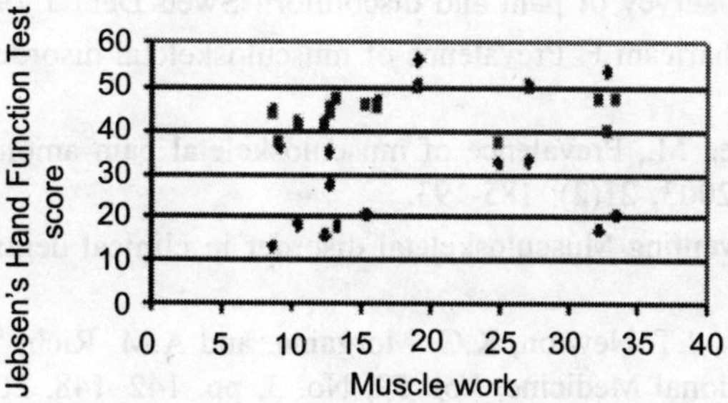

Muscle Work and Hand Dexterity (Pulp Excavation)

Graph 3

Graph 1: Correlation of muscle work vs. hand dexterity in drilling condition
Graph 2: Correlation of muscle work vs. hand dexterity in filling condition
Graph 3: Correlation of muscle work vs. hand dexterity in pulp extraction

30.4 DISCUSSION

The mean pinch grip strength was 16.88±6.373 and the mean Hand grip strength was 48.866 ± 4.926. The mean dexterity assessed using Jebsen Hand Function Test was 48.1176±5.520 for the study population. It was found that though the mean muscle work of ECRL/B was more compared to that of FCU in all the three activities, there was no statistic difference between the muscle work happening in the three activities for both FCU (P-value 0.2059) and ECRL/B(P-value 0.0503). In all the 3 activities the dentists used a precision grip for holding the instruments. This requires proper stabilization of the wrist which is brought about by the extensors and the flexors of the forearm, thus it places more demand on the ECRL/B as well as the FCU muscle. Also the movements of the wrist performed by the subjects consisted mainly of wrist extension (which remained static) along with repetitive ulnar and radial deviation11 which indicates that there is a specific arc of movement that is followed by the wrist which demands stability along with mobility and is achieved by the ECRL/B. that acts not only as a prime mover but also as a stabilizer at the wrist.[11, 12]. This study shows no significant correlation between the muscle work in FCU and ECRL/B with the Hand Dexterity during Drilling (P-value 0.12, 0.914), Filling (P-value 0.2029, 0.3421) as well as Pulp Excavation (P-value 0.4156, 0.9919). There is a possibility that the FCU and ECRL/B do not influence the final score of the Jebsen Hand function test. Also there is a possibility that the motor pattern used could be different for completion of the tasks in Jebsen Hand Function test.[13] There was no significant correlation of the hand dexterity with the pinch grip strength (P-value 0.1129) or the hand grip strength (P-value 0.2379) found in this study. Also there was no significant correlation between the work experience with the hand grip strength and the pinch grip strength as well as with hand dexterity. This could be because the study sample had a work experience which fell in a very narrow range (mean of work experience: 3.911±1.822). The study suggests that in normal subjects, with no pathology, the activity of the ECRL/B is more than FCU in all the above mentioned procedures of drilling, pulp excavation and filling which implicates that the strength of the ECRL/B needs to be more and also due to repetitive use the dentists may develop pain in the elbow. In a study done by Anna Kierklo et al. to find out the "Work-Related Musculoskeletal Disorders Among Dentists –A Questionnaire Survey on 220 polish dentists, it was found that around 15-16% of the respondents presented with pain in the elbow due to improper working habits and deficiency of basic knowledge of ergonomics. Thus it would be ideal to make the dentists aware of the proper ergonomics and also design instruments or splints which would help them relieve the load off the muscles.

30.5 CONCLUSION

There was no correlation between the forearm Muscle Work and Hand dexterity in Dental Surgeons.

REFERENCES

1. Valachi B., Valanchi K. Mechanism leading to musculoskeletal disorder in dentistry. JADA 2003; 134:1344-1350.
2. Rundcrantz B.L., Johnsson B., Moritz U. Cervical pain and discomfort among dentists: epidemiological, clinical and therapeutic aspects. Part 1- A survey of pain and discomfort. Swed Dent J 1990; 14:71-80.
3. Alexopoulos E.C., Stati IC, Charicani F. Prevalence of musculoskeletal disorders in dentists. http:/www.biomedcentral.com/1471-2474/5/16. pp. 1 to 8.
4. Dehghan FM, Amiri Z., Rabiee M., Prevalence of musculoskeletal pain among a group of Iranian dentists (Tehran 1999). Beheshti University Dent J., 2003; 21(2): 185-193.
5. Valanchi B., Valanchi K., Preventing Musculoskeletal disorder in clinical dentistry. J. Am Dent Assoc 2003; 134(12):1604-1612.
6. K.M.S. Ayers, W.M. Thomson, J.T. Newton, K.C. Morgaine, and A.M. Rich, "Self-reported occupational health of general dental practitioners," Occupational Medicine, Vol. 59, No. 3, pp. 142–148, 2009.
7. Sagar Abichandani, Saquib Shaikh and Ramesh Nadiger. Carpal tunnel syndrome – an occupational hazard facing dentistry. Dental Journal. Vol. 63, Issue 5, pp. 230–236, October 2013.
8. Bland J.D. et al. The relationship of obesity, age, and carpal tunnel syndrome: more complex than was thought? Muscle Nerve, 2005; 32: 527–532.
9. Milerad et al. An electromyographic study of dental work. Ergonomics. 1991; Vol. 34, Issue 7.
10. Anna Kierklo et al Work-Related Musculoskeletal Disorders Among Dentists –A Questionnaire Survey. Ann Agric Environ Med 2011, 18, 79–84.
11. Brent Brookbush. Kinesiology by Joint Action – Functional Groups:
12. Medical Discoveries, 2010. Dental Drill.
13. Joan D. Carlson et al. The Effect of Wrist Immobilization on Performance of the Jebsen Hand Function Test. The American Journal of Occupational Therapy. March 1983; Vol. 37, No. 3.

31

Ergonomics and Human Factors at Work

M. K. Pradhan

Safety Directorate of Factories & Boilers, Bhubaneswar, Odisha
Unit-III, Kharavela Nagar, Bhubaneswar–751001

Email: www.dleorissa.in

ABSTRACT

Training and education ensure that employers sufficiently inform workers about ergonomic risk factors at their worksites so they are better able to participate actively in their own protection. A good ergonomics training programme will teach employees how to properly use equipment, tools, and machine controls as well as the correct way to perform job tasks. For example, employers should encourage work methods that allow workers to keep their joints in a neutral position (wrists straight and elbows bent at a right angle) while using tools requiring manual force to prevent excessive force on joints and tendons. To minimize or prevent back disorders, employers should teach workers to avoid long reaches, maintain neutral postures, and use proper lifting techniques. Using correct posture is important whether an employee is sitting, standing, pulling, pushing, lifting, or using tools or equipment.

Keywords: Back disorders, ergonomic risk factors, ergonomics training, neutral postures.

31.1 INTRODUCTION

Ergonomics can be defined simply as the study of work. More specifically, ergonomics is the science of designing the job to fit the worker, rather than physically forcing the worker's body to fit the job. Adapting tasks, work stations, tools, and equipment to fit the worker can help reduce physical stress on a worker's body and eliminate many potentially serious, disabling work related musculoskeletal disorders (MSDs). Ergonomics draws on a number of scientific disciplines, including physiology, biomechanics, psychology, anthropometry, industrial hygiene, and kinesiology.

31.2 THE JOB/TASK BEING DONE

We need to know about ergonomics if you are an employer or an employee in the manufacturing, construction, maritime, and agricultural industries and you or your employees work activities and job conditions include:

- Repeating the same motion throughout your workday,
- Working in awkward or stationary positions,
- Lifting heavy or awkward items,
- Using excessive force to perform tasks, and
- Being exposed to excessive vibration or
- Extreme temperatures.

31.3 IMPORTANCE OF ERGONOMICS

Industries increasingly require higher production rates and advances in technology to remain competitive and stay in business. As a result, jobs today can involve:

- Frequent lifting, carrying, and pushing or pulling loads without help from other workers or devices;
- Increasing specialization that requires the worker to perform only one function or movement for a long period of time or day after day;
- Working more than 8 hours a day;
- Working at a quicker pace of work, such as faster assembly line speeds; and
- Having tighter grips when using tools.

These factors especially if coupled with poor machine design, tool, and workplace design or the use of improper tools create physical stress on workers' bodies, which can lead to injury.

If work tasks and equipment do not include ergonomic principles in their design, workers may have exposure to undue physical stress, strain, and overexertion, including vibration, awkward postures, forceful exertions, repetitive motion, and heavy lifting. Recognizing ergonomic risk factors in the workplace is an essential first step in correcting hazards and improving worker protection. Ergonomists, industrial engineers, occupational safety and health professionals, and other trained individuals believe that reducing physical stress in the workplace could eliminate up to half of the serious injuries each year. Employers can learn to anticipate what might go wrong and alter tools and the work environment to make tasks safer for their workers.

A dramatic increase in MSDs began in the 1970s when these disorders increasingly appeared on companies' injury and illness logs. OSHA cited companies for hazardous workplace conditions that caused problems such as tendinitis, carpal tunnel syndrome, and back injuries.

31.4 MUSCULOSKELETAL DISORDERS (MSDs)

MSDs, or musculoskeletal disorders, are injuries and disorders of the soft tissues (muscles, tendons, ligaments, joints, and cartilage) and nervous system. They can affect nearly all tissues, including the nerves and tendon sheaths, and most frequently involve the arms and back.

Occupational safety and health professionals have called these disorders a variety of names, including cumulative trauma disorders, repeated trauma, repetitive stress injuries, and occupational overexertion syndrome.

These painful and often disabling injuries generally develop gradually over weeks, months, and years. MSDs usually result from exposure to multiple risk factors that can cause or exacerbate the disorders, not from a single event or trauma such as a fall, collision, or entanglement.

MSDs can cause a number of conditions, including pain, numbness, tingling, stiff joints, difficulty moving, muscle loss, and sometimes paralysis. Frequently, workers must lose time from work to recover; some never regain full health.

These disorders include carpal tunnel syndrome, tendinitis, sciatica, herniated discs, and low back pain. MSDs do not include injuries resulting from slips, trips, falls, or similar accidents.

31.5 CAUSES OF WORK RELATED MSDS

Work-related MSDs occur when the physical capabilities of the worker do not match the physical requirements of the job. Prolonged exposure to ergonomic risk factors can cause damage a worker's body and lead to MSDs.

Conditions that are likely to cause MSD problems include the following:

- Exerting excessive force;
- Excessive repetition of movements that can irritate tendons and increase pressure on nerves;
- Awkward postures, or unsupported positions that stretch physical limits, can compress nerves and irritate tendons;
- Static postures, or positions that a worker must hold for long periods of time, can restrict blood flow and damage muscles;

- Motion, such as increased speed or acceleration when bending and twisting, can increase the amount of force exerted on the body;
- Compression, from grasping sharp edges like tool handles, can concentrate force on
- Small areas of the body, reduce blood flow and nerve transmission, and damage tendons and tendon sheaths;
- Inadequate recovery time due to overtime, lack of breaks, and failure to vary tasks
- Can leave insufficient time for tissue repair;
- Excessive vibration, usually from vibrating tools, can decrease blood flow, damage nerves, and contribute to muscle fatigue.
- Whole-body vibration, from driving trucks or operating subways, can affect skeletal
- Muscles and cause low-back pain; and
- Working in cold temperatures can adversely affect a worker's coordination and manual
- Dexterity and cause a worker to use more force than necessary to perform a task.

These risk factors, either alone or in combination, can subject workers' shoulders, arms, hands, wrists, backs, and legs to thousands of repetitive twisting, forceful, or flexing motions during a typical workday. To contribute to MSDs, however, these risk factors must be present for a sufficient duration, frequency, or magnitude.

31.6 CAUSES OF NON-WORK RELATED MSDs

Risk factors not related to the job can cause or contribute to MSDs. These factors include:

- Physical conditioning;
- Medical conditions, such as obesity, diabetes, and arthritis;
- Pregnancy;
- Hobbies that are hand intensive or require manual handling. In these instances, however, because one can control the duration and exposures, hobbies usually are not primary risk factors; and
- Psychological or social workplace stress.

31.7 TYPES OF WORK HAVING ERGONOMIC HAZARDS

MSDs affect workers in almost every occupation and industry in the nation and in workplaces of all sizes. The disorders occur most frequently in jobs that involve:

- Manual handling,
- Manufacturing and production,
- Heavy lifting,
- Twisting movements, and
- Long hours of working in awkward positions.

31.8 SYMPTOMS OF DEVELOPMENT OF MSD

One could have a work-related MSD in case of any of the following:

- Numbness in your fingers,
- Numbness in your thighs,
- Difficulty moving your finger,
- Stiff joints, or
- Back pain.

31.9 BODY PARTS MOST AFFECTED BY MSDs

MSDs can affect nearly all tissues in the human body: the nerves, tendons, tendon sheaths, and muscles. The most frequently affected areas of the body are the arms and the back.

31.10 PARTS OF THE BODY AFFECTED BY MSDs

- Arms
- Hands
- Fingers
- Neck
- Back
- Wrists
- Legs
- Shoulders

30.11 PREVENTION OF MSDs

Many solutions to ergonomic problems in the workplace are simple and inexpensive. For example, awkward and uncomfortable positions can be eliminated by:

- Adjusting the height of working surfaces,
- Providing telephone headsets,
- Supplying anti-fatigue mats,
- Varying tasks,
- Providing short breaks,
- Reducing the weight and size of items workers must lift,
- Putting supplies and equipment within easy reach of the worker,
- Providing ergonomic chairs or stools, and
- Supplying the right tool for the job and the right handle for the worker.

Good ergonomics is good economics.

31.12 ADVANTAGES

Providing a workplace free of ergonomic hazards can do the following:

- Lower injury rates as MSD incidences go down;
- Increase productivity by making jobs easier and more comfortable for workers;
- Improve product quality because fewer errors will be made when using automated processes that demand less physical effort;
- Reduce absences because workers will be less likely to take time off to recover from muscle soreness, fatigue, and MSD-related problems;
- Reduce turnover as new hires are more likely to find an ergonomically designed job within their physical capacity;
- Lower costs as workers' compensation and other payments for illness and replacement;
- workers go down;
- Improve worker safety;
- Increase worker comfort;
- Reduce worker fatigue; and
- Improve worker morale.

31.13 CONTROL OF RISK FACTORS

Employers can prevent MSD hazards by properly designing the job or work station and selecting the appropriate tools or equipment for that job. Based on information from the job analysis, an employer can establish procedures to correct or control risk factors by using:

- Appropriate engineering controls, such as work station, tool, and equipment design or redesign;
- Work practices, such as proper lifting techniques and keeping work areas clean;
- Administrative controls, such as worker rotation, more task variety, and increased rest breaks, and if necessary;
- Personal protective equipment, such as knee pads, vibration gloves, and similar devices.

The National Institute for Occupational Safety and Health recommends using the following guidelines in jobs requiring manual handling:

- Minimize the distance between the load and the body.
- Lift loads from knuckle height.
- Keep the travel distance for the lift to less than 10 feet.
- Minimize twisting.
- Provide good handles for grasping loads.

It is also important that work tools and equipment be ergonomically designed. Most hand tools are designed for only occasional use, not for repetitive use over prolonged periods. When acquiring tools for regular use in an industrial setting, an employer should consider the following ergonomic features:

- Tools should be light-weight and handles designed to allow a relaxed grip so the wrists can remain straight.
- Tools should be designed for use with either hand and be of various sizes so they are appropriate for all workers.
- Tool handles should be shaped so that they contact the largest possible surface of the inner hand and fingers. Avoid tool handles with sharp edges and corners.
- Use power tools to reduce the amount of human force and repetition required.
- Purchase low-vibration tools to reduce tool vibration, and, if necessary, fit absorbent rubber sleeves over the tool handle.

31.14 EXAMPLES OF MUSCULOSKELETAL DISORDERS

Body Parts Affected	*Symptoms*	*Possible Causes*	*Workers Affected*	*Disease Name*
Thumbs	Pain at the base of the thumbs	Twisting and gripping	Butchers, house-keepers, packers, seamstress, cutters	De Quervain's disease
Fingers	Difficulty moving finger; snapping and jerking movements	Repeatedly using the index fingers	Meatpackers, poultry workers, carpenters, electronic, assemblers	Trigger finger
Shoulders	Pain, stiffness	Working with the hands above the head	Power press operators, welders, painters, assembly line workers	Rotator cuff tendinitis
Hands, wrists	Pain, swelling	Repetitive or forceful hand and wrist motions	Core making, poultry processing, meatpacking	Tenosynovitis
Fingers, hands	Numbness, tingling; ashenskin; loss of feeling and control	Exposure to vibration	Chain saw, pneumatic hammer, and gasoline powered tool operators	Raynaud's syndrome (white finger)
Fingers, wrists	Tingling, numbness, severe pain; loss of strength, sensation in the thumbs, index or middle or half of the ring fingers	Repetitive and forceful manual tasks without time to recover	Meat and poultry and garment workers, upholsterers, assemblers, VDT operators, cashiers	Carpal tunnel syndrome
Back	Low back pain, shooting pain or numbness in the upper legs	Whole body vibration	Truck and bus drivers, tractor and subway operators; warehouse workers; nurses aides; grocery cashiers; baggage handlers	Back disability

31.15 MSD MANAGEMENT

MSD management is another important element of an effective ergonomics program. Proper MSD management focuses on early identification and evaluation of signs and symptoms of MSDs and helps eliminate or reduce the risk of developing MSDs. Employers should include the following elements in any MSD management program:

- Injury and illness recordkeeping;
- Early recognition and reporting of MSD symptoms;
- Systematic evaluation and referral to a qualified health care provider;
- Conservative treatment, such as restricted duty jobs, when necessary;
- Conservative return to work;
- Systematic monitoring, including periodic workplace walkthroughs;
- Adequate staffing and facilities where employers provide on-site evaluation;
- Employee training and education;-
- Access to health care providers for each work shift; and
- No barriers to early reporting.

31.16 TRAINING & EDUCATION PROGRAMME

Training and education ensure that employers sufficiently inform workers about ergonomic risk factors at their worksites so they are better able to participate actively in their own protection.

A good ergonomics training programme will teach employees how to properly use equipment, tools, and machine controls as well as the correct way to perform job tasks.

For example, employers should encourage work methods that allow workers to keep their joints in a neutral position (wrists straight and elbows bent at a right angle) while using tools requiring manual force to prevent excessive force on joints and tendons.

To minimize or prevent back disorders, employers should teach workers to avoid long reaches, maintain neutral postures, and use proper lifting techniques. Using correct posture is important whether an employee is sitting, standing, pulling, pushing, lifting, or using tools or equipment.

31.17 CONCLUSION

Work-related MSDs occur when the physical capabilities of the worker do not match the physical requirements of the job. Prolonged exposure to ergonomic risk factors can cause damage a worker's body and lead to MSDs. A good ergonomics training programme will teach employees how to properly use equipment, tools, and machine controls as well as the correct way to perform job tasks. To minimize or prevent back disorders, employers should teach workers to avoid long reaches, maintain neutral postures, and use proper lifting techniques.

REFERENCES

1. R.G. Sell. The human factors section of Bisra, 1950–1964 (The corporate laboratories, British Steel Corporation). Applied Ergonomics, Vol. 2, Issue 4, December 1971, Pages 226–229
2. A Smith. Noise, performance efficiency and safety. Archives of occupational and environmental health, 1990
3. R.G. Gun. The role of regulations in the prevention of occupational injury. Safety Science Vol. 16, Issue 1, January 1993, pp. 47–66
4. Bryon Jr., L. (1999). Educating Engineers on Safety. J. Manage. Eng., 15(2), 30–33.

32

Study of Mechanical Characteristic of Femur and Implants Under Different Loading Pattern

A. K. Dash, D. K. Agarwalla, A. Sahabaz, B. Ritesh, S. S. Sahu, N. Vijoy and S. Nayak

Deparment of Mechanical Engineering

Institute of Technical Education and Research, Bhubaneswar, India

Email: sahabaz.akram@gmail.com

ABSTRACT

The implant plays a vital role for the fracture fixation in bones. But it is observed that due to the failure of implants, the time limit for union of the bone increases extensively and sometimes results in mal-union. The stress and the deformation of the bone as well as the implant are the major parameter which affects the failure of implants. In current analysis, computational analysis has been performed to find out the deformation and stress distribution on the femur bone and the implant to identify the portion subjected to maximum stress and maximum deformation.

Keywords: Implant, fracture, deformation, stress, mal-union.

32.1 INTRODUCTION

The femur, or thigh bone, is the most proximal (closest to the body) bone of the leg invertebrates capable of walking or jumping, such as most land mammals, birds, many reptiles such as lizards, and amphibians such as frogs. In vertebrates with four legs such as dogs and horses, the femur is found only in the rear legs.

32.1.1 Human Anatomy

In human anatomy, the femur is the longest and largest bone. Along with the temporal bone of the skull, it is one of the two strongest bones in the body. The average adult male femur is 48 centimeters (18.9 in) in length and 2.34 cm (0.92 in) in diameter and can support up to 2 times the weight of an adult. It forms part of the hip joint (at the acetabulum) and part of the knee joint, which is located above. There are four eminences, or protuberances, in the human femur: the head, the greater trochanter, the lesser trochanter, and the lower extremity. They appear at various times from just before birth to about age 14. Initially, they are joined to the main body of the femur with cartilage, which gradually becomes ossified until the protuberances become an integral part of the femur bone, usually in early adulthood. The shaft of femur is cylindrical with a rough line on its posterior surface (linea aspera). The intercondylar fossa is present between the condyles at the distal end of the femur. In addition to the intercondylar eminence on the tibial plateau, there is both an anterior and posterior intercondylar fossa (area), the sites of anterior cruciate and posterior cruciate ligament attachment, respectively.

For the present study, the proximal and cortical region have been taken into consideration because of the upper extremity forms the one-third of the total femur length and the tissue composition is much less dense than the lower part. Four curves meet in the upper part making it more vulnerable and prone to fractures and moreover three stresses acts in this region at the same time. 1. Normal stress 2. Torsional shear stress and 3. Bending stress.

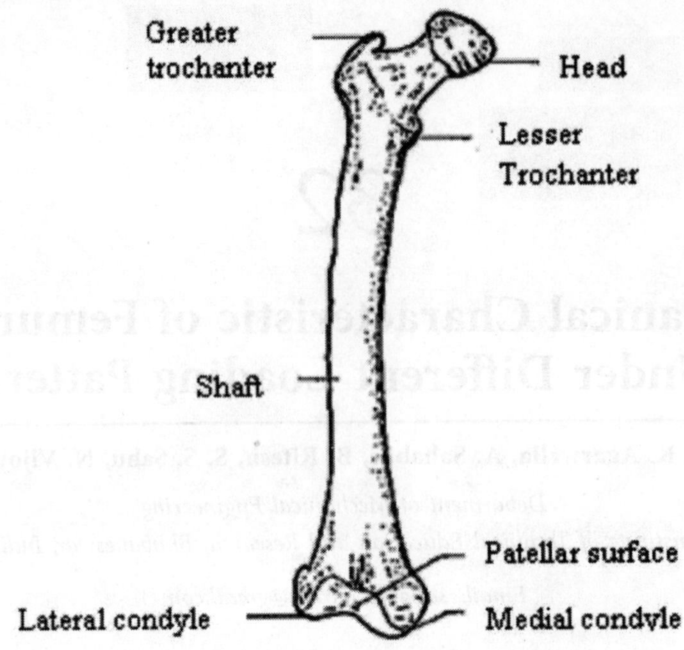

Fig. 32.1 Femur bone

32.1.2 Types of Femur Bone Fractures

The femur is one of the largest and strongest bones in the body. The femur is the thigh bone--it extends from the hip joint down to the knee joint. Because the femur is such a strong bone, it can take tremendous force to cause a femur fracture.

In order for a femur fracture to occur, either a large force must be applied or something is wrong with the bone. In patients with normal bone strength, the most common causes of femur fractures include: 1. Car accidents and 2. Falls from a height. Patients may also have bone that is weakened by osteoporosis, tumor, or infection. These conditions can lead to a so-called pathologic femur fracture.

Femur fractures are generally separated into three broad categories:

32.1.2.1 Proximal Femur Fracture

Proximal femur fractures, or hip fractures, involve the upper- most portion of the thigh bone, just adjacent to the hip joint. A hip fracture is a fracture in the proximal end of the femur (the long bone running through the thigh), near the hip joint.

The term "hip fracture" is commonly used to refer to four different fracture patterns and is often due to osteoporosis; in the vast majority of cases, a hip fracture is a fragility fracture due to a fall or minor trauma in someone with weakened osteoporotic bone.

32.1.2.2 Types

Many subtypes of fractures about the hip joint are colloquially known as 'hip fractures'. Although a true hip fracture involves the joint, the following four proximal femur fractures are commonly referred to as hip fractures. The differences between them are important because each is treated differently.

- *Femoral head fracture* denotes a fracture involving the femoral head. This is usually the result of high energy trauma and a dislocation of the hip joint often accompanies this fracture.
- *Femoral neck fracture* (sometimes Neck of Femur (NOF), sub capital, or intracapsular fracture) denotes a fracture adjacent to the femoral head in the neck between the head and the greater trochanter.
- *Intertrochanteric fracture* denotes a break in which the fracture line is between the greater and lesser trochanter on the intertrochanteric line.

- *Subtrochanteric fracture* actually involves the shaft of the femur immediately below the lesser trochanter and may extend down the shaft of the femur

32.1.2.3 Femur Shaft Fractures

Femoral shaft fractures are among the most common fractures encountered in orthopedic practice. These fractures have age and gender-related bimodal distribution, with injuries occurring most frequently in young males following high-energy trauma.

These fractures can be classified as spiral, oblique, transverse, comminuted or segmental. Of these fractures comminuted and segmental fractures are difficult to treat because of their vertical and rotational instability as compared to other femoral shaft fracture types.

Transverse fracture is a type of fracture in which the break is a straight horizontal line going across the femoral shaft whereas oblique fracture has an angled line across the shaft and spiral fracture line encircles the shaft like the stripes on a candy cane. A twisting force to the thigh causes this type of fracture. In comminuted fracture the bone has broken into three or more pieces. In most cases, the number of bone fragments corresponds with the amount of force required to break the bone.

The treatment of a femoral shaft facture is almost always with surgery. In certain patients, depending on the fracture type and associated injuries, an intramedullary rod may not be an option; in these cases one of the other treatments (plate and screws, external fixator, etc.) will be selected called an intramedullary rod.

Supracondylar Femur Fractures

A supracondylar femur fracture is an unusual injury to the femur just above the knee joint. These fractures often involve the cartilage surface of the knee joint, and must be treated with this cartilage injury in mind. Patients who sustain a supracondylar femur fracture are often at high risk of developing knee arthritis later in life.

Supracondylar femur fractures are more common in patients with severe osteoporosis and in patients who have previously undergone total knee replacement surgery.

The treatment of a supracondylar femur fracture is highly variable, and may utilize a cast or brace, external fixator, plate, screws, or an intramedullary rod.

32.1.3 Literature Survey

This section of the manuscript depicts few of the literatures being published by various authors.

Cicero et al. [1] have investigated the failure of a hip implant, which occurred nine months after the corresponding surgery operation. They have found that the failure has been due to the fatigue process. During the analysis they observed that irregularities surfaced at the interfaces of the base material and the coating leading to notch effect and thereby formation of crack and its propagation. The fatigue failure of a locking compression plate (LCP) fixed across a transverse fracture (8 mm gap) at the mid shaft of femur has been experimentally evaluated by Kanchanomai et al. [2] and they have analyzed and stated that, the fatigue failure of LCP has been possible before the adequate healing of fracture and the full load of walking should not be allowed for the patient. Triantafyllidis et al. [3] have investigated the premature fracture of a stainless steel orthopedic plate implant by alternative episodes of fatigue and cleavage de-coherence. They deduced that the fracture may have been produced due to improper plate fixation and the action of biomechanical forces that resulted cyclic loading and occasionally corrosion assisted cracking. Wallstabe et al. [4] have investigated the failure of an IMHS implant after the rehabilitation period, which required a revision surgery where the upper femur had to be replaced by a hip endoprosthesis. The investigation revealed several causes for the fracture of the lag screw which is the implant component intended to fix the position of the femoral head. These are in descending order of relevance: (1) the screw was placed incorrectly and therefore had a smaller bending stiffness, (2) the femoral fracture zone was not stable, i.e., the lag screw "moved out" of its proper position. Griza et al. [5] have found out that , in cases of nonunion or mal-union of fragments of the bone, the nail can be subjected to an expressive portion of the external loads at full weigh bearing instances, which can promotes short term fatigue fracture of the nail. They have found that the failure has been caused due to severe plastic strain in the locked nails. Schneider [6] et al. have investigated the loads acting in an intramedullary nail during fracture healing in the human femur and found that significant axial forces and bending moments were acting during several activities such as sitting, unsupported leg elevation and partial weight bearing in a standing position. They observed that the forces orthogonal to the nail axis remained small. Maharaja et al. [6] have investigated the numerical analysis of fractured femur bone with prosthetic bone plates. The objective of their study was to compare the bone plates made of different materials like stainless steel, titanium, alumina and nylon and they find out the best material. Out of the several materials used by the author titanium was found out to have less equivalent stress due to loading. Cristofolini et al. [7] have worked on mechanical validation of whole bone composite femur models, compared to cadaveric femur different loading conditions. They also studied the behaviour under axial loading by comparing the vertical displacement of the head as well as the axial strain. Finally, they performed a four point bending test and a torsional test to characterize the whole bone stiffness of the femur. In all these tests, the composite femurs have shown to fall well within the range of cadaveric femur, with no significant differences being detected between the synthetic femurs and the cadaveric femurs. Nassiri et al. [8] have worked on computational modelling of long bone fractures fixed with locking plates. During their work they have found that axial stiffness is also influenced by the working length and plate-bone distance. They observed that the fractured gap had no effect on the construct stability when no bone contact occurred during loading. The stress analysis of the LCP demonstrated that the maximum Von-Mises stresses were found in the innermost screws at the screw-head junction.

32.2 METHODS

Finite element analysis (FEM) is a way to simulate loading conditions on a design and to determine the design response to those conditions. The FEM represents one of the most significant developments in the history of computational methods. It is a computer-based numerical technique for calculating the strength and behaviour of engineering structures. It can be used to calculate deflection, stress, vibration, buckling behaviour and many other phenomena. It can be used to analyze either small or large-scale deflection under loading or applied displacement. It can analyze elastic deformation, or "permanently bent out of shape" plastic. Deformation The computer is required because the astronomical number of calculations needed to analyze a large structure. The power and low cost of modern computers has made Finite Element Analysis available to many discipline and companies.

In the Finite Element Method, a structure is broken down into many small simple blocks or elements. The behavior of an individual element can be described with a relatively simple set of equations. Just as the set of elements would be joined together to build the whole structure, the equations describing the behaviors of the individual elements are joined into an extremely large set of equations. From the solution, the computer extracts the behavior of the individual elements. From this, it can get the stress and deflection of all the parts of the structure. The stresses will be compared to allowed values of stress for the materials to be used, to see if the structure is strong enough.

32.2.1 Steps Adopted in Current Analysis

- Modeling of the upper extremity and cortical zone of the bone.
- Modeling of the plate, intramedullary rod and screw.
- Assembly of the implants with the bone.
- Importing the assembly to the computational software.
- Assigning the material properties and bone properties.
- Meshing the model with relevance center fine.
- For the structural analysis the boundary condition is given by fixing of the femur at the bottom part and different types of loads, i.e., compression, torsion and combined (compression + torsion) is applied at different magnitude.
- The analysis have been done with locking compression plate and then with IM rod.
- Computed the solution model and found the analyses data.

32.3 RESULTS AND DISCUSSION

Table 32.1: Statistics mesh

Nodes	22014
Mess	11773

Table 32.2: Properties of materials

	Structural Steel	*Bone*
Density(kg m^{-3})	7850	1750
Young's modulus (GPa)	200	16
Poisson's ratio	0.3	0.3

32.4 FEMUR WITH LOCKING COMPRESSION PLATE

Table 32.3: Compressive force study

Compressive force (N)	*Equivalent (Von-Mises) Stress (MPa)*		*Total Deformation (mm)*
	Max	*Min*	*Max*
375	73.496	1.672×10^{-4}	1.8848
475	93.095	2.132×10^{-4}	2.3874
575	112.69	2.528×10^{-4}	2.89

Table 32.4: Torsion

***Torsion** (N-mm)*	*Equivalent (Von-Mises) Stress (MPa)*		*Total Deformation (mm)*
	Max	*Min*	*Max*
15	0.0191	3.795×10^{-7}	0.054825
25	0.0426	3.660×10^{-7}	0.000117
35	0.0515	3.150×10^{-7}	0.000164

Table 32.5: Combined force

Compressive force (N) & Torsion (N-mm)	*Equivalent (Von-Mises) Stress (MPa)*		*Total Deformation (mm)*
	Max	*Min*	*Max*
(375,15)	20.799	2.067×10^{-5}	0.16082
(475,25)	26.342	7.346×10^{-6}	0.20368
(575,35)	31.885	1.217×10^{-5}	0.24655

32.5 FEMUR WITH INTRAMEDULLARY NAIL

Table. 32.6: Compressive force

Compressive Force (N)	*Equivalent (Von-Mises) Stress (MPa)*		*Total Deformation (mm)*
	Max	*Min*	*Max*
375	185.91	1.018×10^{-4}	3.597
475	235.49	1.284×10^{-4}	4.556
575	285.07	1.383×10^{-4}	5.516

Table 32.7: Torsion

Torsion (N-mm)	*Equivalent (Von-Mises) Stress (MPa)*		*Total deformation (mm)*
	Max	*Min*	*Max*
15	0.0774	7.58×10^{-6}	0.000740
25	0.1291	4.109×10^{-6}	0.001234
35	0.1807	7.087×10^{-6}	0.001727

Table. 32.8: Combined force

Force(N) & Torsion (N-mm)	*Equivalent (Von-Mises) Stress (MPa)*		*Total Deformation (mm)*
	Max	*Min*	*Max*
(375,15)	182.92	1.19×10^{-4}	3.598
(475,25)	235.5	1.259×10^{-4}	4.557
(575,35)	285.08	1.336×10^{-4}	5.517

32.6 ILLUSTRATIONS, DIAGRAMS AND PHOTOGRAPHS

32.6.1 Femur with Locking Compression Plate

(a)

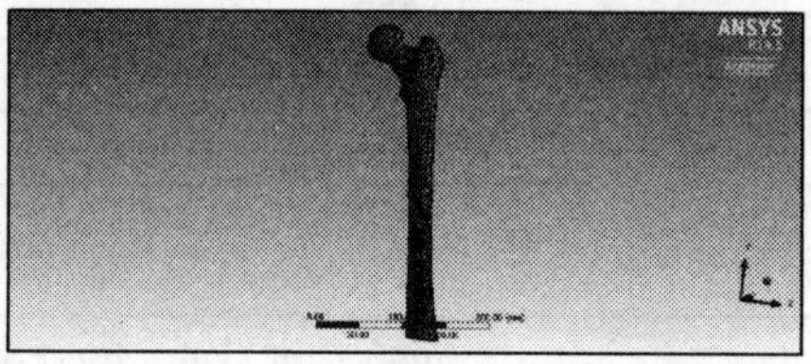

(b)

Fig 32.1 (a) Complete assemble model (b) Image after meshing

32.6.1.1 Compressive Load Distribution in Femur and Implant

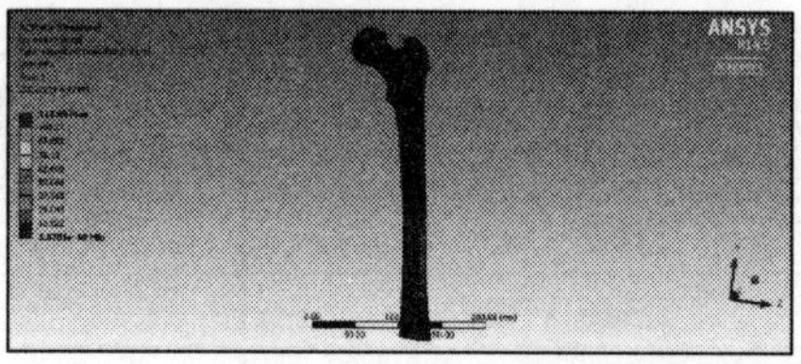

(a) Equivalent stress under compression

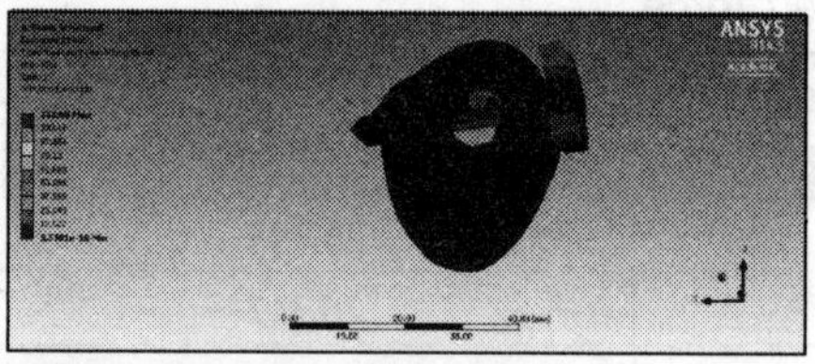

(b) Cross-section view of model under compression

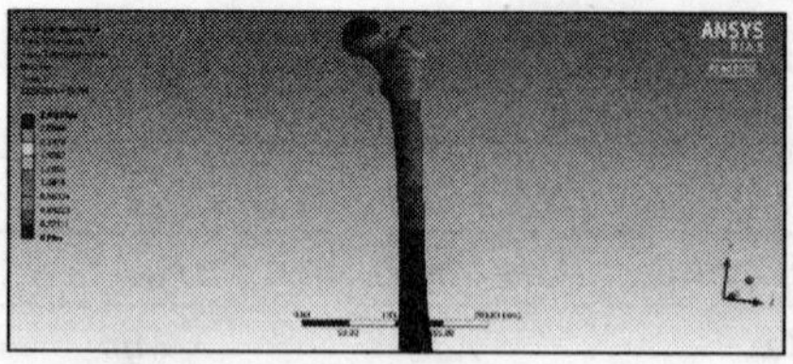

(c) Total deformation under compression

Fig. 32.2 (a) and (c) depicts the maximum equivalent stress and maximum total deformation at 575 N compressive force to be 112.69 MPa and 2.89 mm respectively

32.6.1.2 Torsional Force in Femur and Implant

(a) Equivalent stress under torsional force

(b) Cross-section view of model under torsional force

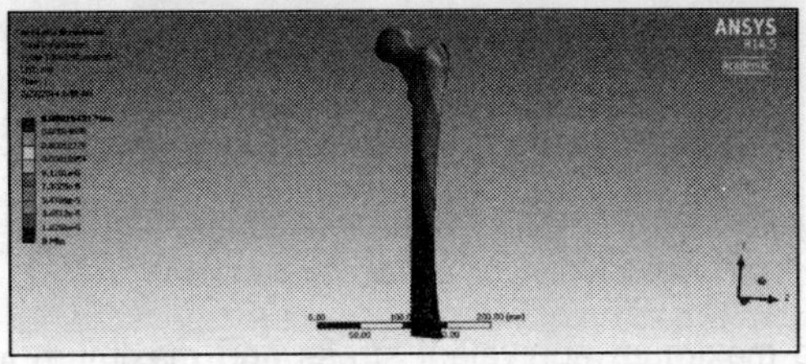

(c) Total deformation under torsional force

Fig. 32.3 (a) and (c) depicts the maximum equivalent stress and maximum total deformation at 35 N.mm bending moment to be 0.0515 MPa and 0.000164 mm respectively. Figures. 32.2(a) and (c) depicts the maximum equivalent stress and maximum total deformation at 35 N.mm torsional force to be 0.0515 MPa and 0.000164 mm respectively

32.6.1.3 Effect of Combined Force in Femur and Implant

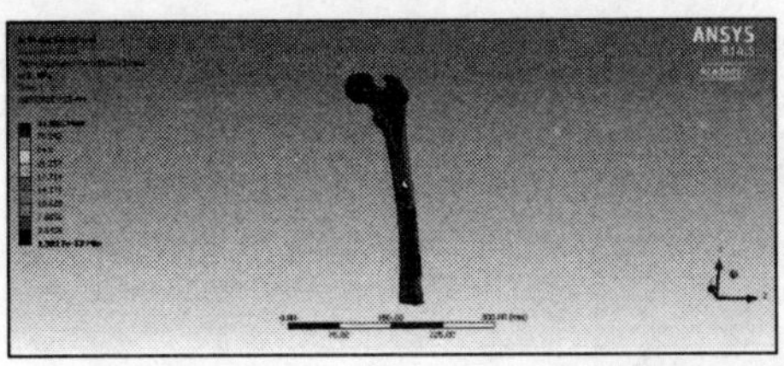

(a) Equivalent stress under combined force

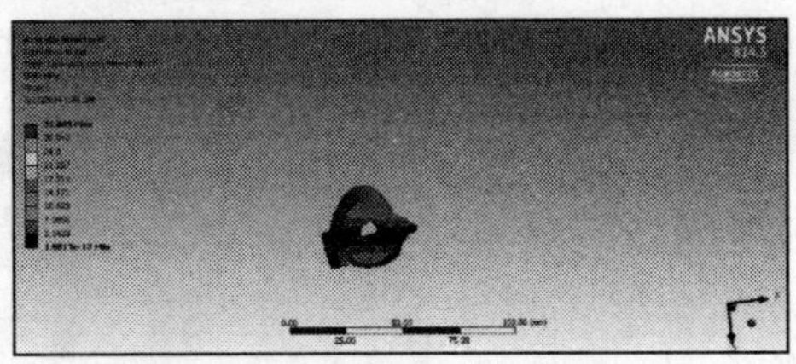

(b) Cross-section view of model under combined force

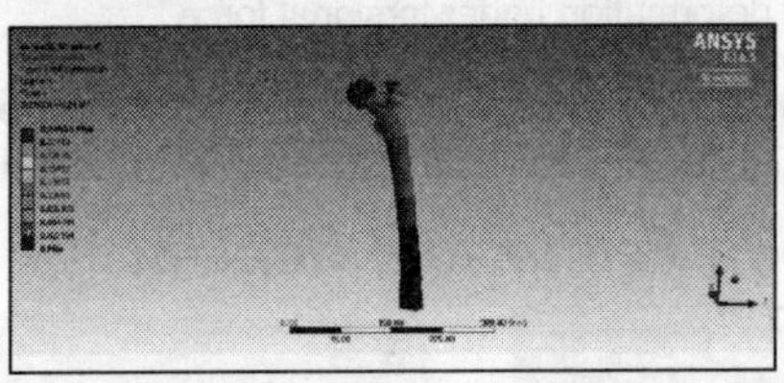

(c) Total deformation under combined force

Fig. 32.4 (a) and (c) depicts the maximum equivalent stress and maximum total deformation at combined force of 575 N compressive force and 35 N.mm torsional force to be 31.885 MPa and 0.2465 mm respectively

32.6.2 Femur with Intramedullary Nail

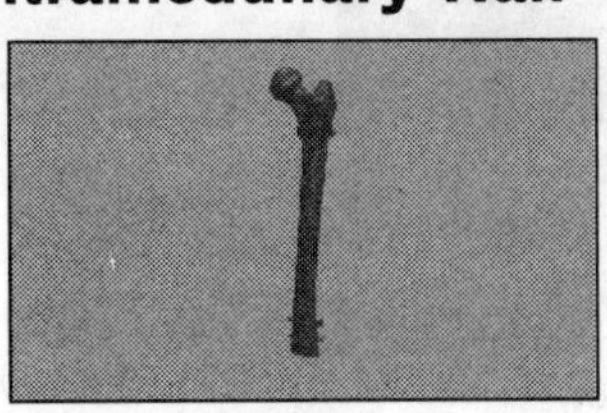

(a)

(b)

Fig 32.5 (a) Complete assemble model (b) Image after meshing

32.6.2.1 Compressive Load Distribution in Femur and Implant

(a) Equivalent stress under compression

(b) Cross-section view under compression

(c) Total deformation under compression

Fig. 32.6 (a) and (c) depicts the maximum equivalent stress and maximum total deformation at 575 N compressive force to be 285.07 MPa and 5.516 mm respectively

32.6.2.2 Torsional Force in Femur and Implant

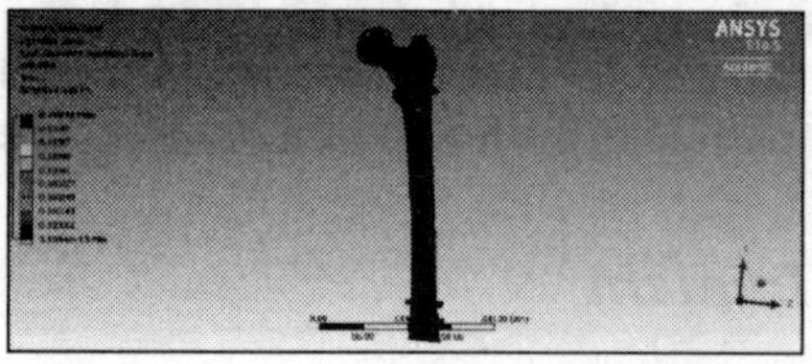

(a) Equivalent stress under torsional force

(b) Cross-section view of model under torsional force

(c) Total deformation under torsional force

Fig. 32.7 (a) and (c) depicts the maximum equivalent stress and maximum total deformation at 35 N.mm torsional force to be 0.18074 MPa and 0.001727 mm respectively

32.6.2.3 Effect of Combined Force in Femur and Implant

(a) Equivalent stress

(b) Cross section view

(c) Total deformation

Fig. 32.8 (a) and (c) depicts the maximum equivalent stress and maximum total deformation at combined force of 575 N compressive force and 35 N.mm torsional force to be 285.08 MPa and 5.517 mm respectively

32.7 CONCLUSION

The result from the analyses revealed that

- For the locking compression plate, for the same compressive force, the total deformation and the stress induced is less as compared to IM rod.
- For the locking compression plate, for the same torsional force, the total deformation and the stress induced is less as compared to IM rod.
- For the locking compression plate, for the same combined force, the total deformation and the stress induced is less as compared to IM rod.
- Top portion of the femur head and upper trochanter is subjected to maximum deformation when subjected to compressive, torsional and combined forces.
- Maximum stress is induced at the implant and bone interface.
- The performance of the locking compression plate under various loading condition is found to be better than the intramedullary rod.

REFERENCES

1. S. Cicero, F. Gutie´rrez-Solana, J.A. A ʻlvarez, L. Sànchez. (2007). Failure analysis of a hip implant by using the FITNET fitness for service procedure. Engineering Fracture Mechanics, 24:688–702.
2. C. Kanchanomai, V. Phiphobmongkol, P. Muanjan. (2008). Fatigue failure of an orthopedic implant – A locking compression plate. Engineering Failure Analysis, 15: 521-530.
3. G.K. Triantafyllidis, A.V. Kazantzis, K.T. Karageorgiou. (2007). Premature fracture of a stainless steel 316L orthopaedic plate implant by alternative episodes of fatigue and cleavage decoherence. Engineering Failure Analysis, 14, 1346-1350.
4. R. Wallstabe, C. Klinger. (2012). Cause of an implant fracture: Material- or fabrication failure. Engineering Failure Analysis, 25, 261-270.
5. S. Griza, C.G. Zimmer, A. Reguly, T.R. Strohaecker. (2009). A case study of subsequential intramedullary nails failure. Engineering Failure Analysis, 2009, 16, 728–732.
6. Erich Schneider, Markus C. Michela, Martin Gengea, Kurt Zubera, Reinhold Ganz. (2001). Loads acting in an intramedullary nail during fracture healing in the human femur. Engineering Failure Analysis.
7. Luca Cristofolini, Marco Viceconti, Angelo Cappello, Aldo Toni. (1996). Mechanical validation of whole bone composite femur models. Journal of Biomechanics, 29 (4), 525–535.
8. M. Nassiri, B. MacDonald, J.M. O'Byrne. (2013). Computational modelling of long bone fractures fixed with locking plates- How can the risk of implant failure be reduced?. Journal of orthopaedics, 10, 29-37.

33

To Compare Cervical Spine and Glenohumeral Range of Motion & Scapular Stability Between Patients with Levator Scapulae Syndrome (LSS) and Matched Normal Subjects

Ajit Dabholkar, Priyanka Shukla and Sujata Yardi

Department of Physiotherapy, Padmashri Dr. D.Y. Patil University, Nerul, Navi Mumbai–400706

Email: ajitdabholkar78@yahoo.co.in

ABSTRACT

Aim of the study is to investigate shortened levator scapulae can pull the ipsilateral cervical vertebrae laterally causing painful and tender cervical joints and possible causing referred pain from the joints to the head. The scapulae and the cervical spine are anatomically linked through the cervicoscapular muscles, upper trapezius and levator scapulae. Changes of the alignment of either the scapulae or the cervical spine can potentially influence the biomechanics of the other by altering the tension at cervico-scapular muscles. There is affection of cervical extension and rotation in patients with LSS. There is reduced internal rotation of the glenohumeral joint in patients with LSS. Scapular stability is affected at rest in patients with LSS.

Keywords: Levator scapulae, glenohumeral range, scapular stability.

33.1 INTRODUCTION

Cervical spine pain affects nearly 70% of the population at some time in their lives. It can arise from unspecified causes[1,2]. It is often associated with a variety of complications, such as headache and loss of range of motion in the cervical spine.[3] This neck pain may also results from long periods of overstretching of Levator Scapulae muscle which results in inflammation and pain within the muscle which is often referred as "Levator Scapulae Syndrome"(LSS). This pain and stiffness which worsens on attempting everyday motions results in restriction of motion in neck and shoulder[4]. Shortened levator scapulae also can pull the ipsilateral cervical vertebrae laterally causing painful and tender cervical joints and possible causing referred pain from the joints to the head[5]. This condition typically co-exists with cervical, glenohumeral or scapular dysfunction rather than an over-use condition in its own right[11]. Compensatory elevation of the scapula in the symptomatic shoulder as it strives to avoid impingement predisposes the levator scapula to hypertonicity and shortening. This tightness becomes painful in its own right and presents as tender on palpation. The neural connections of this muscle allow referred pain proximally into the cervical spine and distally into the upper arm. This can be due to neural irritation arising from altered posture and mechanics of the scapula and cervical spine[6]. According to Harpa et al., there was an altered stability of the scapula in patients with cervical spine disorders. There exists very limited research studying the influence of levator scapulae syndrome on cervical and shoulder motions along with the scapular stability and the strength of scapular muscles. The present study is therefore an attempt to compare cervical spine and glenohumeral range of motion, scapular stability and trapezius strength between patients with levator scapulae syndrome and matched normal subjects. To compare cervical spine & glenohumeral range of motion with scapular stability between patients with levator scapulae syndrome and matched normal subjects.

33.2 MATERIALS AND METHODOLOGY

Marker, Measuring tape, Universal inclinometer (BASELINE), Universal goniometer, Pressure algometer, Baseline PUSH – PULL dynamometer, Flexible ruler, Weighing scale, Height scale

33.2.1 Study Design

Cross-sectional study

33.2.2 Study Setting

Physiotherapy OPD, No. 27, Padmashree Dr. D.Y. Patil Hospital and Research Centre.

33.2.3 Sample Size

60: 20 subjects with levator scapulae syndrome, 20 subjects with trapezius spasm and tightness, 20 matched normal subjects.

33.2.4 Study Duration

6 months

33.2.5 Inclusion Criteria

Group 1: Patients with levator scapulae syndrome showing following signs and symptoms, Sharp pain in and around the neck, superior angle of scapula or between the scapula, Tightness of levator scapulae muscle, Presence of trigger points, Radiating pain to the head, Patient presenting with any of the above the symptoms were included. Group 2: Patients with upper trapezius spasm, tightness and with presence of trigger points. Group 3: Age and gender matched controls.

33.2.6 Exclusion Criteria

Cervical myelopathy, Cervical radiculopathy, Congenital deformity of head and neck, Fracture in and around neck and shoulder, Any other musculoskeletal injury to cervical spine and shoulder joint, Hypermobility of shoulder.

33.2.7 Procedure

Approval for the study was gained from institutional Ethics Committee. All subjects were explained the purpose of the study and consent was taken from all of them prior assessment. Patients meeting the inclusion criteria for Levator Scapulae Syndrome (LSS) were selected formed group 1. Age and gender matched healthy subjects formed the control group. In total 60 subjects were recruited; 20 in group 1 with Levator Scapulae Syndrome, 20 in group 2 with trapezius spasm and tightness, and 20 age and gender matched healthy controls.

33.2.8 Outcome Measures

After meeting the inclusion criteria all the subjects in all the groups were assessed for the following: Tightness of Upper Trapezius[3] and Levator Scapulae, Presence of trigger points[7,8]. Following measures were assessed: Cervical ROM: Cervical ROM was assessed using inclinometer (BASELINE), Shoulder ROM: Shoulder ROM was taken using universal goniometer, Scapular stability: The current study uses the method of assessing scapular stability proposed by Kibler.[9]

33.3 RESULTS AND OBSERVATIONS

Table 33.1: Comparison of cervical ROM in all the three groups

ROM	Group	Mean	SD	Sum of ranks	Mean of ranks	P value
Flexion	1	66.45	7.2	532.5	26.62	0.3496
	2	69	7.49	607	30.35	
	3	70.55	7.43	690.5	34.52	

Extension	1	64.9	9.9	741.5	37.07	
	2	69.85	8.53	622.5	31.12	0.042*
	3	73.95	12.4	466	23.3	
Lt. S.F.	1	48.9	6.84	539	26.95	
	2	49.65	4.91	581	29.05	0.2623
	3	53.15	8.02	710	35.5	
Rt. S.F.	1	51.2	7.69	596.5	29.82	
	2	50.75	7.39	554	27.7	0.5059
	3	54.65	10.4	679.5	33.97	
Lt. Rotation	1	83.15	7.18	421.5	21.07	
	2	87.9	4.05	643	32.15	0.0039**
	3	89.85	4.46	765.5	38.27	
Rt. Rotation	1	82.65	5.52	432	21.6	
	2	85.4	4.76	585.5	29.27	0.0017**
	3	88.4	3.53	812.5	40.62	

Table 33.2: Comparison of glenohumeral ROM between all the three groups

ROM	*Group*	*Mean*	*SD*	*Sum of ranks*	*Mean of ranks*	*P value*
LT. Extension	1	43.00	8.8	519.5	25.97	
	2	47.10	7.2	688.5	34.42	0.2964
	3	45.95	9.8	622	31.1	
RT. Extension	1	44.90	8.96	577.5	28.87	
	2	47.50	8.35	695	34.75	0.3933
	3	44.25	8.78	557.5	27.87	
Lt. I.R.	1	59.40	7.24	457	22.84	
	2	64.25	11	629	31.45	0.03*
	3	66.75	8.47	744	37.2	
Rt. I.R.	1	58.60	10.8	437	21.85	
	2	63.75	9.5	635	31.75	0.0123*
	3	67.25	6.58	758	37.9	
Lt. E.R.	1	87.50	5	590.5	29.52	
	2	86.75	8	606	30.3	0.8881
	3	87.25	8.03	633.5	31.6	
Rt. E.R.	1	87.75	5.5	582	29.1	
	2	87.50	5.5	561.5	28.07	0.3606
	3	89.50	3.94	686.5	34.32	

Table 33.3: Comparison of scapular stability using Kibler method.

Kibler Method	*Group*	*Mean*	*SD*	*Group comparison*	*P Value*
Ki1	1	0.59	0.45	1 VS 3	0.0025**
	2	0.50	0.52	2 VS 3	0.0410*
	3	0.26	0.34	1 VS 2	0.4121
Ki2	1	0.86	0.54	1 VS 3	0.0241*
	2	0.70	0.55	2 VS 3	0.0643
	3	0.45	0.51	1 VS 2	0.2431

Ki3	1	0.80	0.59	1 VS 3	0.2374
	2	0.99	0.79	2 VS 3	0.0976
	3	0.57	0.42	1 VS 2	0.5237
km1	1	0.53	0.36	1 VS 3	0.3545
	2	0.54	0.44	2 VS 3	0.6147
	3	0.46	0.38	1 VS 2	0.9891
Km2	1	0.57	0.44	1 VS 3	0.34
	2	0.43	0.28	2 VS 3	0.8269
	3	0.51	0.51	1 VS 2	0.5757
Km3	1	0.44	0.26	1 VS 3	0.0359*
	2	0.51	0.37	2 VS 3	0.0539
	3	0.28	0.19	1 VS 2	0.7637

P value:*significant, ** very significant,*** extremely significant

33.4 DISCUSSION

In current study, we found that there was reduced cervical range of motion in patients with levator scapulae syndrome compared to patients with trapezius spasm and matched normal subjects (Table 33.1). We found a significant difference between patients with levator scapulae syndrome and matched normals for cervical extension and cervical rotations but not for cervical flexion and side flexion (Table 33.1). This findings correlate with the study done by ***Lee et al.,*** who stated that between-groups differences were observed as lower neck muscle endurance time, reduced left rotation, relatively reduced extension at second test, but greater range of retraction, when the subclinical neck pain and normal groups were compared. These data suggest that there are early range changes associated with the development of neck pain. According to **Janda** long period of overactivity of the levator scapulae along with upper trapezius leads to concomitant tightness of pectorals and sternocledomastoid and this will lead to inhibition of deep neck flexors and scapular stabilizers (lower trapezius and serratus anterior). It is postulated that because of these muscular imbalances there is a tendency for forward head posture. According to ***Rene Cailliet*** this assumed position, decreases the normal active and passive rotational ROM. A person can normally decrease the ability to turn the head to the right or the left by assuming an exaggerated forward head posture. The present study also reveals significant decrease in internal rotation at the glenohumeral joint in patients with levator scapulae syndrome compared to control group. No significant results were obtained for the other glenohumeral range of motion between the groups (Table 33.2). ***Bersen et al,*** concluded from their study that levator scapulae contracts concentrically during the first half of shoulder abduction and eccentrically during second half of shoulder abduction. A shortened levator scapula shall therefore be unable to contract eccentrically and generate the force required for abduction in the second half which is accompanied by external rotation leading to shortening of the external rotators. As a result these shortened muscles will reduce the internal rotation range of motion. In current study, lateral scapular slide test measured from inferior angle of scapula to the thoracic spine at 0° shoulder abduction showed very significant results between patients with levator scapulae syndrome and control group, significant results between patients with trapezius spasm and tightness and control group, but non-significant results between patients with levator scapulae syndrome and patients with trapezius spasm and tightness. Significant results also found between patients with LSS and control group at 45° shoulder abduction but non-significant at 90° shoulder abduction and internal rotation (Table 33.3). The adaptive shortening of levator scapulae and rhomboid muscle as both being the upward rotator of the scapula, results in elevation, adduction and downward rotation of the scapula. This position turns the glenoid fossa downward and causes anterior tilting of the scapula, thus lowering and roundening the shoulder. This three-dimensional positional fault of the scapulae due to muscle imbalance alters the force couple formed by the rhomboids and teres major muscles. In levator scapulae syndrome, patient uses levator scapulae and rhomboid muscles to control the way the scapula is positioned on the rib cage, instead of using the correct muscles, being the lower trapezius and serratus anterior. This will leads to concomitant tightness of pectorals and sternocledomastoid and this will lead to inhibition of deep neck flexors and scapular stabilizers (lower trapezius and serratus anterior). The reduced scapular upward rotation reflects either long, inefficient upper trapezius and serratus anterior muscles or short overactive levator scapula and rhomboid muscles which would simultaneously elevate and retract the clavicle and downwardly rotate the scapula leading to alteration in length tension relationship of axioscapular muscles and scapular dyskinesia. Thus, levator scapulae syndrome can influence the cervical extension, and side rotations; glenohumeral internal rotation; scapular stability at rest. Therefore, evaluation and management strategies needs to be addressed in patients with Levator Scapulae Syndrome and its influence on various factors needs to be undertaken clinically.

REFERENCES

1. Cote P., Cassidy J.D., Carrol L., Kristman V. The annual incidence and course of neck pain in the general population: A population based-cohort study. Pain. 2004; 112: 267-273.
2. Cote P., Cassidy J.D., Carrol L., The Saskatchewan health and back pain survey: The prevalence of neck pain and related disability in Saskatchewan adults. Spine. 1998; 23: 1689-1698.
3. Croft P.R., Lewis M., Papageorgiou A.C., Thomas E., Jayson M.I., Macfarlane G.J., Silman A.J: Risk factors for neck pain: a longitudinal study in the general population. Pain 2001, 93: 317-325.
4. Joseph D., Kurnik, D.C. Levator scapulae syndrome. The chiropractic news source. Dynamic chiropractic. June 15, 1998, Vol. 16, Issue 13.
5. Levator Scapulae Syndrome- Muscular Injuries. Guide to sports medicine.
6. Maria Zuluga. The shoulder complex. Sports Physiotherapy: Applieed science and practice. Churchil Livingstone. 1995
7. Takal E.P. Pain pressure threshold on upper trapezius and levator scapulae muscles. Scand J Rehab. Med. 1990, 22(2): 63-68.
8. Giburm Park, Chan Woo Kim, Si Bog Park, Mi Jung Kim, Seong Ho Jang, Reliability and Usefulness of the Pressure Pain Threshold Measurement in Patients with Myofascial Pain. Ann Rehabil Med. 2011 June; 35(3): 412-417.
9. Leen T' Jonck et al. Measurement of scapular position and rotation: A reliability study. Proceedings of the first conference of the ISG.

34

Effect of Fatigue on Hand Dexterity and Reaction Time

Tejashree A. Dabholkar, Bhakti Meghraj and Sujata Yardi

Padmashri D.Y. Patil University Nerul, Navi Mumbai

Email: kamble.teju@yahoo.co.in

ABSTRACT

Hand dexterity is a pre-requisite for various skillful activities reactive decides the speed of performance. Fatigue can influence reaction time and dexterity both the components. Hence, this study aims at assessing effect of fatigue on these two parameters. In this Experimental Design Postgraduate students and interns of physiotherapy faculty were included. All subjects were evaluated for 40% base line maximum contraction of grip strength, jebsen score for dexterity, reaction time, subjects were randomly divided into experimental and control group. Expermental group were induced with using 40% of maximum contraction and 10% of MVC for control group. Reassessment of outcome measures done immediately. Study shows statistically significant difference for jebson hand function ($p < 0.001$), reaction time ($p < 0.0001$), grip strength ($p < 0.001$) in experimental group whereas no significant difference found for control group. Fatigue has effect on hand dexterity and reaction time.

Keywords: Reaction time, dexterity, fatigue.

34.1 INTRODUCTION

Skeletal muscles have to contract to produce force necessary in everyday life. Muscle fatigue, defined as the failure to maintain the required or expected power output[1]. In short it decreases muscle performance and productivity. Psychosocial work characteristics are significant predictors for the onset of fatigue in the working population[2]. Community surveys have shown that fatigue is common in developed countries[3]. Fatigue and need for recovery were found to be independent risk factors for being injured in an occupational accident. This means that in the push back of occupational accidents, fatigue, and even more importantly need for recovery, need special attention[4]. Workers with jobs in which they work with tools or machinery had a fivefold increased risk for being injured in an occupational accident[5]. Hence, it is very important to know whether fatigue alone has any influence on the working ability[6]. Journal of sports sciences issued an article in 2004 with a conclusion that volitional fatigue of 35-40 minutes caused around 69% decreased in accuracy of ground stroke hitting and 30% decreased in service accuracy in the tennis players. The results of this study suggest that fatigue was accompanied by a decline in tennis skills in tennis palyers.[7] Hand Dexterity is defined as fine, voluntary movements used to manipulate small objects during a specific task[8]. An examination of dexterity provides a unique way of evaluating the neuro motor function of the entire hand because sensation and intrinsic hand strength combine to produce the manipulative skills that facilitate dextrous movements[9]. Hand dexterity test are often used to pre-screen employees because dexterity of hand is very important factor in most of the work places and even in day to day activities. They measure how well a person uses their fingers, arms, and hands while performing certain tasks. Fine motor skills efficiency significantly influences the quality of the task outcome as well as speed of the task performance. Speed of task performance reflected on reaction time. Reaction time is defined as a time required for muscle action after a particular stimulation[9]. A good dexterous hand requires fast acting muscles. Reaction time test is valid measure of cognitive function. Mental fatigue leads to increased reaction time[10]. Since there are plethora of information on fatigue and task performance but very scanty literature on fatigue and its effect on hand dexterity and reaction time in normal. Hence, the aim of the study is to find out reaction time and hand function under the influence of fatigue.

34.2 METHODOLOGY

Institutional ethics committee approval was taken. Subjects were selected according to the inclusion and exclusion criteria. Female subjects between the age of 18-30 years were selected whereas those with past upper and lower limb injuries, diabetes and high blood pressure were excluded from study. A random division of the subjects into two groups was done, i.e Experimental group & controlled group. Basic evaluation was carried out for each subject of both groups. Name, age, height, weight was recorded. Grip strength, reaction time and hand dexterity of the dominant hand of each person was calculated. A study was found with results-dominant hands of the right-handers worked significantly faster than the non-dominant hands for all ($p < 0.05$)[11]. Hence, all the components were calculated for dominant hand.

34.2.1 Grip Strength

The subjects are comfortably seated on a chair and the subjects' upper limb is positioned according to the test position given by American society of hand surgery. Maximum voluntary contraction (MVC), i.e grip strength of the dominant hand is recorded with the aid of hand held dynamometer.

34.2.2 Reaction Time

to measure a reaction time, an application called reaction time calculator was used. Subjects were seated on the chair and the reaction time calculator was kept in front of the subjects (within the reach).subjects were told to press start button on screen and followed by which to press green circle as soon as it appears on screen. The reaction time between the visual stimulus and the motor response is directly displayed on the screen in milliseconds which is noted down. To eliminate the effects of mental fatigue subjects are given trials for 2-3 times with adequate rest periods.

34.2.3 Hand Dexterity

To measure the hand dexterity jebsens hand function test (JHFT) is used which consist of 7 different activities. It was performed using the dominant hand. The subjects were explained verbally and what has to be done was shown by the investigator to the subject. Time was calculated for each activity one after the other. The score of the JHFT was recorded in seconds.

34.2.4 Inducing Fatigue

For the experimental group: 40% of MVC is calculated and noted down which is said to be the submaximal value of the subject. Then 90% of the sub maximal value is also calculated. Then fatigued is induced with the dynamometer by asking the subjects to hold the contraction at the submaximal readings. Visual feedback is given to the subjects by holding the mirror in front of the dynamometer. Subjects have to hold the contraction at submaximal level until they get the feeling of giving away or tiredness, this shows that fatigue has set in. And if the reading has fall down below 90% of submaximal value then also it is considered that hand is fatigued.

Immediately after the fatigue has set in, reaction time was recorded followed by post readings for jebsens hand function and MVC in the same way as above.

For controlled group: 10% of MVC is calculated and noted down. Since fatigue is not to be induced in this group, only 10% of MVC is to be held by the subject for duration of 19 one min only. Immediate after 1 min post reading for reaction time, hand function and MVC were taken.

For both the groups subjects were told to rate the fatigue activity on the BORG scales. For experimental group it was expected to have very hard and for control group a very light after inducing fatigue.

34.3 DATA PRESENTATION AND INTERPRETATION

34.3.1 Analytical Statistics

For outcome measures data were collected on standardized forms and encoded for computerized analysis using Graph Pad PrismVersion 6.03. Within the group Paired *t* test was used and between the group Unpaired test was used. ANOVA test was used for multiple group comparisons.

Table 34.1: Effect of fatigue on reaction time

	Control	*Experimental*
P value	0.5259#	<0.0001 ****
95% CI	6.267 to 11.95	47.35 to 83.37
	**** -Extremely significant # - not significant	

Table 34.2: Effect of fatigue on hand dexterity

	Control	*Experimental*
P value	0.007#	< 0.0001 ****
95% CI	1.237 to –0.2102	11.08 to 14.94

Table 34.3: Effect of fatigue on individual component of dexterity test

	Control		*Experimental*	
	P value	*95% CI*	*P value*	*95% CI*
Writing	0.5847#	1.540 to 0.8779	0.0009***	0.4568 to 1.537
Card turning	0.1194#	0.1458 to 0.01780	0.0005***	0.4028 to 1.244
Picking up small objects	0.2152#	0.5033 to 0.1193	< 0.0001****	0.6648 to 1.126
Stacking checkers	0.2557#	0.2750 to 0.07664	< 0.0001****	0.6648 to 1.126
Simulated feeding	0.0606#	0.3859 to 0.009112	< 0.0001****	0.8765 to 1.770
Moving light objects	0.4692#	–0.08984 to 0.04264	0.0002***	0.4627 to 1.283
Moving heavy objects	0.0677#	0.09728 to 0.003681	0.0304*	0.05008 to 0.9211

34.4 DISCUSSION

The age group selected was from 18-30 years where the mean of age was 22.6 SD±2.22 for experimental group and 21.92 SD±1.93 for control group. BMI mean was 21.92±3.2 for experimental subjects and for control group was 21.69±2.46 thus both the groups have homogeneous distribution of the samples. The mean of grip strength for control group is 17.92 kgs ±3.673 and 18.2±4.255 for experimental group. Mean of reaction time was is 605 milliseconds (ms) and JHFT was 55.50 seconds(s) for control group and for experimental group reaction time was 564 ms, and JHFT was 50.5 sec. In this study it was found that fatigue has affected the hand muscles. Fatigue has prolonged reaction time and JHFT score. This study also showed reduction in MVC post fatigue.

Study shows significant difference between pre and post score of JHFT (P-value < 0.0001), the possible reason for reduced hand function may be changes in the local metabolism at muscle[15]. Pedersen et al. reported that increased intra muscular concentrations of lactic acid, bradykinin, arachidonic acid and serotonin after fatiguing contractions may affect the muscle spindle system and proprioceptive acuity and thus may alter the functional activity[12]. Djupsjobacka et al. in separate study reported that increased intramuscular concentrations of several contractile substances alter the muscle spindle output as measured through reflex arc. Since local blood flow and metabolic changes are more pronounced at the muscle than in the articular structures, muscle mechanoreceptors may be more affected than articular mechanoreceptors[13-15]. Peripheral fatigue occurs at the level of sarcomere and involves failure at neuromuscular junction, sarcolemma and transverse tubules[16]. This study finding supports the "dysfunctional mechanoreceptor" theory proposed by Voight et al. – muscle fatigue desensitized muscle spindle threshold thereby decreasing the afferent feedback to the central nervous system[17]. Barens (1980) found that forearm blood flow was maximum with contractions of 20-25% MVC and declined thereafter[18]. Further in 1939 Barcroft and Millen and Edward et al. in[9] 1972 found that partial occlusion occurred in the leg muscles at forces greater than 20% MVC[19]. This shows that blood supply reduces in the muscles which are fatigue resulting in less nutrition to the muscle. Blood supply is important component for muscle contraction; hence muscle contraction is compromised in fatigued muscles which may be the possible reason for increased JHFT score. As fatigue progresses, the number of active motor unit decreases, muscle fibre conduction velocity decreases[20]. Motor unit fires more slowly[21] and the motor units become more asynchronized[22]. This leads to decreased mean frequency of electromyogram (EMG) signal and eventually to task failure[18]. Muscle fatigue may impair a person's ability to properly execute a task. MRI studies showed increased activity in the prefrontal areas of the brain after muscle fatigue, which may explain increase in processing time[23]. At the muscle level, fatigue causes an increase in muscle response time or "electromechanical delay"[24], possibly due to decrease in muscle fibre velocity. Muscle fatigue may also affect the body's ability to successfully reproduce a movement. In a study of rapid elbow flexion/extension, fatigue of extensor muscle caused an undershoot of final position during extension[25], however other studies showed no effect of fatigue on the end point trajectories in multi joint tasks[26], in each of these studies, it was presumed that subjects changed their neural control (muscle activation) or co-ordination strategies to achieve the same over all task goal. The central nervous system reduces the alpha-motor nerve activation frequency as fatigue develops which may also be the possible reason for increased reaction time. Jochen Baumeister and his colleagues studied the effects of induced fatigue on brain activity during sensorimotor control. EEG gave results of significant decrease in frontal theta, alpha-1 and alpha-2 frequencies after inducing fatigue these components are important for processing information thus may lead to increased reaction time.[27] This study has shown significant difference in MVC (P-value 0.0001). MVC has decreased drastically post fatigue. Decline in isometric force production is characteristic of skeletal muscle fatigue. Force decline can be due to reduction of cross-bridges ability to generate force which comes early and reduction of sarcoplasmic reticulum Ca^{2+} release, which develops late in fatigue. The reduction of force due to impaired cross-bridge function would be caused by accumulation of inorganic phosphate ions, which results from phosphocreatine breakdown.

34.5 CONCLUSION

Our study shows that fatigue can influence reaction time and hand dexterity and this finding can have its clinical implication in various occupations requiring fine dexterity to get maximum outcome.

REFERENCES

1. American Physiological Society: Cellular mechanisms of muscle fatigue
2. Swaen, G.M.H; L.G.P.M van A., Bultmann, U; Kant, I.J. Fatigue as a risk factor for being injured in an occupational accident. Occupational and Environmental Medicine, suppl. Fatigue at Work (2003): 88-92.
3. Population based study of fatigue and psychological distress. BMJ 1994 (Published 19 March 1994)
4. Swaen, G.M.H; L.G.P.M van A., Bultmann, U; Kant, I.J. Fatigue as a risk factor for being injured in an occupational accident. Occupational and Environmental Medicine, suppl. Fatigue at Work (2003): 88-92.
5. Mc. Donald C. Epidemiology of work related diseases. London BMJ Books, (2002). J Sports Sciences 20: 4.
6. Backman C., Cork S., Gibson G., Parsons J. Assessment of hand function: the relationship between pegboard dexterity and applied dexterity. Can J Occup Ther. 1992; 59: 208-13.
7. Robert Langner, Michael B. Steinborn, Anjan Chatterjee, Walter Sturm, Klaus Willme. (2010). Mental fatigue and temporal preparation in simple reaction-time performance .Acta Psychologica, 133: 64-72.
8. Zeliha Tulum, Lamia pinar and ferdi baskurt Dokuz Eylul. Comparison of Pressure Pain Threshold, Grip Strength, Dexterity and Touch Pressure of Dominant and Non-Dominant Hands within and Between Right- and Left-Handed Subjects university of physical Therapy anf Rehabilitation, 35340, Izmir-Turkey.
9. Pedersen J., Lonn J., hellstrom F., djupsjobacka M., Johansson H. Localized muscle fatigue decreases the movement sense in human shoulder. Med sci sports Exerc. 1999 31: 1047-1052.

10. Djupsjobacka M, Johnnson H, Bergenhein M. Influences on gamma- muscle-spindle system from muscles afferents stimulated by increased intramuscular concentration of achidonic acid. Brain Res. 1994; 663; 293-302.
11. Djupsjobacka M., Johnnson H., Bergenhein M., Sjolander P. Influences on the gamma-muscle-spindle system from contra lateral muscles afferents stimulated by KCI and lactic acid. Neurosci Res. 1995; 21; 301-309.
12. Djupsjobacka M., Johnnson H., Bergenhein M, Wenngren BI. Influences on gamma- muscle-spindle system from muscles afferent stimulation by increased intramuscular concentrations of bradykinnin and 5- HT. Neurosci Res 1994; 663; 293-302.
13. Powers S.K, Howley E.T. Exercise physiology. Teippett S., Canner G.C. The effects of fatigue on and the relationship of arm dominance to shoulder proprioception. J orthop Sports Med 1998; 23; 348-352.
14. Voight M.L., Hardin J.A., Blackburn T.A., Tippet S., canner G.C. The effects of muscle fatigue on and the relationship of arm dominance to shoulder proprioception. J orthop Sports phys ther. 1998; 26; 262-265.
15. Barnes W.S., (1980). The relationship between maximum isometric strength and intramuscular circulatory occlusion. Ergonomics 23: 351-357.
16. Edwards R.H.T., Hill DK, MacDonnell M. (1972). Myothermal and intramuscular pressure measurements during isometric contractions of the human quadriceps muscle. J. Physiol (Lond) 224: 58-59.
17. Farina D., Fattorini L., Felici F., Filligoi G. (2002). Nonlinear surface EMG analysis to detect changes of motor unit conduction velocity and synchronization. J. Appl Physiol 93:1753–1763.
18. Bigland-Ritchie B., Woods J.J. (1984). Changes in muscle contractile properties and neural control during human muscular fatigue. Muscle Nerve 7: 691-699.
19. Arihara M., Sakamoto K. (1999). Contribution of motor unit activity enhanced by acute fatigue to physiological tremor of finger. Electromyogr Clin Neurophysiol 39:235–247. van Duinen H, Renken R., Maurits N., Zijdewind I (2007) Effects of motor fatigue on human brain activity, an fMRI study. NeuroImage 35: 1438-1449.
20. Wilder D.G., Aleksiev A.R., Magnusson M.L., Malcolm PH, Spratt K.F., Goel V.K. (1996). Muscular response to sudden load: a tool to evaluate fatigue and rehabilitation. Spine 21: 2628-2639.
21. Jaric S, Blesic S., Milanovic S., Radovanovic S, Ljubisavljevic M., Anastasijevic R. (1999). Changes in movement final position associated with agonist and antagonist muscle fatigue. Eur J. Appl Physiol 80: 467-471.
22. Lucidi C.A., Lehman S.L. (1992). Adaptation to fatigue of long duration in human wrist movements. J. Appl Physiol 73: 2596-2603.
23. Cote JN, Mathieu P.A., Levin MF, Feldman A.G. (2002). Movement reorganization to compensate for fatigue during sawing. Exp Brain Res 146: 394-398.
24. Heuer H, Schulna R., Luttmann A. (2002). The effects of muscle fatigue on rapid finger oscillations. Exp Brain Res 147: 124-134.
25. Selen LPJ, Beek P.J., van Dienn J.H. (2007). Fatigue-induced changes of impedance and performance in target tracking. Exp Brain Res 181: 99-108. 44
26. R.H fits Cellular mechanisms of muscle fatigue Department of Biology, Marquette University, Milwaukee, Wisconsin, American physiological society.
27. Glenn Elert Fingertip Reaction Time, the physics fact book 2003. H. westerblad, D.G. Allen, J.D. Bruton, F.H. Andrade, J. Lannergren. Mechanisms underlying the reduction of isometric force in skeletal muscle fatigue 4 JAN 2002.

35

Modeling of Lung Diseases: A Review

Ajay Kumar and V. K. Katiyar

Department of Mathematics, Indian Institute of Technology Roorkee–247667, India

Email: ajayiitr85@gmail.com

ABSTRACT

The lungs are paired organs in the chest that execute respiration. Lungs are important because they supply oxygen to body cells and release carbon dioxide to the atmosphere. Lung disease refers to disorders that affect the lungs, the organs that allow us to breathe. The main causes of lung diseases are smoking, air pollution, radon and asbestos. Particle pollution – especially fine particles-contains microscopic solids or liquid droplets that are so small that they can get deep into the lungs and cause serious health problems. Therefore, the objective of present work is to give a brief review on modeling of lung diseases: types, causes and effects on health. There are many lung diseases, but in our study we will concentrate mainly on Asthma, Pneumonia, Tuberculosis and Lung Cancer.

Keywords: Lungs, asthma, pneumonia, tuberculosis and lung cancer.

35.1 INTRODUCTION

The lungs are paired organs in the chest that execute respiration. Each human has two lungs (right and left). Each lung is between 10 and 12 inches long. The right lung composed of three lobes whereas the left lung is lightly smaller consisting of only two lobes. The right lung accounts for 55% of overall gas volume and the left lung for 45%. Mechanical changes in lung structure during respiration, is known as lung mechanics. It is a general term to describe all those things that determine the relationships between pressures, flows and volumes in the lung.

Lung diseases are some of the most common medical conditions worldwide. Infections, genetics and smoking are responsible for most lung diseases. Lung disease can result from problems in any part of body system. The term lung disease refers to many disorders affecting the lungs, such as asthma, tuberculosis, pneumonia and lung cancer.

Asthma is a chronic (long-term) disease that makes it hard to breathe. Asthma causes recurring periods of wheezing (a whistling sound when you breathe), chest tightness, shortness of breath, and coughing. The coughing often occurs at night or early in the morning. Asthma affects people of all ages, but it most often starts during childhood. Asthma can't be cured, but it can be managed with proper treatment. People with asthma can lead normal, active lives.

Tuberculosis (TB) is an infectious disease that is caused by a bacterium called Mycobacterium tuberculosis. TB primarily affects the lungs, but it can also affect organs in the central nervous system, lymphatic system, and circulatory system among others. The disease was called "consumption" in the past because of the way it would consume from within anyone who became infected.

Pneumonia is an inflammation of the lungs. Usually, the inflammation is caused by an infection, but it can also because by breathing certain chemical fumes. It's a more common problem than most people think. Usually pneumonia is a mild disease, but some forms are very dangerous. In all cases, you'll need to seek care from your health care provider.

Lung cancer is a disease characterized by uncontrolled cell growth of the tissues in the lung. If left untreated, this growth can spread beyond the lung in a process called metastasis into nearby tissue or other parts of the body. Most cancers that start in lung, known as primary lung cancers, are carcinomas that which causes 80-90% of lung cancers. Nonsmokers account for 10-15% of lung cancer cases.

The objective of this study is to provide a brief review on four lung diseases discussed above.

35.2 LITERATURE REVIEW

35.2.1 Asthma

The acute and/or chronic inflammation in lungs can be associated with structural changes which are the causes of various lung diseases. In the chronic obstructive pulmonary, fibrotic disorders and acute lung injury this complex and variable mechanism occurs.

The first mathematical model describing the aerosol deposition in the lungs was introduced by Findeisen (1935) which was based on very straightforward pulmonary morphologies. After that, Altshuler (1959) presented a model of aerosol removed by respiratory tract. For any breathing condition, a theory was developed by Taulbee and Yu (1975) to predict the particle deposition and its distribution in human respiratory tract. In which a convection-diffusion equation for the particle concentration with a loss term was used to describe the transport and deposition of particles. Heyder et al. (1984) provided a review in which the general concepts of primary models and their numerical verification and on characteristics of secondary models were discussed. For the computation of particle deposition in two real bifurcation geometries Balashazy and Hofmann (1993a) presented a computer model under different realistic air flow fields, Brownian diffusion and interception, gravitational settling and simultaneous effects of inertial impaction. Further, the particle deposition patterns in an asymmetric bifurcation were analyzed. This pattern was based on a numerical method, which was used for the calculation of aerosol particle trajectories and air velocity fields in a three-dimensional bifurcation model. In the above mentioned model, for the computation of airflow numerically, a finite difference method was used by solving the Navier-Stokes equations. Balashazy and Hofmann (1993b) extended their work to study aerosol deposition efficiencies and deposition patterns in bifurcating airways under expiratory flow conditions. Particularly they studied the effects of particle size, flow rate, inlet boundary flow profile and bifurcation geometry on particle deposition. The effects of asymmetry in branching angle and airway diameter were described by Balashazy and Hofmann (1995). Further, for several particle flow rates and sizes, the flow divisions on both expiratory and inspiratory particle deposition patterns were discovered.

In asthmatic condition, Martonen et al. (2003) developed mathematical model for the study of the changes in human lungs. They considered the issue of how in silico modeling can be integrated with aerosol therapy. The effects of asthma-induced airway constriction on particle deposition patterns and aerosol dynamics was investigated by Longest et al. (2006) in bifurcating models of pediatric airways. The airway inflammation, airway hyper-responsiveness and reversible airway obstruction can describe the Asthma (Faffe and Zin, 2009). Tissue damage and consequential structural changes are the causes of asthmatic airway inflammation in various chronic inflammatory disorders. To examine the effects of acute asthma on the airflow patterns, realistic airway geometry and pressure drop, Inthavong et al. (2010) discussed two models with six generations of the airway tree from an acute asthma episode. Ross et al. (2012) examined the relationships among SDB (sleep-disordered breathing), obesity and asthma in a cohort of children, those who referred for specialty asthma care. Further they hypothesized that both SDB and obesity were related with severe asthma, defined after one year of care in a specialty asthma clinic. Walker et al. (2013) investigated assess bronchial and alveolar NO (Nitric Oxide) levels primarily comparing healthy children to PCD (primary ciliary dyskinesia) and secondarily to cystic fibrosis (CF) and asthmatic children. In asthmatics bronchial NO was found to be significantly higher than in healthy controls and in children with CF alveolar NO was significantly lower.

35.2.2 Tuberculosis

Frequency of meeting with infectious persons, duration and the proximity are the generic reasons for the spreading of TB. In the history, TB has been used as a main example of "social disease" and the smoke, become completely curable disease when the impressive chemotherapy for TB became available in the end of the 1940 and beginning of 1950s (Lonnroth et al., 2009). In the effective treatments of TB, the slogan “Prevention starts with cure” played an important role. Further, (Chisholm, 1948; Rosen, 1974) reported that for reducing inequity and improving living conditions, the need to control TB has usually been used as a debate. The existing universal tuberculosis (TB) control paradigm mostly focuses on cutting transmission during early case detection and effective treatment. Medical interventions are at the core of the global strategy (Lonnroth & Raviglione, 2008; WHO, 2006). On the other hand, several recent observations have specified that the impact of the present TB control strategy has been less than expected (WHO, 2008a).

A detail study of the role of TB risk factors and social determinants and the historical and recent progress in TB control has been presented by Lonnroth et al. (2009). Pienaar et al. (2010) discussed the model (applied to different communities) as a tool to help identify the persons most likely to transmit TB and the interventions which are most likely to reduce the number of infected people in that setting. Mishra and Srivastava (2013) proposed susceptible-exposed-infectious-quarantine cured-susceptible with vaccination (SEIQRS-V) model captures both the spatial and temporal dynamics of TB disease spread process. Okuonghae (2013) presented a novel mathematical model that distinguishes between susceptibility amongst the population. The effect of treatment under the heterogeneity was also examined.

35.2.3 Pneumonia

Pneumonia is a leading cause of death and a major source of human morbidity. The initial immune response plays a central role in determining the course and outcome of pneumococcal disease. More than 2 million children in the developing countries still die from pneumonia each year.

A mathematical model was proposed by Rudnev and Romanyukha (1995) which involved "averaged dynamics variables" characterizing of lung resistance. Coopera et al., (1997) derived the computer models for predicting mortality of hospital patients with pneumonia by describing the application of eight statistical and machine-learning methods. Palmer et al. (2000) assessed a critical pathway designed to manage community-acquired pneumonia more efficiently than its management with conventional therapy.

For predicting community acquired pneumonia among patients with respiratory complaints, Heckerling et al. (2004) used genetic algorithms to search for optimal hidden layer architectures, connectivity, and training parameters for ANN for predicting pneumonia among patients with respiratory complaints. Rudan et al. (2008) provided a report on mortality and morbidity burden of childhood pneumonia, mainly among children aged less than 5 years.

To assisting ICU clinicians in diagnosing ventilator-associated pneumonia, Charitos et al., (2009) developed a probabilistic model. A mathematical model was developed by Otieno et al. (2012) for study pneumonia among children under five years of age. To predict the exposure to pneumococcal bacteria, Shrestha et al. (2013) developed a mathematical model and they found that antiviral treatment was expected to be efficacious only when administered during the early phase of influenza infection. Carloni et al. (2013) reported a mathematical approach to evaluate abnormal remodeling in IPFI (idiopathic pulmonary fibrosis). In that study, distribution of IPF lesions was compared with the hypothetical distribution of maximal mechanical stress which was obtained by a simplified mathematical model.

35.2.4 Lung Cancer

Lung cancer is the primary cause of cancer mortality in developed countries. Lung cancer kills more than 1.3 million people every year. Lung cancer ranks top in both incidence and mortality in most part of the world. Scientists strive to explore biomarkers and their possible role in the diagnosis, treatment and prognosis of lung cancer. The prognostic information provided by the CA 125 antigen assay was analyzed by Diez et al. (1994). Further, by using a multivariate analysis, the influence of CA 125 serum level on postoperative outcome was studied. Further, to provide the information on recurrence and the risk of death in patients, Diezl et al. CA125 and SCC (squamous cell carcinoma antigen) antigens. The patients were considering here those had undergone curative surgical treatment for NSCLC. Kulpa et al. (2000) presented the diagnostic usefulness of tumor markers in lung cancer by using reference group workers of a chemical plant producing chromites and chromate pigments. Balashazy et al. (2003) computed local deposition patterns in central human airway bifurcations, quantify their in homogeneities at the cellular level, and point to the possible consequences of the inhomogeneity regarding the health effects of inhaled aerosols by computational fluid dynamics models. Holdenrieder et al., (2004) investigated the potential of circulating, nucleosomes DNA for the early prediction of the efficacy of chemotherapy in patients with advanced lung cancer. Maciel et al. (2005) investigated whether serum low molecular weight proteins (LMW) can discriminate lung cancer patients from healthy donors. This work shows the usefulness of 2D-gel electrophoresis and mass spectrometry proteomic techniques for the identification of up- and down-regulated proteins in serum from adenocarcinoma lung cancer patients. Utley et al. (2006) described the modeling exercise, which was concerned with augmenting the information that was available to clinicians and patients.

Chi-Shing Cho (2007) provided a report which summarizes the pre-clinical and translational researches against lung cancer in relation to biomarker discovery and validation. Reich and Kim (2011) modeled lung cancer growth dynamics to show early diagnosis of lethal lung cancer by means of radiographic or CT screening. To obtain a complete picture of factors influencing potential lung cancer risk, Chung-Min Liao et al. (2011) emphasized the need to consider the Nano and ultrafine particle-bound PAHs (polycyclic aromatic hydrocarbons) data in additional to genetic susceptibility and respiration data. Hazeltona et al. (2012) developed a longitudinal biological based model of the relationship between individual smoking histories and the probability for lung cancer incidence, lung cancer mortality, CT screen detection and other cause mortality. Further the model related to these different outcomes to an underlying lung cancer disease pathway and an effective other cause mortality pathway, which were both influenced by the individual smoking history. Koleva et al. (2013) investigated a mathematical model that describes lung cancer regression in patients treated by radiotherapy and chemotherapy. A good correlation between the predicted and clinical data concluded that the mathematical model has potential to describe lung cancer regression.

REFERENCES

1. Altshuler, B. (1959). Calculation of regional deposition of aerosol in the respiratory tract. Bulletin of Mathematical Biophysics, 21, 257-270.
2. Balashazy, I. and Hofmann, W. (1993a). Particle deposition in airway bifurcations: I. Inspiratory flow. Journal of Aerosol Science, 24, 745-772.
3. Balashazy, I. and Hofmann, W. (1993b). Particle deposition in airway bifurcations: II. Expiratory flow. Journal of Aerosol Science. 24, 773-786.
4. Balashazy, I. and Hofmann, W. (1995). Deposition of aerosols in asymmetric airway bifurcations. Journal of Aerosol Science, 26(2), 273-292.
5. Balashazy, I.; Hofmann, W. and Thomas H. (2003). Local particle deposition patterns may play a key role in the development of lung cancer. Journal of Applied Physiology, 94, 1719-1725.
6. Carloni, A.; Poletti, V.; Fermo, L.; Bellomo N. and Chilosi M. (2013). Heterogeneous distribution of mechanical stress in human lung: A mathematical Approach to evaluate abnormal remodeling in IPF. Journal of Theoretical Biology, 332, 136-140.
7. Charitos, T.; Gaag, V.L.C.; S. Visscher, Schurink, K.A.M. and Peter Lucas, J.F. (2009). A dynamic Bayesian network for diagnosing ventilator-associated pneumonia in ICU patients. Expert Systems with Applications 36, 1249-1258.
8. Chi-Shing Cho, W. (2007). Potentially useful biomarkers for the diagnosis, treatment and prognosis of lung cancer. Biomedicine & Pharmacotherapy, 61, 515-519.

36

Analysis of Breast Thermograms Using Wavelet Based Structural Texture Features

S. Prabha[1], C. M. Sujatha[1] and S. Ramakrishnan[2]

[1]*Dept. of Electronics and Communication Engineering, CEG Campus, Anna University, India*
[2]*Biomedical Engineering Group, Department of Applied Mechanics, IIT Madras, Chennai, India*
Email: harprabha@gmail.com

ABSTRACT

Breast cancer diagnosis using thermograms entirely depends on the identification of asymmetrical thermal pattern. In this work, wavelet based structural texture features are used for the analysis breast tissues which include normal, carcinoma, nodule and fiber adenoma conditions. Breast tissues are extracted from background tissue by multiplying ground truth masks with raw images. The corresponding left and right breasts are delineated. Normal and abnormal groups are categorized based on the healthy and pathological conditions of the separated breast tissues. Two levels of discrete wavelet decomposition is performed on these tissues. Texture features such as coarseness, contrast and directionality are determined from the extracted wavelet coefficients. The results show that the considered features are able to differentiate normal and abnormal conditions. Further, it is observed that the directionality feature is highly significant ($p < 0.005$) in discriminating carcinoma and nodule tissues. Hence, it appears that this can be used in the automated detection of asymmetry in breast thermograms for efficient diagnosis of breast cancer.

Keywords: Breast thermogram, discrete wavelet transform, directionality, asymmetry analysis.

36.1 INTRODUCTION

Breast thermography is a non-invasive, non-contact and radiation free procedure which records and displays thermal patterns emitted by human skin (Ng et al., 2001). Statistics on incidence of breast cancer among women reveals that there is a need for early detection of the disease. Mammography which is the current gold standard technique has high false positive rate and less sensitivity in dense breast tissue. Thermography is functional imaging technique that is reported as an adjunct tool and has high sensitivity and specificity (Ng et al., 2009 and Minikina et al., 2009).

As human body is thermally symmetrical, any deviation from ideal health condition results in asymmetrical thermal patterns. Therefore, asymmetry analysis is significant in the identification of abnormality. Several features that contribute in the detection of asymmetry have been reported in review paper recently (Borchartt et al., 2013). Basic statistical features (Lipari et al., 1997, Wiecek et al., 1999), histogram based features (Kuruganti et al., 2002), gray level co-occurrence matrix based features (Acharya et al., 2010) and fractal measures (Mahnaz et al., 2010 and Serrano et al., 2010) were found to be useful features in identification of pathological conditions of breast tissues. It is reported that the higher order statistics such as skewness, kurtosis, variance and difference variance were most significant in the asymmetry analysis of breast thermograms (Qi et al., 2002 and Koyal et al., 2004). However, it is observed that such textural features which are derived based on local properties are not very prominent in improving the sensitivity of breast cancer diagnosis (Borchartt et al., 2013). Hence, there is a need for identification of significant features that could effectively extract local and global properties of the image.

Structural texture analysis involves extraction of texture elements based on local and global primitives (Chen et al., 1998). These features are found to be effective in asymmetry analysis of MR and mammography breast images (Lu 2003 and Ferrari 2008). Wavelet produces multiresolution representation of texture image and extracts variations at different scales. Tamura et al., and Abbadeni et al. have reported the computational measures of structural texture features for natural images. Wavelet based structural texture features have been reported as a significant texture features for image retrieval and classification applications (Muwei et al., 2009).

In this work, wavelet based structural texture features such as coarseness, contrast and directionality are extracted to analyze the asymmetry between normal and abnormal tissues. Three different abnormal breast conditions such as carcinoma, fiber adenoma and nodule are considered in this analysis.

36.2 METHODS AND MATERIALS

Breast thermal images for this study are obtained from online database of the project PROENG. (http://visual.ic.uff.br/proeng/). FLIR ThermaCam S45 camera was used to capture images at University Hospital of Federal University of Pernambuco. The acquisition procedure and protocol details have already been reported elsewhere [Motta].

In this analysis, 20 abnormal breast images are considered for. These images are selected such that the pathologies are either in left or right region. Breast tissues are extracted from background tissue by multiplying ground truth masks with raw images. The corresponding left and right breasts are delineated. Normal and abnormal groups are categorized based on the healthy and pathological conditions of the separated breast tissues

36.3 WAVELET BASED STRUCTURAL TEXTURE FEATURES

The grouped normal and abnormal breast tissues are subjected to discrete wavelet transform. Further analysis is carried out on the wavelet coefficient to quantify the microstructural changes caused due to increased metabolic activities in the affected tissues.

Wavelet represents image variations in different scales. At each level decomposition, four sub-bands, namely, low-frequency (LF), the low-high frequency (LH), the high-low frequency (HL) and high-high frequency (HH) are formed. Most of the low frequency information is contained in the LL frequency sub-band. Structural texture features which include coarseness, contrast and directionality are extracted from LLfrequency sub-band of both first and second level of decomposition.

Coarseness is the granularity measurement of texture. The moving average A_Z *(s, t)* over the neighborhood of size $2^Z \times 2^Z$ ($z = 0, 1, 2, 3, 4, 5$) at the point *(s, t)* is computed as

$$A_z\ (s,\ t) = \sum\nolimits_{j=s-2^{z-1}}^{j=s+2^{z-1}} \sum\nolimits_{k=t-2^{z-1}}^{k=t+2^{z-1}} x(i,j) \ / \ 2^{2z} \qquad ...(1)$$

where $x(j, k)$ is the coefficient of the LL sub-band at *(j, k)*. The differences between pairs of non-overlapping moving averages in the horizontal and vertical directions for each pixel are computed as

$$E_{z,h}\ (s,\ t) = |A_z\ (s + 2^{z-1},\ t) - A_z\ (s - 2^{z-1},\ t)| \qquad ...(2)$$

$$E_{z,h}\ (s,\ t) = |A_z\ (s,\ t + 2^{z-1}) - (s,\ t - 2^{z-1})| \qquad ...(3)$$

The best neighborhood size that maximizes *E* in either direction is determined as $S_{best}\ (s, t) = 2^z$. The global coarseness is calculated by averaging S_{best} over the entire LL sub-band:

$$\text{Coarse} = \frac{1}{MN} \sum\nolimits_{j=1}^{M} \sum\nolimits_{k=1}^{N} S_{\text{best}}\ (j, k) \qquad ...(4)$$

Contrast is varied by stretching or shrinking gray scale. The local contrast of the image over the neighborhood of size $W \times W$ is computed as:

$$l\_\text{con}\ (i,\ j) = \frac{\max_{x \in W \times W}(x) - \min_{x \in W \times W}(x)}{\max_{x \in W \times W}(x) + \min_{x \in W \times W}(x)} \qquad ...(5)$$

where $\max_X$ and $\min_X$ represent the maximum and minimum magnitude of LL sub-band coefficients. The global contrast is defined as the mean of all the local contrast values:

$$\text{Con} = \frac{1}{MN}\sum_{j=1}^{M}\sum_{k=1}^{N} l\_\text{con}(i, j) \qquad ...(6)$$

Directionality: The frequency distribution of orientation of local edges against their directional angle is a measure of degree of directionality. The edge strength $e(x, y)$ and the directional angle $a(x, y)$ are computed using

$$e(x, y) = 0.5\,(|\Delta_X(x, y)| + |\Delta_Y(x, y)|) \qquad ...(7)$$

$$a(x, y) = \tan^{-1}(\Delta_Y(x, y)/\Delta_X(x, y)) \qquad ...(8)$$

where $\Delta_X(x, y)$ and $\Delta_Y(x, y)$ are the horizontal and vertical grey level differences between the neighboring pixels, respectively.

A histogram of quantized direction values $H_{dir}(a)$ is constructed by counting numbers of the edge pixels with the corresponding directional angles and edge strength greater than a predefined threshold. The histogram is relatively uniform for images without strong orientation and exhibits peaks for highly directional images. The degree of directionality relates to the sharpness of the peaks:

$$F_{dir} = 1 - rn_p \sum_{p=1}^{n_p} \sum_{\alpha \in w_p} (a - a_p)^2 H_{dir}(a) \qquad ...(9)$$

where n_p is the number of peaks, a is the quantized directional angle (cyclically in modulo 180°), a_p is the position of the pth peak, w_p is the range of the angles attributed to the pth peak (that is, the range between valleys around the peak) and r denotes a normalizing factor related to quantizing levels of the angles a.

36.4 RESULTS AND DISCUSSION

A typical gray scale breast thermogram is shown in Fig. 36.1(a). This thermogram depicts asymmetry pattern which is due to the presence of carcinoma condition in the left breast. The breast tissues are segmented from background tissues using ground truth mask to extract quantitative information about asymmetry. Figure 36.1(b) and (c) show the ground truth mask and the segmented breast tissues.

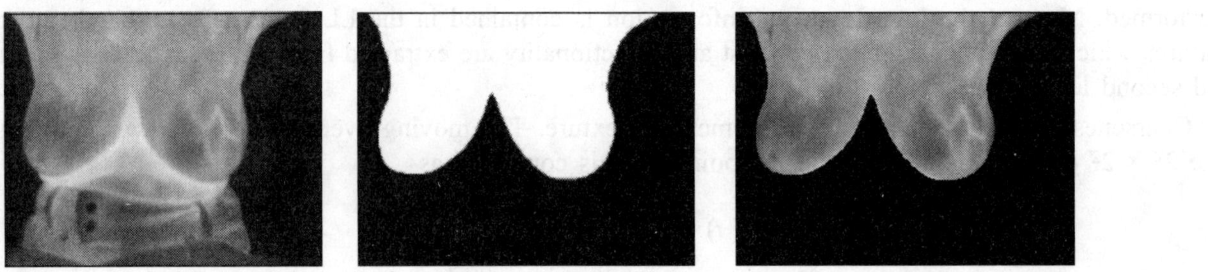

(a) (b) (c)

Fig. 36.1 Segmentation of breast tissues (a) Typical breast thermogram (b) Ground truth mask (c) Segmented breast tissue

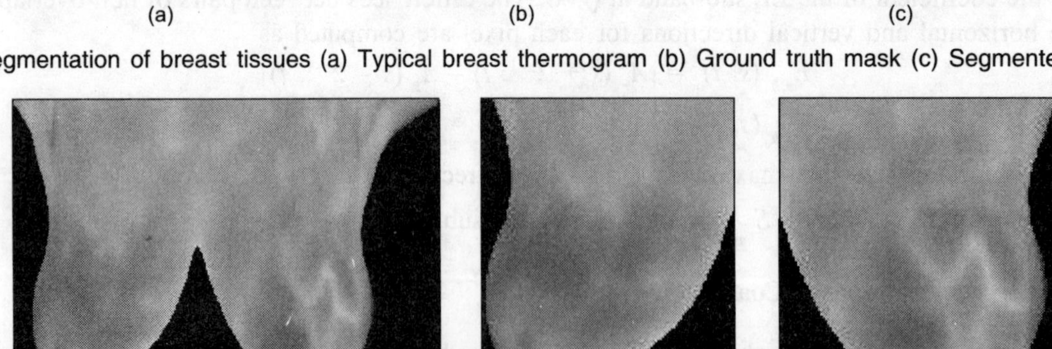

(a) (b) (c)

Fig. 36.2 Different stages in the separation of breast regions (a) Non-breast regions eliminated (b) Right breast tissue (c) Left breast tissue

The breast tissues after the removal of non-breast regions are shown in Fig. 36.2(a). Left and right regions are separated using midpointof inframammary folds and the obtained results are shown in Fig. 36.2(b) and (c).

The normal and abnormal breast regions are subjected to discrete wavelet transform up to two levels of decomposition. Structural texture features such as coarseness, contrast and directionality are extracted from the wavelet coefficients. The normalized average values of these features are shown in bar plot representation in Fig. 36.3.

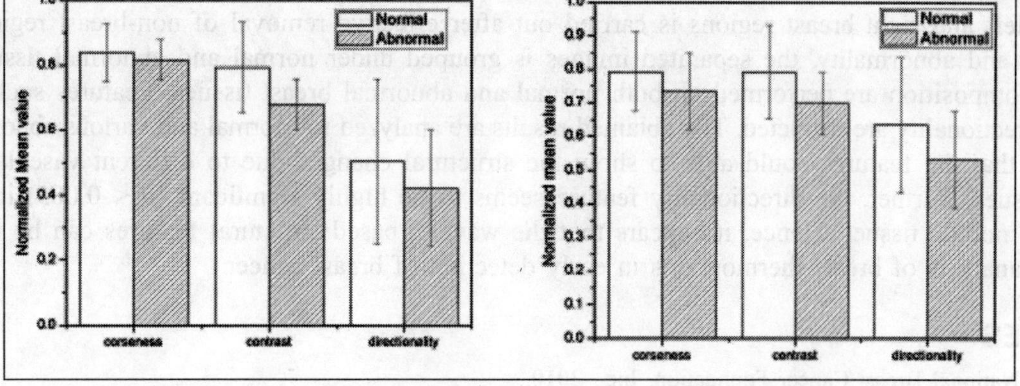

Fig. 36.3 Bar plot representation of the normalized mean value of structural features at (a) First level decomposition and (b) Second level decomposition

This plot shows that the extracted features such as the coarseness, directionality and contrast are able to differentiate the normal and abnormal conditions significantly in first and second level of decomposition.

The bar plot of normalized mean value of structural features derived from various pathological conditions are shown in Fig. 36.4. The structural features are able to significantly differentiate pathological conditions in both the levels of decomposition. The p-test values of features derived from abnormal breast tissues are presented in Table. 36.1.

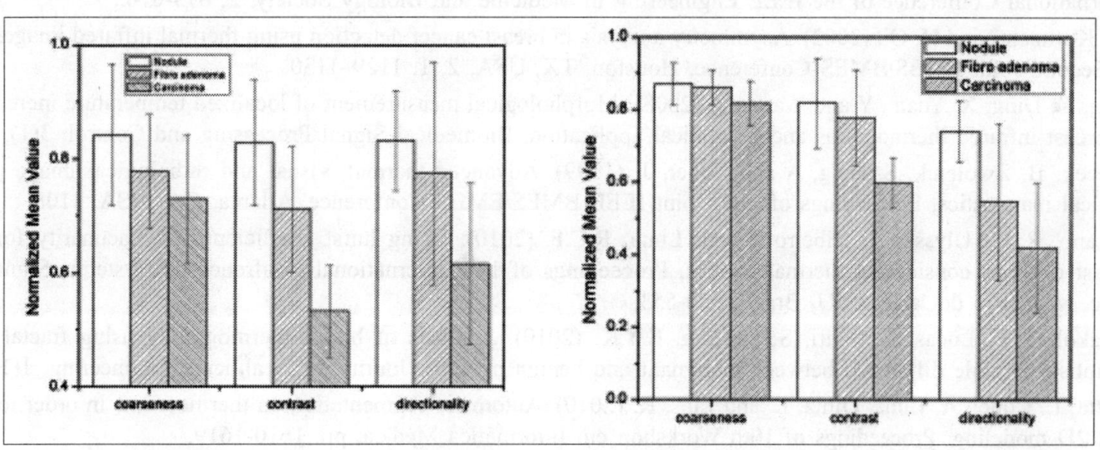

Fig. 36.4 Bar plot representation of normalized mean value for different abnormal conditions at (a) First level decomposition and (b) Second level decomposition

Table 36.1: P-test values for different pathological conditions

Parameters	*Scale 1*		*Scale 2*	
	Carcinoma and Nodule	*Carcinoma and Fibro adenoma*	*Carcinoma and Nodule*	*Carcinoma and Fibro adenoma*
Coarseness	0.0925	0.4392	0.7897	0.1988
Contrast	0.0007	0.0094	0.6027	0.005
Directionality	0.0006	0.223	0.0014	0.0713

The contrast and directionality features are found to be significant in differentiating the abnormal breast tissues. High statistical significance is observed with directionality feature in identifying the different abnormal conditions.

36.5 CONCLUSION

In this work, an attempt is made to analyze the breast infrared images using wavelet based structural features. For this study, normal and abnormal conditions such as carcinoma, fibro adenoma and nodule are considered. Before extracting features for the analysis, the breast tissues are segmented using ground truth mask and raw image. Separation of left and right breast regions is carried out after effective removal of non-breast regions. Depending on the healthy and abnormality, the separated images is grouped under normal and abnormal tissues. Two levels of wavelet decomposition are performed on both normal and abnormal breast tissues. Features such as coarseness, contrast and directionality are extracted. The obtained results are analyzed for normal and various abnormal conditions. It is observed that the features could able to show the structural changes due to different vascularized nature of the various tissues. Further, the directionality feature seems to be highly significant ($p < 0.005$) in differentiating carcinoma and nodule tissues. Hence, it appears that the wavelet based structural features can be used effectively in asymmetry analysis of breast thermograms in early detection of breast cancer.

REFERENCES

1. NBCF, National Breast Cancer Foundation, Inc., 2010.
2. Ng E and Sudharsan N. (2001). Numerical computation as a tool to aid thermographic interpretation. International Journal of Medical Engineering and Technology, 25 (2), 53-60.
3. Ng EYK. (2009). A review of thermography as promising non-invasive detection modality for breast tumor. International Journal of Thermal Sciences, 48(5), 849-859.
4. Minikina W. and Dudzik S. (2009). Infrared thermography: Error and Uncertainities. 1st ed. A john wiley and sons Ltd. Publication, 15.
5. Borchartt, T.B. Conci, A. Lima, R. Resmini and Sanchez, A. (2013). Breast thermography from an image processing viewpoint: A survey, Signal Processing, 93(10), 2785–2803.
6. Lipari and Head, J. (1997). Advanced infrared image processing for breast cancer risk assessment, Proceedings for 19th International Conference of the IEEE Engineering in Medicine and Biology Society, 2, 673-676.
7. P.T. Kuruganti and H. Qi, (2002). Asymmetry analysis in breast cancer detection using thermal infrared images, Proceedings of Second Joint EMBS/BMES Conference, Houston, TX, USA, 2, 1, 1129–1130.
8. Tang, X. Ding, X. Yuan . Y and Wang, Q. (2008). Morphological measurement of localized temperature increase amplitudes in breast infrared thermograms and its clinical application, Biomedical Signal Processing and Control, 3(1), 312-318.
9. Wiecek, B. Zwolenik, S. Jung, A and Zuber, J. (1999). Advanced thermal, visual and radiological image processing for clinical diagnostics, Proceedings of First Joint IEEE BMES/EMBS Conference, Atlanta, GA, USA, 1108
10. Serrano, R. C. Ulysses, J. Ribeiro S. and Lima, R.C.F. (2010). Using hurst coefficient and Lacunarity for diagnosis of breast diseases considering thermal images, Proceedings of 17th International Confrence on Systems, Signals and Image Processing, Rio de Janeiro, RJ, Brazil, 550-553.
11. Tavakol, M.E. Lucas, C. Sadri, S. and Ng, E.Y.K. (2010). Analysis of breast thermography using fractal dimension to establish possible difference between malignant and benign patterns, Journal of Healthcare Engineering, 1(1), 27-43.
12. Motta, L. Conci, A. Lima, Diniz, E. and Luís, R. (2010). Automatic segmentation on thermograms in order to aid diagnosis and 2D modeling. Proceedings of 10th Workshop em Informática Médica, pp. 1610-1619.
13. Acharya, U.R..Ng, E.Y.K. Tan, H. Sree, V. (2010) Thermography based breast cancer detection using texture features and support vector machine, Journal of Medical Systems 01–08.
14. Abbadeni, N. Djemel, Z. and Shengrui, W. (2000). Autocovariance-based perceptual textural features corresponding to human visual perception., Proceedings of 15th International Conference on Pattern Recognition Vol. 3.
15. Muwei, J. Guo, H and Lei, L. (2009). Texture Image Classification Using Visual Perceptual Texture Features and Gabor Wavelet Features, Journal of Computers 4.8.
16. Tamura, H. Shunji, M and Takashi, Y. (1978). Textural features corresponding to visual perception. Systems, Man and Cybernetics, IEEE Transactions 8.6: 460-473.
17. Chen, H. Pau, F. (1998). The Handbook of Pattern Recognition and Computer Vision, World Scientific Publishing Co., 2nd ed, pp. 207-248.

18. Mahnaz, T. Caro, L. Saeed, S and Ng, E. (2010). Analysis of Breast Thermography Using Fractal Dimension to Establish Possible Difference between Malignant and Benign Patterns, Journal of Healthcare Engineering , Vol. 1, No. 1, pp. 27–43.
19. Lu, S and Bottema, J. (2003). Structural image texture and early detection of breast cancer, Proceedings of the 2003 APRS Workshop on Digital Image Computing.
20. Ferrari, Ricardo J. (2008). Can bilateral asymmetry analysis of breast MR images provide additional information for detection of breast diseases, Computer Graphics and Image Processing, SIBGRAPI'08. XXI Brazilian Symposium on IEEE.

37

Hip Implantation: Modeling & Analysis

Subhomoy Chatterjee, Sreerup Banerjee, Santanu Majumder and Amit Roy Choudhury

Department of Aerospace Engineering & Applied Mechanics
Bengal Engineering & Science University
Shibpur, Howrah–711103, West Bengal, India

Email: dbtbesus@gmail.com

ABSTRACT

Loosening, due to stress shielding, is among the main factors behind the failure of the Total Hip Replacement (THR). Off-the-shelf implants are made in standard sizes and shapes that serves as a good mean of the individual geometric requirements but does not care for the individual deviations. This results in miss-matching in the exact required geometry. This accumulates stress on some particular zones leaving other zones of stress shielding. This study aims to customize the implant geometry for a particular subject by means of image data of the boney structures. Stress analysis has been performed in Finite Element Analysis (FEA) package ANSYS 14.

Keywords: ANSYS 14, finite element analysis (FEA), implant geometry, off-the-shelf implants, stress analysis, total hip replacement (THR).

37.1 INTRODUCTION

Total Hip Replacement (THR) was started with cemented implants [1]. Presently it is an established procedure of management of arthritis of hip joint with an average of 500,000 cases in the USA alone [2]. Among the main factor of concern in THR is loosening of implant mainly caused by bone resorption due to stress shielding [3-8]. Loosening occurs because the prefabricated implants have standard shapes and sizes. These are fabricated according to average human geometry and serves as a good mean of requirements but fail to take care of individual deviation of need leading to a miss-match between the geometry of the implant and the one that is exactly required. This, in turn, results in the difference in the stress accumulation, leaving some stress shielding zones. Different regions of the femur are subjected different strains as a result of dissimilar stress distribution. Bone, being a living entity, responds to this mechanical stimuli and resorps at the stress shielding regions. So the implant loosens over time creating the requirement for revision surgery.

Stiffer Implants generate lower cement interface stress but higher prosthesis stress [9-12]. Cementless implants showed better performance over cemented ones in terms of preservation of bone quality [13]. However, tendency of intra-operative crack formation and fractures of the bone, early stem loosening and secondary proximal bone loss were observed here, possibly due to stress shielding and femoral osteolysis. This is supposed to be related to a geometric mismatch between the stem and the endosteal bone due to variability of upper femoral geometry [14]. The probable existence of dimensional and material property difference between actual situation and simulated situation is still a matter of concern [15].

Optimized implant geometry is being attempted by researchers [16-23]. The manufacturing aspect deserves serious importance for the intricate 3D shapes. Marked development of manufacturing process is adding an edge to favor geometric customization. Laser Engineered Net Shaping (LENS), Rapid prototyping, electron beam machining and 3D Computer Numerical Control (CNC) milling machining techniques are preferred for developing the intricate implant shapes.

In this study customization of implant geometry is attempted for the geometric requirement of a particular subject based on the individual geometry of the upper femoral part. This anatomical geometry is obtained from imaging data as obtained by CT-scan and then processing the image to obtain the exact bone model. The implant shape and dimensions are deduced from this model. Stress analysis has been performed in Finite Element Analysis (FEA) package ANSYS 14.

37.2 METHODOLOGY

CT-Scan data of cadaveric bone has been taken in DICOM format and these image slices are processed in a medical image processing package, MIMICS (Materialise NV, Leuven, Belgium) to generate the 3D reconstructed geometry of the femur [Fig. 37.1]. The dense boney part, in the model, is identified as density of 450 Hounsfield Unit as lower threshold. The marrow cavity of the bone is measured at various levels to obtain the dimensions of the implant stem. The implant model is designed in the commercial package CATIA (Dassault Systemes, Vélizy-Villacoublay, France). Finite element analysis (FEA) of stress and strain is done using commercial FE package ANSYS 14 (ANSYS Inc., Cecil Township, Pennsylvania, USA). Before performing FEA, the implant and bone models are aligned to perform virtual implantation using ANSYS Workbench and a frictional contact is established at the interface.

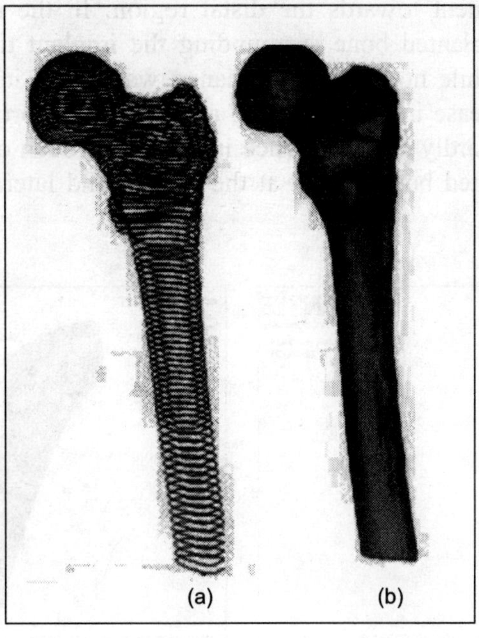

Fig. 37.1 Solid modeling from CT scan by piling the scan slices (a) Piling of the scan slices (b) Constructed 3 D model

10 noded, tetrahedral element (SOLID 187) has been chosen for both the bone and the implant for FEA. The material property of the bone is assigned from MIMICS, as obtained by an empirical equation using the image intensities obtained from the scan data, while that of the implant has been assigned as 117 GPa approximating surgical grade titanium alloy (Ti6Al4V). Physiological load (joint force and selected muscle pull force) is applied at the implant head as according to Bitsakos et al. (2005) [Table 37.1] [24]. A comparison of the strain profiles obtained from FEA for implanted and un-implanted femur has been performed. The customized implant, designed for the cadaveric femur is fabricated by CNC cutting method and inserted into the same cadaveric bone by orthopaedic surgeons to test for proper geometric fit.

Table 37.1: Muscle forces and joint loading considered in the FEA. Data taken from Bitsakos et al. (2005) [24].

Muscle Name/Joint	*Force component (N)*		
	X	Y	Z
Glutes Medias	42.2	4	-81.7
Glutes Minimus	21.5	-7	-45.2
Piriformis	113.4	61.6	-38
Hip Joint	-861.3	250.8	2056.9

Table 37.2: Values within which Von-Mises strain of the different zones of the coronal section of the femur lies.

Particulars	*Natural*	*	*Implanted*	*
	Proximal	Implant Tips	Proximal	Implant Tips
Middle	0.0025	0.0220	0.0025	0.0170
Middle	0.0050	0.0050	0.0025	0.0220
Lateral	0.0025	0.0220	0.0025	0.0170

37.3 RESULTS

It was found that the Von-Mises strain in the proximal region of the un-implanted femur was within 0.005 [Fig. 37.2(a)]. But there is marked increment towards the distal region. In the region of the un-implanted bone, that corresponds to the region of the implanted bone surrounding the implant tip, the strain in the medial and lateral portions reached as high as 0.022 while no significant change was found in the middle region. But for implanted case, the bone showed the same increase in strain (0.022) at the middle portion at this region. The strain at medial and lateral sides was within 0.017. Hardly any difference in strain has been observed along the medio-lateral aspect in the proximal region of the implanted bone except at the medial and lateral edges of the implant [Fig. 37.2(b)].

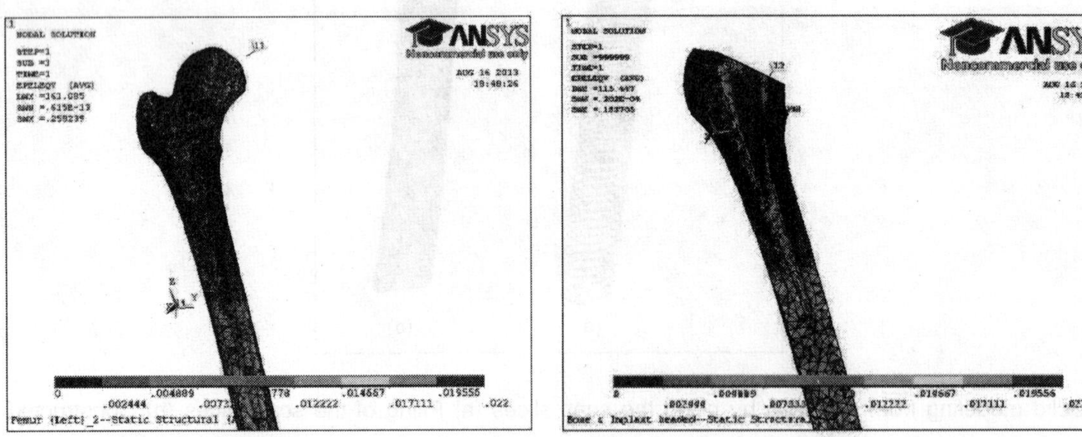

(a) (b)

Fig. 37.2 Strains generated in (a) Un-implanted femur, (b) Implanted femur

37.4 CONCLUSION

The proximal region of the femur is mostly important in osteointegration with the implant because of the presence of the cancellous bone. Customization of implant geometry has been found to eliminate the difference in strain, as compared to the un-implanted femur, in the medio-lateral aspect of the proximal region. The importance of medio-lateral aspect for strain equalization is more important than antero-posterior aspect because the joint reaction force has a greater component in this direction. So it can be concluded that customization of implant geometry produces congenial atmosphere for osteointegration at the medio-lateral region of the interface.

37.5 DISCUSSION

Loosening is the main factor of failure for the case of THR [25]. Loosening is directed by the stress behavior at the bone-implant interface [13, 26-29]. Implantation changes the location and the direction of loads, to which the bone is not adapted to [30]. Homogenizing the strain, at the interface, would result in more even distribution of stress and thus preventing stress-shielding. Customization of the geometry of the implant, for the exact requirement of the subject according to the individual geometry and density profile of the bone, is anticipated to promote uniformity of stress and strain profile at the interfacial region. Here in this study the marrow lumen of the femur is identified by considering an upper density threshold of 450 Hu. Further extensive study is required that will consider not only the geometry but the stiffness property of the bone and to monitor the stiffness of the implant and customize it as well.

REFERENCES

1. R. Huiskes, A survey of finite element analysis in orthopaedic biomechanics: the first decade, Journal of Biomechanics, 16, 385-409, (1983).
2. B. Latham and T. Goswami, Effect of geometric parameters in the design of hip implants paper IV. Materials & Design, 25, 715-722, (2004).
3. I. Jonkers, N. Sauwen, G. Lenaerts, M. Mulier, G.V. Pere and S. Jaecques, Relation between subject-specific hip joint loading, stress distribution in the proximal femur and bone mineral density changes after total hip replacement Journal of Biomechanics, 41, 3405-3413, (2008).
4. J.J. Jacobs, D.R. Sumner and J.O. Galante, Mechanisms of bone loss associated with total hip replacement. Orthopedic Clinics of North America, 24, 583-590, (1993).
5. O. Caglar, B. Atilla, M. Tokgozoglu, M. Alpaslan, The effect of collar on aseptic loosening and proximal femoral bone resorption in hybrid total hip arthroplasty. Orthopedics, 31(3), (2008).
6. T.A. Gruen, G.M. McNeice and H.C. Amstutz, Modes of failure of cemented stem-type femoral components: a radiographic analysis of loosening, Clin Orthop Relat Res, 141, 17-27 (1979).
7. K.L. Markolf and H.C. Amstutz, The effect of calcar contact on femoral component micromovement: a mechanical study, J Bone Joint Surg Am, 62(8), 1315-1323, (1980).
8. I. Oh and W.H. Harris, Proximal strain distribution in the loaded femur: an in vitro comparison of the distributions in the intact femur and after insertion of different hip replacement femoral components, J Bone Joint Surg Am, 60(1), 75-85, (1978).
9. M. Fagan, Material selection in the design of the femoral component of cemented total hip replacements, Clinical Materials, 1, 151-167, (1986).
10. J. Lewis, M.J. Askew, R.L. Wixson, G.M. Kramer and R.R. Tarr, The influence of prosthetic stem stiffness and of a calcar collar on stresses in the proximal end of the femur with a cemented femoral component, Journal of Bone & Joint Surgery (Am), 66A, 280-286, (1984).
11. P. Prendergast, J. Monaghan, D. Taylor, Materials selection in the artificial hip joint using finite element stress analysis, Clinical Materials, 4(4), 361-376, (1989).
12. A. Rohlmann, U. Mossner, G. Bergmann, G. Hees and R. Kolbel, Effects of stem design and material properties on stresses in hip endoprostheses, Journal of Biomedical Engineering, 9, 77-83, (1987).
13. P.J. Prendergast, Finite element models in tissue mechanics and orthopaedic implant design, Clinical Biomechanics, 12(6), 343-366, (1997).
14. P.C. Noble, J.W. Alexander, L.J. Lindahl, D.T. Yew, W.M. Granberry and H.S. Tullos, The anatomic basis of femoral component design, Clinical Orhopaedics & Related Research, 235, 148-165, (1988).
15. A. Roy Chowdhury and S. Saha, Application of Finite Element Method in Orthopaedic Implant Design- A Review, Journal of Long term Effects of Orthopaedic Implant. (Accepted).
16. P.R. Fernandes, J. Folgado and R.B. Ruben, Shape optimization of a cementless hip stem for a minimum of interface stress and displacement, Computer methods in biomechanics and biomedical engineering, 7(1), 51-61, (2004).
17. R.J. Yang, K.K. Choi, R.D. Crowninshield and R.A. Brand, Design sensitivity analysis: a new method for implant design and a comparison with parametric finite element analysis, Journal of Biomechanics, 17, 849-854, (1984).
18. R. Huiskes, R. Boeklagen, Mathematical shape optimization of hip prosthesis design, Journal of biomechanics, 22, 793-804, (1989).
19. H. Katoozian, D.T. Davy, Three-dimensional shape optimization of femoral components of total hip prostheses, Bioeng Conf. ASME, 24, 552-555, (1993).

20. H. Katoozian, D.T. Davy, Effects of loading conditions and objective function on three-dimensional shape optimization of femoral components of hip endoprostheses, Medical Engineering & Physics, 22, 243-251, (2000).
21. P. Kowalczyk, Design optimization of cementless femoral hip prostheses using finite element analysis, Journal of Biomechanical Engineering, 123, 396-402, (2001).
22. D.P. Nicolella, B.H. Thacker, H. Katoozian and D.T. Davy, The effect of three-dimensional shape optimization on the probabilistic response of a cemented femoral hip prosthesis, Journal of biomechanics, 39, 1265-1278, (2006).
23. Y.S. Yoon, G.H. Jang and Y.Y. Kim, Shape optimal design of the stem of a cemented hip prosthesis to minimize stress concentration in the cement layer, Journal of Biomechanics, 22, 1279-1284, (1989).
24. C. Bitsakos, J. Kerner, I. Fisher and A.A. Amis, The effect of muscle loading on the simulation of bone remodelling in the proximal femur, Journal of Biomechanics, 38, 133-139, (2005).
25. H. Malchau, P. Herberts, Prognosis of total hip replacement. revision and re-revision rate in THR: a revision-risk study of 148,359 primary operations, Scientific Exhibition Presented at the 65th Annual Meeting of the AAOS., (1998).
26. P.B. Chang, K.A. Mann, D.L. Bartel, Cemented femoral tem performance effects of proximal bonding geometry, and neck length, Clinical Orthopaedics and Related Research, 355, 57-69, (1998).
27. R. Huiskes, W. Vroemen, A standardized finite element model for routinecomparative evaluations of femoral hip prsotheses, Acta Orthopaedica Belgica, 52, 258-261, (1986).
28. S.A. Maher, P.J. Prendergast, Discriminating the loosening behavior of cemented hip prostheses using measurements of migration and inducible displacements, Journal of Biomechanics, 35, 257-265, (2002).
29. A.M. Rohlmann, U. Ossner, G. Bergmann, G.K. Hees and R. Obel, Effects of stem design and material properties on stress in hip endoprostheses, Journal of Biomedical Engineering, 9, 77-83, (1987).
30. N. Goetzen, F. Lampe, R. Nassut and M.M. Morlock, Load-shift - numeric evaluation of a new design philosophy for uncemented hip prostheses, Journal of Biomechanics, 38, 595-604, (2005).

38

Autonomic Reactivity Variables of Indian Sportspersons Withdrawn from Power Events

Dhananjoy Shaw[1] and Pawan Kumar[2]

[1]*Indira Gandhi Institute of Physical Education & Sports Sciences, University of Delhi, New Delhi–110018*

[2]*Bhaskaracharya College of Applied Sciences, University of Delhi, Dwarka, New Delhi–110075, India*

Email: dhananjoy_shaw@yahoo.com

ABSTRACT

Autonomic nervous system (ANS) in human plays significant role for maintenance of balance of several vital functions before and after activities. In sports, it occupy a very important parameter for enhancing performance and endurance. A pilot study has been done to evaluate autonomic reactivity variables of Indian sportspersons withdrawn from power events. Respiratory, circulatory and other sympathetic and parasympathetic (S&PS) body indices have been measured in 235 male (Age Group: 30-35 years). Results showed 9 to 32 score (Percentile range: i.e., 1 as percentile 1 and 32 as 100 percentile) at Deep Breathing Test. Expiratory inspiratory ratio ranged from 1.09 to 1.63 score as percentile range and Valsalva manoeuvre ratio ranged from 1.13 to 2.01 score as percentile range. Other S&PS parameters i.e. hand grip test ranged from 24 to 60 score as percentile range. Inces like Cold Pressor Response (CPR) test showed a ranged from 2 to 14 score as percentile. Tilt Table Response (Lying to standing test) found variation between 2 and 14 score as percentile and the variable, namely, 30:15 ratio ranged from 1.11 to 2.01 score as percentile range. Results of study suggest that ANS indices can be used for the assessment of performance and endurance related activities in sports persons.

Keywords: Autonomic nervous system (ANS), autonomic reactivity variables, deep breathing test, expiratory inspiratory ratio, cold pressor response (CPR), tilt table response.

38.1 INTRODUCTION

The Autonomic Nervous System (ANS) regulates the caliber of blood vessels to vary the distribution of blood, the organs whilst also maintaining the correct blood pressure. It regulates body temperature by varying the blood flow to the skin and by the control of sweating [1]. The ANS has two components sympathetic and parasympathetic. The sympathetic nervous system controls many important functions of the autonomic nervous system, i.e., increasing heart rate, force of contraction, excitability, conductivity of muscles. Parasympathetic nervous system works opposite to sympathetic nervous system, i.e. stimulate digestive secretion, slow the heart rate, constrict the pupils and dilate blood vessels. [2]. There are many test battery available (activity and reactivity) to measure the disturbance of autonomic function clinically. It is seen that as soon as the sportsmen stop participating actively in competitive sports they encounter with lots of problem like hypertension, diabetes, etc. There are numerous studies of normative studies on autonomic function [3-7] with clinical norms. But, no such norms are available to measure the autonomic function of the sportspersons (retired or not actively participating in the competitive sports) as, they might not be the same presently used for clinical purpose. Seeing the above context, researcher has taken sports dominating by power events as the delimitation of the study and made effort to form new norms.

38.2 OBJECTIVE

Objective of the study is to develop percentile norms of autonomic sympathetic and parasympathetic reactivity variables of Indian sportspersons withdrawn from competitive games and sports from power events.

38.3 SAMPLE

The study was conducted on 215 males. Based on the nature of game and sports the sportspersons were classified into power events (throwers, judo players, wrestlers, short distance swimmers, cricket fast bowlers and power lifters). Ages of the sample ranged from 30 to 35 years of same socio-economic status.

38.4 METHODOLOGY

Subjects were asked to come with two hours fasting before the test. No medication was taken before 48 hours of the testing. Subjects were make to rest for 30 minutes before the commencement of the test and then heart rate variability (HRV) was performed, which quantifies autonomic drive to the myocardium. The ECG analog were filtered and quantified using the software, namely 1) AUTONOMIC FUNCTION TEST HRV_Soft version 1.1, 2) HRV Software, Biomedical Signal Analysis Group, Department of Applied Physics, University of Kupio, Finland. Both sympathetic and parasympathetic drives to myocardium were assessed by Hand Grip Test (HGT), Cold Pressor Test (CPT), Lying to Standing Test (LST), Deep Breathing test (DBT), Expiratory Inspiratory ratio (E: I ratio), Valsalva maoeuvre ratio (VM ratio) with regard to HRV variables (sympathetic and parasympathetic reactivity). This was achieved by simultaneous measurement of ECG and respiration on a digital polygraph (Medicaid Company, Chandigarh, India).

38.5 STATISTICS

Collected data was computed with percentile from one to hundred. The findings have been presented with table.

38.6 RESULTS

Table 38.1 showing percentile norms of Indian sportspersons withdrawn from competitive games and sports from power events pertaining to autonomic sympathetic and parasympathetic reactivity variables.

Table 38.1: Autonomic variables data obtained after study

P	*DBT2*	*E: I ratio 2*	*VM2*	*HGT1*	*CPT1*	*LST1*	*30:15 ratio*
1	9	1.09	1.13	24	2	2	1.11
10	12	1.14	1.19	28	4	4	1.17
20	13	1.21	1.23	30	4	5	1.21
30	15	1.22	1.27	34	6	6	1.29
40	16	1.23	1.30	36	6	6	1.35
50	17	1.26	1.39	38	6	6	1.41
60	18	1.27	1.43	40	8	8	1.50
70	19	1.29	1.53	44	10	8	1.59
80	21	1.34	1.59	48	10	10	1.67
90	22	1.47	1.72	52	12	12	1.81
100	32	1.63	2.01	60	14	14	2.01

Subject No. = 215 (Age: 3035 yrs.)

P-Percentile: 1 = Sympathetic Reactivity, 2 = Parasympathetic Reactivity

ANS (Sympathetic & Parasympathetic) Functions/Reactivity Index Studied

1. Deep Breathing Test (DBT)
2. Change in Heart Rate (CHR)

3. Expiratory Inspiratory Ratio (EIR)
4. Valsalva Manoeuvre Ratio (VMR)
5. Hand Grip Test (HGT)
6. Cold Pressor Response Test (CPRT)
7. Lying to Standing Test (LST)

In Table 38.1 the variable, namely, deep breathing test having 9 to 32 score as percentile range, i.e. 1 as percentile 1 and 32 as 100 percentile. In the same table it is noted that, the variable, namely, expiratory inspiratory ratio ranged from 1.09 to 1.63 score as percentile range. The variable namely, valsalva manoeuvre ratio ranged from 1.13 to 2.01 score as percentile range. The variable namely, hand grip test ranged from 24 to 60 score as percentile range. The variable namely, cold pressor test ranged from 2 to 14 score as percentile. The variable, namely, lying to standing test ranged from 2 to 14 score as percentile and the variable namely, 30:15 ratio ranged from 1.11 to 2.01 score as percentile range.

38.7 CONCLUSION

Autonomic nervous system variables were studied in Indian sportspersons withdrawn from power events and to assess the causes for decline in performance and endurance in sports events. The study concluded with the percentile norms of Indian sportspersons withdrawn from competitive games and sports from power events pertaining to autonomic sympathetic and parasympathetic reactivity variables for versatile use and evaluation. Our study suggests that ANS indices can be used for the evaluation and understanding of sportspersons performances in sports events.

REFERENCES

1. http://www.answers.com/topic/Autonomic_nervous_system.http://en.wikipedia.org/wiki
2. David A. Gelber, Michael Pfeifer, Beth Dawson, Mary Schumer. (1997) Cardiovascular Autonomic Nervous System Tests: Determination of Normative Values and Effect of Confounding Variables, Journal of the Autonomic Nervous System, 62:40-44.
3. Walsh N.E., Schoenfeld L., Ramamurthy S., et al. (1989) 'Normative Model for Cold Pressor Test', American Journal of Physical Medicine & Rehabilitation 68: 6-11.
4. Ward K.D., Sparrow D., Landsberg L., Young JB, Vokonas P.S., Weiss S.T. (1996) Influence of Insulin, Sympathetic Nervous System Activity and Obesity on Blood Pressure: The Normative Aging Study. Journal of Hypertension 14: 301-308.
5. Jong-Chyou Denq, Peter C O'Brien and Phillip A Low. (1998) Normative Data on Phases of the Valsalva Maneuver, Journal of Clinical Neurophysiology 15: 535-540.
6. Freeman R., S.T. Weiss, M. Roberts, S.M. Zbikowski and D. Sparrow. (1995) The Relationship between Heart Rate Variability and Measures of Body Habitus', Clinical Autonomic Research, 5: 261-268.

39

Role of Mechanical Factor in Cause Production of Avascular Necrosis of Capital Femoral Epiphysis in Perthes' Disease of Hip

Jaganath Sahoo, Binod C. Raulo, Mahesh C. Sahu, Rabiteja Patra and Amiya K. Dash

Department of Orthopaedics, IMS and SUM Hospital
and
Department of Mechanical Engineering Institute of Technical Education & Research
Siksha O Anusandhan Univeristy. Bhubaneswar–751003, Odisha, India
Email: drjsahoo@gmail.com

ABSTRACT

Avascular necrosis of capital femoral epiphysis in case of Perthes' disease of hip is of ediopathic nature. Many a speculation is there, but nothing stood the test of time. However, the analysis of different forces acting around hip joint depicts a pertinent etiology for cause production of avascular necrosis of capital femoral epiphysis.

Keywords: Perthes' disease, avascular necrosis, capital femoral epiphysis.

39.1 INTRODUCTION

Hip joint is a diarthrodeal joint of human body where stability is considered to be an important aspect in addition to mobility. The anatomical configuration of head of femur to the acetabular socket is such that it imparts more stability as recurrent dislocation of hip is hardly encountered with. However, avascular necrosis of head femur is one of common complication seen in orthopaedic practice following fracture neck of femur, dislocation of hip, in association with systemic blood discariasis such as sickle cell anemia or even following heavy consumptions of steroids and in case of Perthes' disease of hip which is idiopathic in nature seen in children.

No doubt the cause behind avascular change is compromisation of vascular supply to head of femur in adult and capital femoral epiphysis in case of children. The peculiarity observed on radiological examination that the spherocity of head of femur is not at all maintained following Perthes' disease of hip in contrast to other conditions leading to avascular change. Thus, a mechanical analysis was thought to play a substantial role in perpetuating the avascular change of capital femoral epyphysis to a peculiar shape which is not usually seen with other conditions leading to avascular necrosis.

39.2 VASCULAR SUPPLY OF CAPITAL FEMORAL EPIPHYSIS

The capital femoral epiphysis gets its nutrition from three sources: 1 Epiphysial vessels, 2. Metaphysial vessel 3. Synovial fluid of hip joint. The lateral epiphysial vessels supply the poster superior part of capital epiphysis arising from sub-synovial vascular ring. The medial epiphysial vessels supply antero medial part of capital epiphysis which passes through ligamentum teres femoris. The vessel through the ligamentum teres femoris supply nutrition to capital femoral epiphysis up to age of 5-7 years after which it recedes. The metaphysial vessel proceeds to supply nutrition at this age to the epiphysis. The metaphysial vessels arise from extra capsular arterial ring.

The synovial fluid being contained in a blind pouch covers the capital femoral epiphysis. The nutritient element passes from synovial fluid into epiphysis by a squeezing phenomena following compression due to body weight transmission.

39.3 WEIGHT TRANSMISSION THROUGH CAPITAL FEMORAL EPIPHYSIS

The head of the femur or the capital femoral epiphysis along with the neck is designed in such a way that when the weight is borne; the force due to load is exerted almost at the center of head of femur and passes through calcar femorale. The similar situation with the capital femoral epiphysis where the metaphysical cartilage remains 150 angulated to horizontal, the weight transmission passes through the center of the metaphysial plate exerting uniform pressure through out the metaphysial cartilage plate as well as epiphysis (Fig 39.1). Thus, creating squeezing phenomena to distribute the nutrients throughout the capital femoral epiphysis. Over and above the shape of the capital femoral epiphysis is maintained in a spherical shape being molded inside the acetabular socket due to compression force following weight transmission since form follows the pressure (Woulfs law). This position of femoral head inside the acetabulum is influenced by balance of adductor and abductor forces acting in coronal plane. Thus, to have this physiological state of contained femoral epiphysis in the actabulum, the forces of adductor and abductor are to equalize each other in coronal plane to maintain the equilibrium.

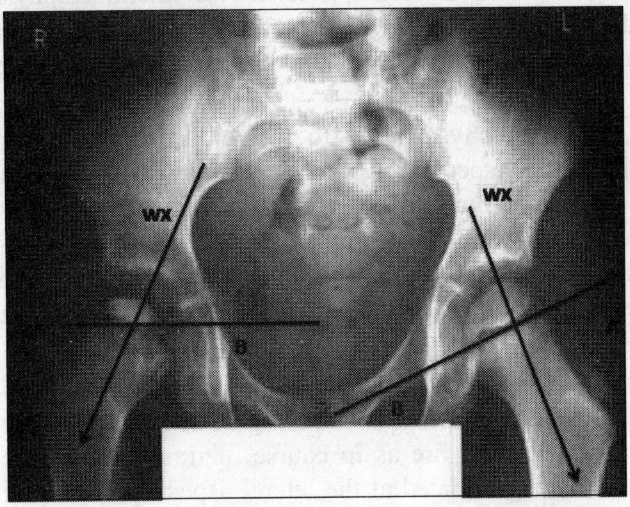

Fig. 39.1 Weight transmission axis through capital femoral epiphysis normal (left) and Perthes' disease (Right)

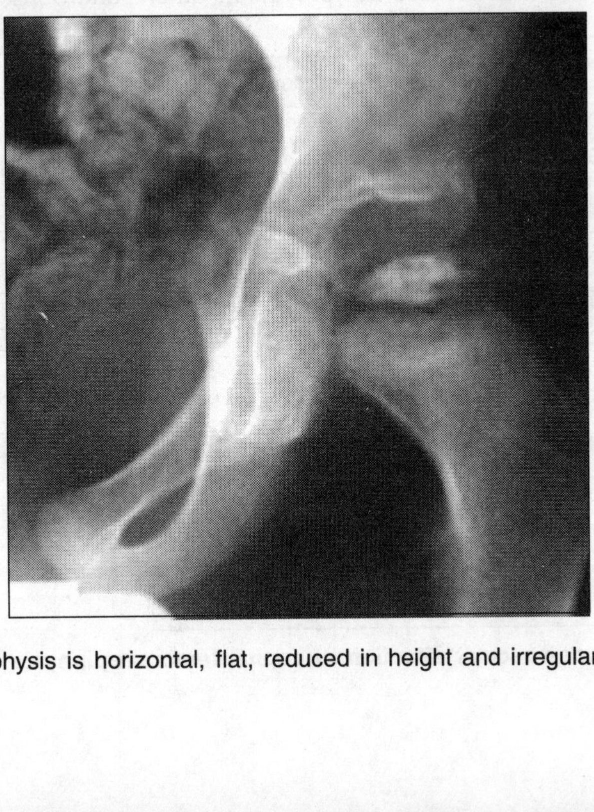

Fig. 39.2 The epiphysis is horizontal, flat, reduced in height and irregular in Perthes' disease

39.4 MECHANICAL SITUATION FOLLOWING PERTHES' DISEASE IN HIP

The adductor force being more than the abductor force tends to push the capital femoral epiphysis laterally, there by the metaphysial cartilage line (which is about 150 to horizontal) becomes horizontal (Fig 39.2). In this adverse situation the weight bearing axis shifts from center of metaphysial plate to midial side. As a result the medial aspect of metaphysial plate along with epiphysis is moulded inside the acetabular socket where as the lateral aspect becomes flat without any compression effect and having no molding cover of acetabulum.

39.5 ANALYSIS OF VASCULAR SITUATION AND NUTRITIONAL STATUS OF CAPITAL FEMORAL EPIPHYSIS IN PERTHES' DISEASE

Perthes' disease is seen usually at the age of 5-7 years of a child. This is the age when compromisation of vascular supply to the capital femoral epiphysis is seen as the vessels through ligamentum teres femoris recedes and metaphysial vessels do not proceed fully to supply the epiphysis. Thus, it is obvious that the epiphysis along with metaphysial plate is relatively under nourished with blood supply during this period.

The mechanical scenario again is precarious at the capital femoral epiphysis as the adductor force pushes the epiphysis laterally and metaphysial plate becomes horizontal and in this situation the capital epiphisis along with the neck tends to adopt a valgus attitude. With valgus possiton the lateral epiphyseal vessels get kinked and the medial epiphysial vessels in side ligamentum teris femoris get tithered. The lateral part of epiphysis is devoid of adequate blood supply to start with and is followed by the same situation on the medial aspect of epiphysis later on.

The lateral displacement of capital epiphysis leads to weight transmission compression force to be exerted more on the medial aspect and the lateral aspect of metaphyseal plate is devoid of any compression as a result the lateral part of the epiphysis receives poor nutrients from synovial fluid. Thus, the metaphyseal plate along with the epiphysis is not uniformly nourished which may perpetuate the lateral aspect of the epiphysis along with the metaphysial plate to undergo avascular change due to death of bone generating cells. Over and above the kinking of lateral epiphysial vessel jeopardize the nutrition supply to lateral part of epiphysis.

So this could be a pertinent cause for avascular necrosis of capital femoral epiphysis which eventually starts from the latral aspect due to lack of adequate nutrition, becomes flat, depress with reduction of height and looks sclerosed in the X-ray (Fig. 39.2). In addition to the tethering of medial epiphyseal vessel in ligamentum teres femoris leads the situation from frying pan to fire as in course of time medial side of epiphysis is effected. This denotes that the pathological process being initiated at the lateral aspect, proceeds in due course engulfing the total epiphysis which becomes irregular and sclerosed and looses the spheroidal shape since epiphysis is peripheral part of spherical head. Such mechanical situation does not prevail in other conditions leading to avascular necrosis of head of femur.

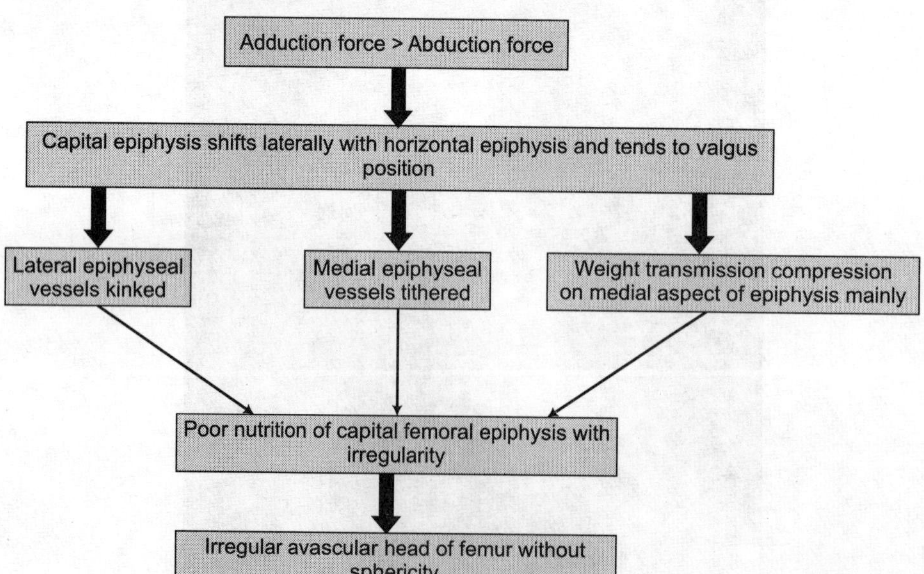

Flow chart of mechanico-vascular situation of capital femoral epiphysis in Perthes' disease

Compromisation of vascular supply to capital femoral epiphysis at the age of 5-7 years, the mechanical situation due to imbalance of adductor and abductor forces at hip joint leading to a valgus attitude affecting categorically the vascular supply of capital femoral epiphysis by epiphysial vessels and due to medialisation of weight transmission force cumulatively affect the nutritional status of epiphysis culminating to avascular changes with irregularity.

39.6 CONCLUSION

In case of Perthes' disease the capital femoral epiphysis remains in a dual adverse situation, i.e. Compromisation of vascular supply and unequal compression force leads to non-uniform nutritional supply from synovial fluid. Observing the mechanical situation it seems logical that the capital epiphysis being pushed laterally, becomes flat, avascular and looses the spherical shape (Fig. 39.3). Such situation does not prevail in other conditions leading to avascular necrosis of head of femur as a result even if the head becomes totally sclerosed but retains the spherical shape (Fig 39.4). Thus the adverse mechanical situation due to disparity between adductor and abductor forces seems to play as the prime cause for such type of avascular change in Perthes' diseas of hip.

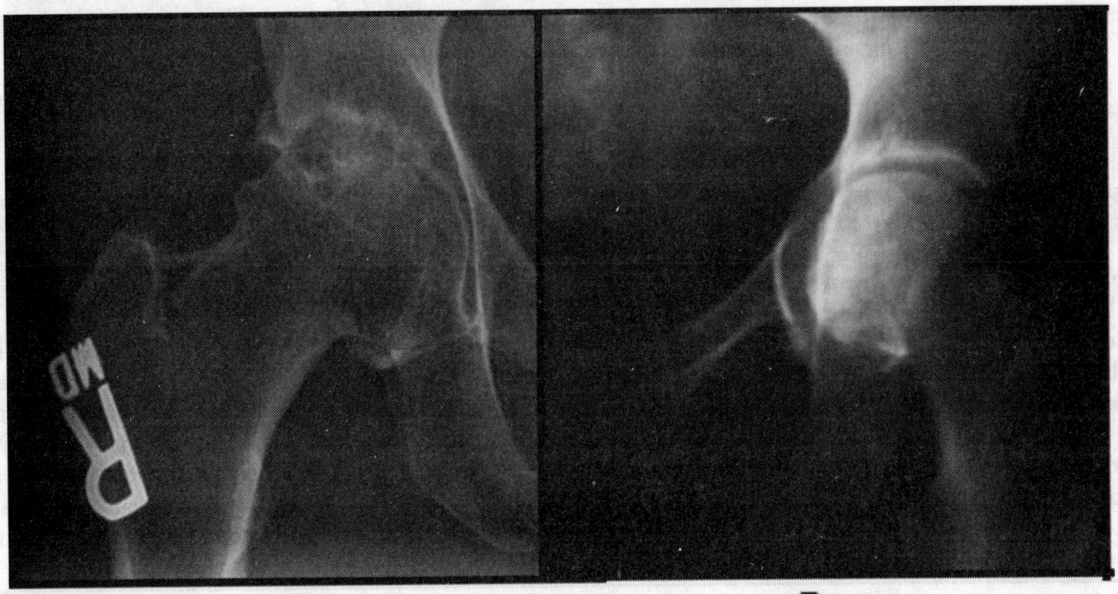

A B

Fig. 39.3 (a) Avascular head in Perthes' disease (b) Avascular head in other than Perthes'

REFERENCES

1. Dutta D and Dutta SK. (2013) Elementary Orthopedic Biomechanics. New Central Book Agency (P) Ltd; London 1998. ISBN: 978817318004.
2. Clifford R. Wheeless. (1994) Wheeless' Textbook of Orthopaedics, Vol III, North Carolina.
3. Kim, HK. (2012) Pathophysiology and new strategies for the treatment of Legg-Calve-Perthes disease. Journal of Bone and Joint Surgery (A) 94:659-669.
4. Perry DC, Bruce CE, Pope D, Dangerfield P, Platt MJ, Hall A. (2012). "Legg-Calvé-Perthes disease in the UK: Geographic and temporal trends in incidence reflecting differences in degree of deprivation in childhood". Arthritis & Rheumatism 64:1673-1679.
5. Daniel AB, Shah H, Kamath A, Guddettu V, Joseph B. (2012) Environmental tobacco and wood smoke increase the risk of Legg-Calvé-Perthes disease. Clin Orthop Relat Res. 470:2369-2375.

40

Cricket Ball Impact on Human Brain Simulation Using Finite Element and Impact Biomechanics Methods

G. Gauri, H. Anand, D. Anees, D. Naresh, Y. Pragya, K. Prashant and M. Rajesh

Biomedical Engineering Department, SGSITS, Indore
Indore Institute of Computer Aided Engineering (IICAE), Indore

Email: Gaurig1995@gmail.com

ABSTRACT

Human brain is an imperative organ of complete body. It controls every part of human body. Human brain is divided into different lobes and each lob manages different body parts and different behavior, i.e. frontal lobes are mainly responsible for decision making and planning. Even with a small injury to the brain can lead to total dysfunction in terms of behavior. Sports related injuries especially in cricket; head are often vulnerable to the severe impact of hard ball. Aim of the study is the analysis of effect of hard cricket ball impact to the human brain and measurement of brain injury like stress, strain and pressure. Finite Element Analysis (FEM) is a very advance tool to analyze stress-strain and acceleration response behavior of the structures. Finite Element Analysis (Hypermesh) is used to prepare simulation of ball impact to head scenario. Simplified Finite Element Model of human brain was developed with appropriate viscoelastic material properties. Finally, a Cricket Helmet was developed to reduce brain injury usually occurs during cricket match to the sportspersons. It was also found that newly developed helmet may provide protection from head injury by cricket ball.

Keywords: Brain injury, cricket helmet, frontal lobes, human brain, hypermesh, finite element analysis (FEM), simplified finite element model.

40.1 INTRODUCTION

The anatomy of the brain is complex due its intricate structure and function (Fig. 40.1). This amazing organ acts as a control center by receiving, interpreting, and directing sensory information throughout the body. The brain and spinal cord are the two main structures of the central nervous system. There are three major divisions of the brain. They are the forebrain, the midbrain, and the hindbrain.The forebrain is responsible for a variety of functions including receiving and processing sensory information, thinking, perceiving, producing and understanding language, and controlling motor function. The midbrain and the hindbrain together make up the brainstem. The midbrain is the portion of the brainstem that connects the hindbrain and the forebrain. This region of the brain is involved in auditory and visual responses as well as motor function.The hindbrain extends from the spinalcord and is composed of the metencephalon and myelencephalon. The metencephalon contains structures such as the pons and cerebellum. These regions assist in maintaining balance and equilibrium, movement coordination, and the conduction of sensory information. The myelencephalon is composed of the medulla oblongata which is responsible for controlling such autonomic functions as breathing, heart rate, and digestion.

Brain damage or brain injury is the destruction or degeneration of brain cells. Brain damage may occur due to a wide range of conditions, illnesses, or injuries. Possible causes of widespread (diffuse) brain damage include prolonged hypoxia (shortage of oxygen), poisoning, infection, and neurological illness. Common causes of focal or localized brain damage are physical trauma (traumatic brain injury), stroke, aneurysm, or neurological illness.

The extent and effect of brain injury is often assessed by the use of neurological examination, neuro-imaging, and neuropsychological assessment. Brain injury does not necessarily result in long-term impairment or disability, although the location and extent of damage both have a significant effect on the likely outcome.

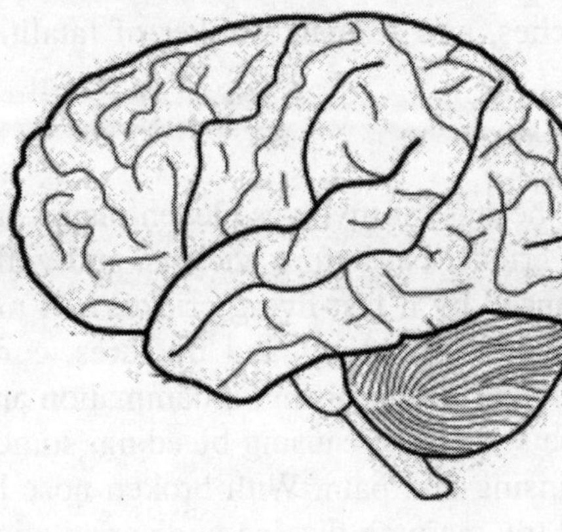

Fig. 40.1 Human brain

Any injury that results in trauma to the scalp, skull, or brain can be classified as a head injury. The terms traumatic brain injury and head injury are often used interchangeably in medical literature. This broad classification includes neuronal injuries, hemorrhages, vascular injuries, cranial nerve injuries, and subdural hygromas, among many others. These classifications can be further categorized as open (penetrating) or closed head injuries. This depends on if the skull was broken or not. Because head injuries cover such a broad scope of injuries, there are many causes—including accidents, falls, physical assault, or traffic accidents—that can cause head injuries. Many of these are minor, but some can be severe enough to require hospitalization.

All traumatic brain injuries are head injuries. But head injury is not necessarily brain injury. There are two types of brain injury: traumatic brain injury and acquired brain injury. Both disrupt the brain's normal functioning. Traumatic Brain Injury (TBI) is caused by an external force, such as a blow to the head and that causes the brain to move inside the skull or damages the skull. This in turn damages the brain.

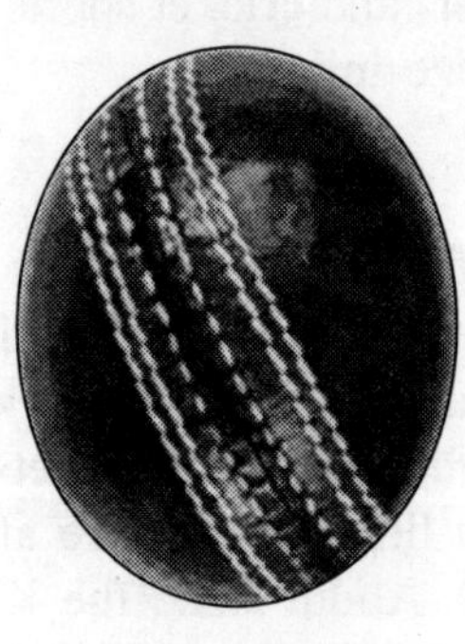

Fig. 40.2 Cricket ball

An Acquired Brain Injury (ABI) occurs at the cellular level. It is most often associated with pressure on the brain. This could come from a tumor. Or it could result from neurological illness, as in the case of a stroke. The severity of brain damage can vary with the type of brain injury. A mild brain injury may be temporary. It causes headaches, confusion, memory problems, and nausea. In a moderate brain injury, symptoms can last longer and be more pronounced. In both cases, most patients make a good recovery although even in mild brain injury 15% of people will have persistent problems after one year.

Cricket balls, weighing between 155.9 and 163.0 grams, are known for their hardness and for the risk of injury involved when using them. The danger of cricket balls was a key motivator for the introduction of protective equipment. A cricket ball is a hard, solid ball used to play cricket. A cricket ball consists of cork covered by leather, and manufacture is heavily regulated by cricket law at first class level. The manipulation of a cricket ball, through employment of its various physical properties, is the staple component of bowling and dismissing batsmen –movement in the air, and off the ground, is influenced by the condition of the ball and the efforts of the bowler, while working on the cricket ball to obtain an optimum condition is a key role of the fielding side. Injuries due to cricket balls are often recorded in matches, and a small number of fatalities have been recorded or attributed to cricket balls.

40.1.1 Cricket Injuries

Cricket is a direct contact sport, in a w ay, because it involves eleven players a side with a bowler bow ling a cricket ball, with batsman swinging a paddle-like cricket bat before wickets, and with players running and crouching. Most injuries to the face, head, and neck are caused by a fast-flying cricket ball making forceful contact Typical injuries to the head and neck during cricket are facial lacerations and fractures, concussion, spinal injuries and stenos is, cervical disk and eye injuries. With all these injuries, bruising, inflammation and swelling will appear with feelings of pain. A facial laceration is a cut to the face soft tissue causing bleeding, some bruising and swelling. Bone fractures to the face will cause severe swelling, bruising and pain. With broken nose bones, the nose will be deformed with bleeding Concussion is mild brain or head trauma with dizziness, disorientation, nauseousness, and swelling at point of contact (haematoma). Batsman and bowlers can develop spinal stenos is and injuries causing pain. When spinal or cervical vertebrae of the neck are hurt, players feel neck and arm pain with numbness or tingling sensations.

40.1.2 Causes of Head and Neck Injuries

The sources of head and neck injuries to cricket players are usually direct contact by a cricket ball at breakneck speed. Overstretching muscles in the neck is another cause. Spinal injuries can occur from repetitive bow ling or bat swinging with force. Other factors causing inflammation of tendon or damage to bone can be w hen neck muscles are pulled from not warming up properly before a cricket match, dehydration or poor diet, illness and incorrect batting or bow ling technique.

40.1.2.1 Preventing Head and Neck Injuries

Wearing batting helmets is key to reduce risk of head trauma. All players should learn the right cricket exercise technique for their role and also adequately w arm up before playing. Having medical assistance on-hand is important and players can seek advice from physiotherapists and cricket sports professionals about fluid in-take, diet, protective gear, and cricket sporting techniques to minimize injury.

40.1.2.2 Dangers of Cricket Balls

Cricket balls are very hard and potentially lethal; hence most of today's batsmen and close fielders often wear protective equipment. In 1998, Indian cricketer Raman Lamba died when a cricket ball hit his head in a club match in Dhaka. Lamba was fielding at forward short-leg without wearing a helmet when a ball struck by batsman Mehrab Hossain hit him hard on the head and rebounded to wicket-keeper Khaled Mashud. Other cricketers known to have died as a result of on field injuries in a first-class fixture after being hit while batting: George Summers of Nottinghamshire on the head at Lord's in 1870, Abdul Aziz, the Karachi wicket-keeper, the ball hit over the heart in the 1958-59 at Quaid-e-Azam Trophy final. Ian Folley of Lancashire (playing for Whitehaven) aws hit on the face in 1993. Glamorgan player Roger Davis was almost killed by a ball in 1971 when he was hit on the head while fielding. The Indian batsman Nariman Contractor had to retire from the game after being hit by a ball on the head in the West Indies. A cricket umpire, Alcwyn Jenkins, died in 2009 in Swansea, Wales after being hit on the head by a ball thrown by a fielder.

40.1.3 Cricket Helmet

A helmet is a form of protective gear worn to protect the head from injuries. Helmets attempt to protect the user's head by absorbing mechanical energy and protecting against penetration. Their structure and protective capacity are altered in high-energy impacts. Beside their energy-absorption capability, their volume and weight are also important issues, since higher volume and weight increase the injury risk for the user's head and neck. Anatomical helmets adapted to the inner head structure were invented by neurosurgeon sat the end of the 20th century. Helmets used for different purposes have different designs. For example, a bicycle helmet must protect against blunt impact forces from the wearer's head striking the road. A helmet designed for rock climbing must protect against heavy impact, and against objects such as small rocks and climbing equipment falling from above. Practical concerns also dictate helmet design: a bicycling helmet should be aerodynamic in shape and well ventilated, while a rock climbing helmet must be lightweight and small so that it does not interfere with climbing. A helmet has two major parts, namely:

1. The outer shell and
2. The energy absorbing liner.

The outer shell, the main purpose of which is to prevent penetration of the helmet by sharp objects, is made from polymers reinforced with fibres (glass, carbon or Kevlar) or thermoplastics like ABS or polycarbonate. In the FE analysis the outer shell is assumed to be made of glass fibre, reinforced plastic and modeled as a homogeneous isotropic elastic material.As cricket ball impact injuries are very crucial, it has to be analyzed carefully. It is almost impossible to assess the ball impact head injury experimentally. Finite Element Methods is a key player to solve complex experimental methodology be converting them in the form of simulation. Hypermesh is the product of Altair Engineering is the preprocessor which is widely used to create human body finite element modeling. This research work includes the ball impact head injuries and helmet performance during impact.

40.2 METHODOLOGY

FEA consists of a computer model of a material or design that is stressed and analyzed for specific results. It is used in new product design, and existing product refinement. A company is able to verify a proposed design will be able to perform to the client's specifications prior to manufacturing or construction. Modifying an existing product or structure is utilized to qualify the product or structure for a new service condition. In case of structural failure, FEA can be used to help determine the design modifications to meet the new condition. There are generally two types of analysis that are used in industry: 2-D modeling, and 3-D modeling. While 2-D modeling conserves simplicity and allows the analysis to be run on a relatively normal computer, it tends to yield less accurate results. 3-D modeling, however, produces more accurate results while sacrificing the ability to run on all but the fastest computers effectively. Within each of these modeling schemes, the programmer can insert numerous algorithms (functions) which may make the system behave linearly or non-linearly. Linear systems are far less complex and generally do not take into account plastic deformation. Non-linear systems do account for plastic deformation, and many also are capable of testing a material all the way to fracture.

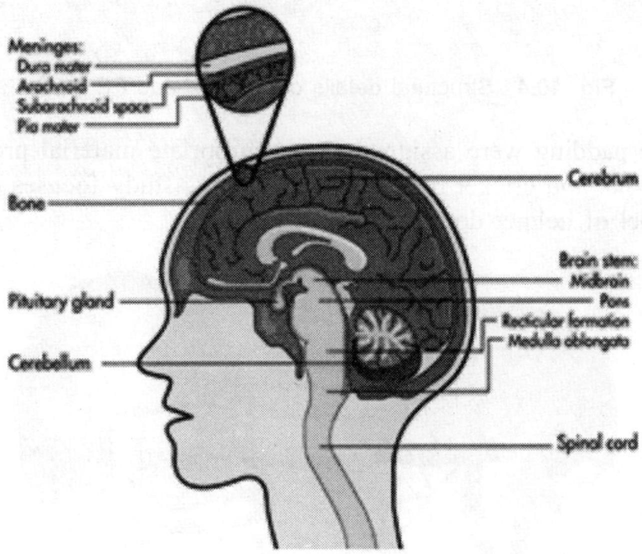

Fig. 40.3 Simplified finite element model of human brain

FEA uses a complex system of points called nodes which make a grid called a mesh. This mesh is programmed to contain the material and structural properties which define how the structure will react to certain loading conditions. Nodes are assigned at a certain density throughout the material depending on the anticipated stress levels of a particular area. Regions which will receive large amounts of stress usually have a higher node density than those which have been predicted in earlier studies. Finite element method provides a greater flexibility to model complex geometries than finite difference and finite volume methods do. It has been widely used in solving structural, mechanical, heat transfer, and fluid dynamics problems as well as problems of other disciplines.

The advancement in computer technology enables us to solve even larger system of equations, to formulate and assemble the discrete approximation, and to display the results quickly and conveniently. This has also helped the finite element method become a powerful tool. Simplified finite element human head model was developed using Hyper-mesh under Radioss solver. As shown in Fig. 40.3 below the simplified model contains all the necessary structure of real complex human brain like scalp, skull and brain Fig. 40.4. Material properties of skull, scalp and brain were taken from the study done by Zhant et al., in 2001.

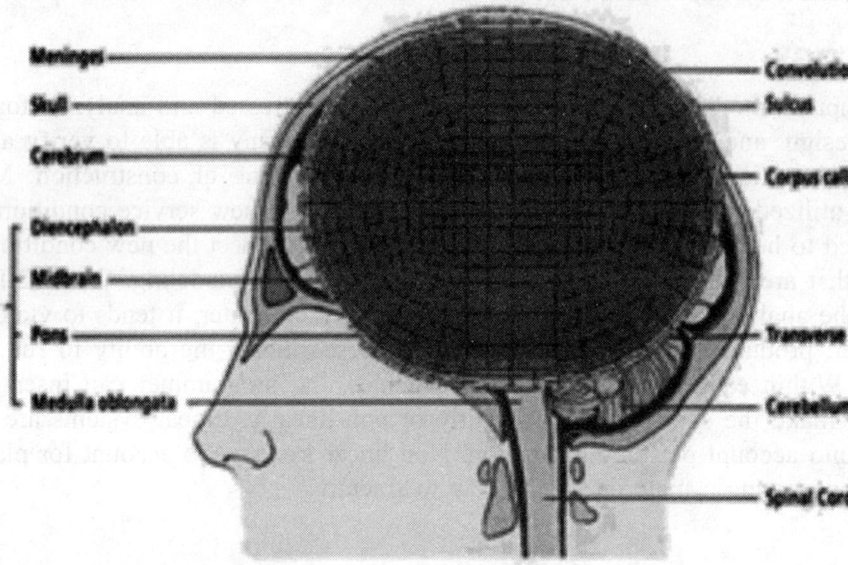

Fig. 40.4 Structural details of human head FE model

The helmet shell and foam padding were assigned with appropriate material properties Fig. 40.5 below shows the CAD model of Cricket helmet and cricket ball Fig. 40.6. As this study focuses only on head injury we did not consider facial injury, FE model of helmet doesn't include that part.

Fig. 40.5 Cricket helmet

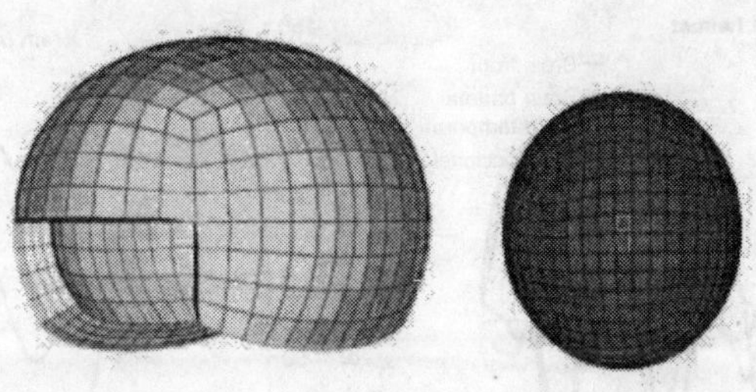

Fig. 40.6 Showing the CAD FE model of cricket helmet and cricket ball

Any cricket ball has radius of about 35 mm. Figure below shows the FE model of cricket ball. As ball is significantly harder as compare to head, it was assumed as a rigid structure (Fig. 40.7). The maximum average velocity of cricket balls observed between 130 to 150 kmph. We simulated the ball with impact velocity of 140 kmph. Two cases were considered with and without helmet as shown in figure below.

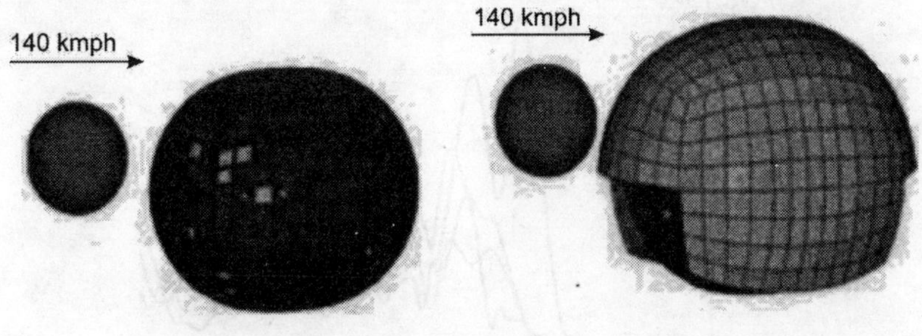

Fig. 40.7 Simulation of cricket ball impact on human brain and hemet

40.3 RESULTS

Results were plotted on frontal, parietal, temporal and occipital locations of skull and brain. stress and pressure time history of each locations were plotted for brain and skull. skull deformation was also captured in both the cases. Fig. 40.8(a) shows the brain pressure time history at different locations of without helmet case. As shown in figure below, maximum brain pressure at frontal location was found about 2000 kPa. Maximum brain pressure at parietal location was found about 1000 kPa. Maximum brain pressure at temporal location was found about 500 kPa. Maximum brain negative pressure at occipital location was found about 1000 kPa. Figure 40.8(b) shows the brain pressure time history at different locations of with helmet case. As shown in figure below, maximum brain pressure at frontal location was found about 100 kPa. Maximum brain pressure at parietal location was found about 50 kPa. Maximum brain pressure at temporal location was found about 40 kPa. Maximum brain negative pressure at occipital location was found about 200 kPa. Figure 40.8(c) shows the brain stress time history at different locations of without helmet case. As shown in figure below, maximum brain stress at frontal location was found about 0.7 kPa. Maximum brain stress at parietal location was found about 0.4 kPa. Maximum brain stress at temporal location was found about 3 kPa. Maximum brain negative stress at occipital location was found about 0.35 kPa.

Figure 40.9 (a) shows the brain stress time history at different locations of with helmet case. As shown in figure below, maximum brain stress at frontal location was found about 0.1 kPa. Maximum brain stress at parietal location was found about 0.15 kPa. Maximum brain stress at temporal location was found about 0.1 kPa. Maximum brain negative pressure at occipital location was found about 0.12 kPa. Figure 40.9(b) shows the skull pressure time history at different locations of without helmet case. As shown in figure below, maximum skull pressure at frontal location was found about 0.017 GPa. Maximum skull pressure at parietal location was found about 0.005 GPa. Maximum skull pressure at temporal location was found about 0.002 GPa. Maximum skull negative pressure at occipital location was found about 0.01 GPa. Figure 40.9(c) shows the skull pressure time history at different locations of with helmet case. As shown in figure below, maximum skull pressure at frontal location was found about 0.0004 GPa. Maximum skull pressure at parietal location was found about 0.0002 GPa. Maximum skull pressure at temporal location was found about 0.001 GPa. Maximum skull negative pressure at occipital location was found about 0.001 GPa.

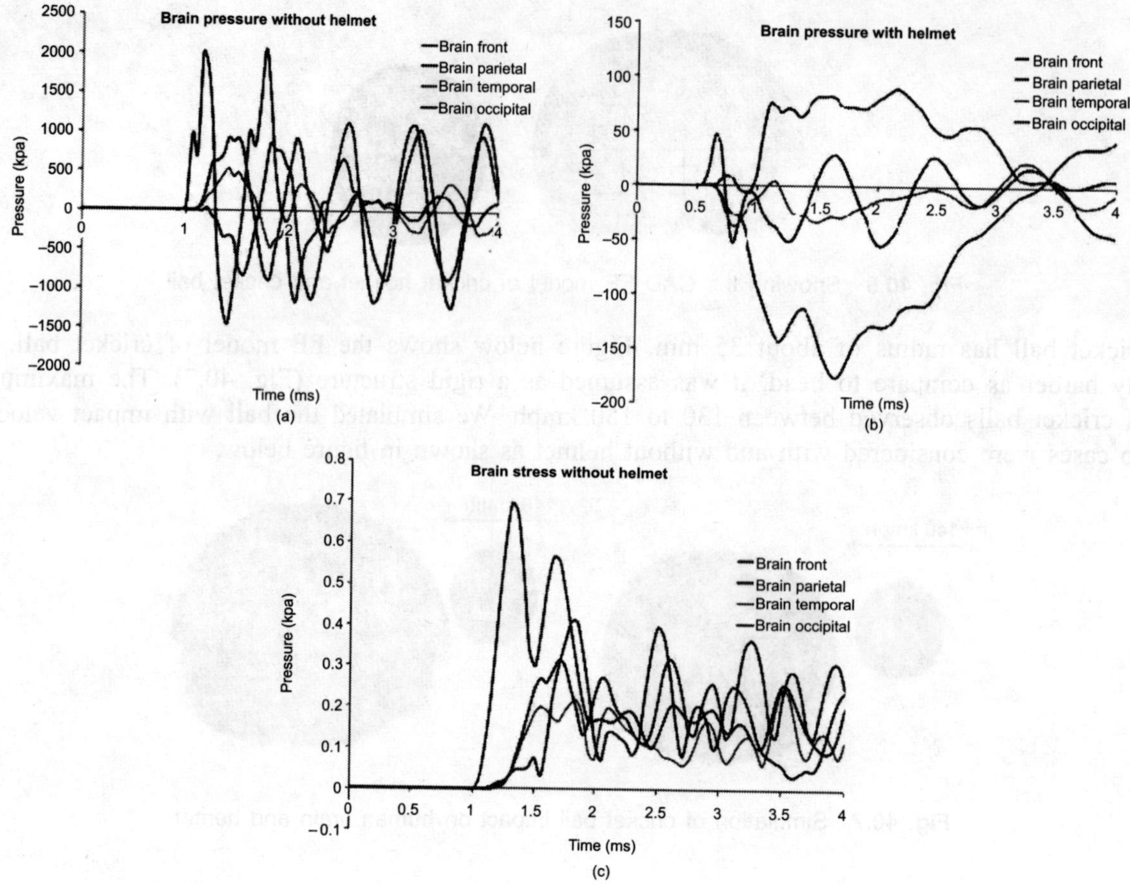

Fig. 40.8 (a) Brain stress time history without helmet (b) Brain stress time history with helmet (c) Skull pressure without helmet

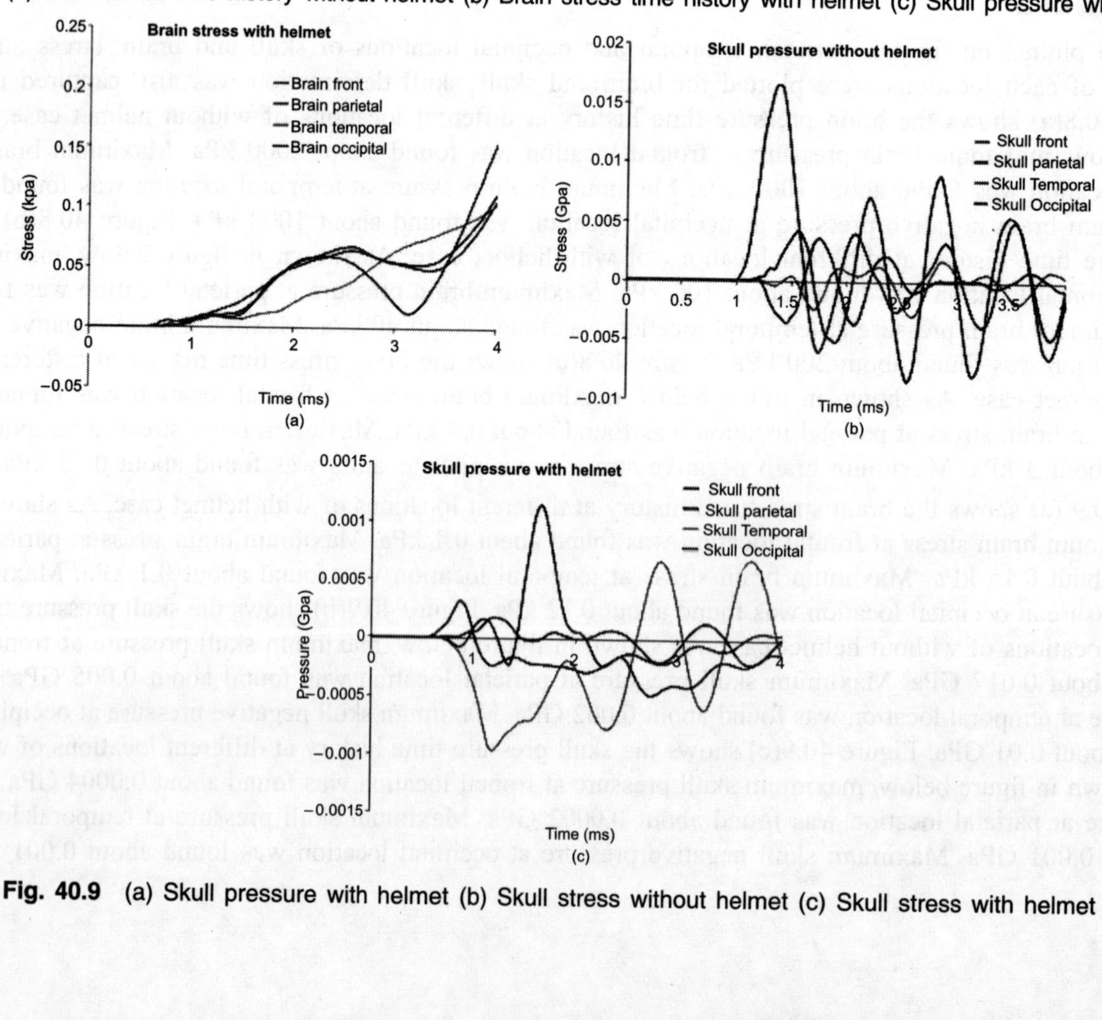

Fig. 40.9 (a) Skull pressure with helmet (b) Skull stress without helmet (c) Skull stress with helmet

Figure 40.10(a) shows the skull stress time history at different locations of without helmet case. As shown in figure below, maximum skull stress at frontal location was found about 40 MPa. Maximum skull stress at parietal location was found about 20 MPa. Maximum skull stress at temporal location was found about 15 MPa. Maximum skull negative stress at occipital location was found about 8 MPa. Figure 40.10(b) shows the skull stress time history at different locations of with helmet case. As shown in figure below, maximum skull stress at frontal location was found about 2 MPa. Maximum skull stress at parietal location was found about 1 MPa. Maximum skull stress at temporal location was found about 2 MPa. Maximum skull negative stress at occipital location was found about 2 MPa. Figure 40.10(c and d) shows the performance of helmet during cricket ball impact to the head. It is clear from the figure that cricket helmet design is capable enough to reduce the pressure and stress at different locations of brain and skull. significant reduction was also found in skull deformation because of helmet.

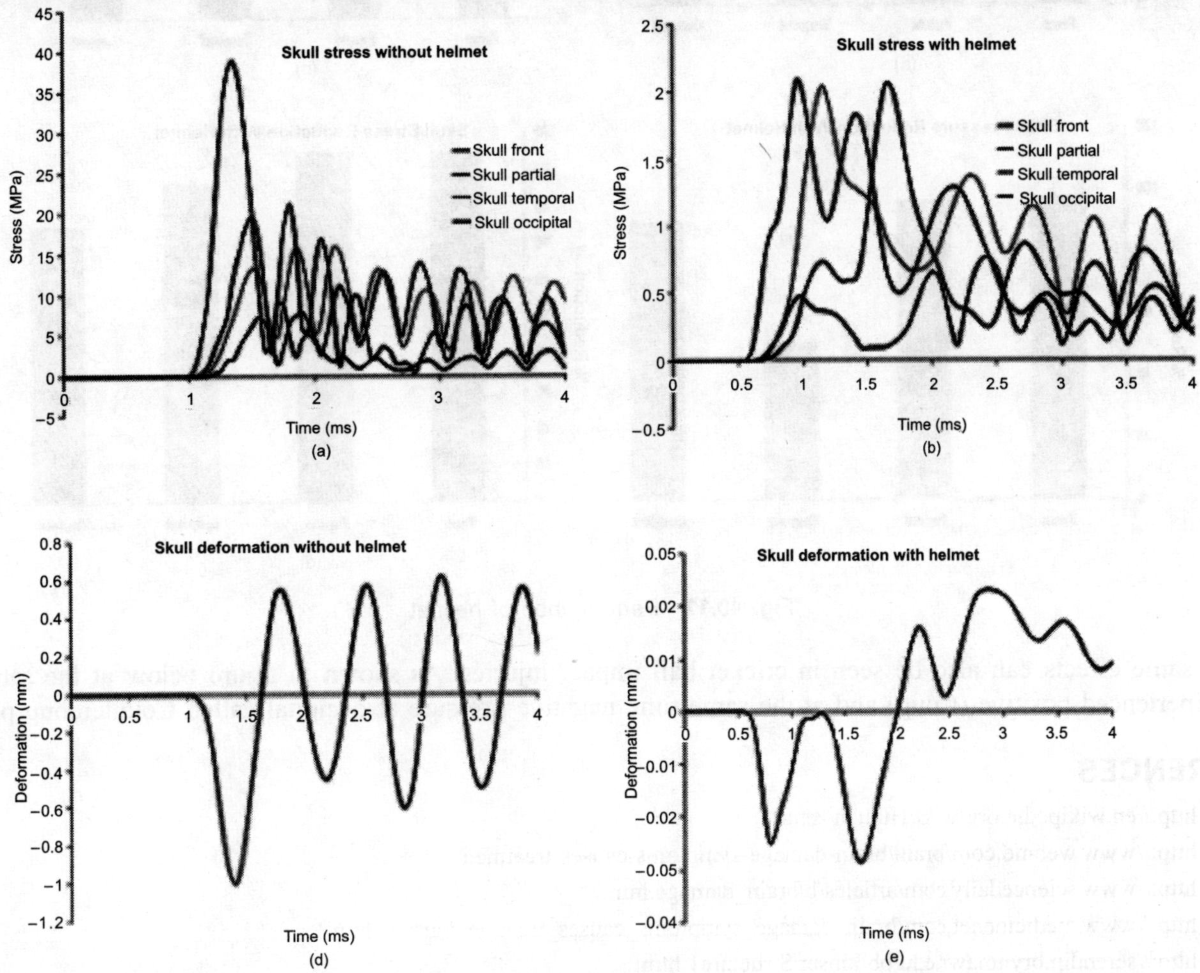

Fig. 40.10 (a) Skull deformation time history of without helmet and max. skull deformation was found about 1.1 mm; (b) Skull deformation time history of with helmet-max skull deformation was found about 0.03 mm and Minimum skull deformation was found about 03 mm; (c) Skull deformation without helmet; (d) Skull deformation with helmet

40.4 CONCLUSION AND DISCUSSION

Brain injury can be at the site of impact, but can also be at the opposite side of the skull due to a countercoup effect (the impact to the head can cause the brain to move within the skull, causing the brain to impact the interior of the skull opposite the head-impact). If the impact causes the head to move, the injury may be worsened, because the brain may ricochet inside the skull causing additional impacts, or the brain may stay relatively still (due to inertia) but be hit by the moving skull (both are countercoup injuries).

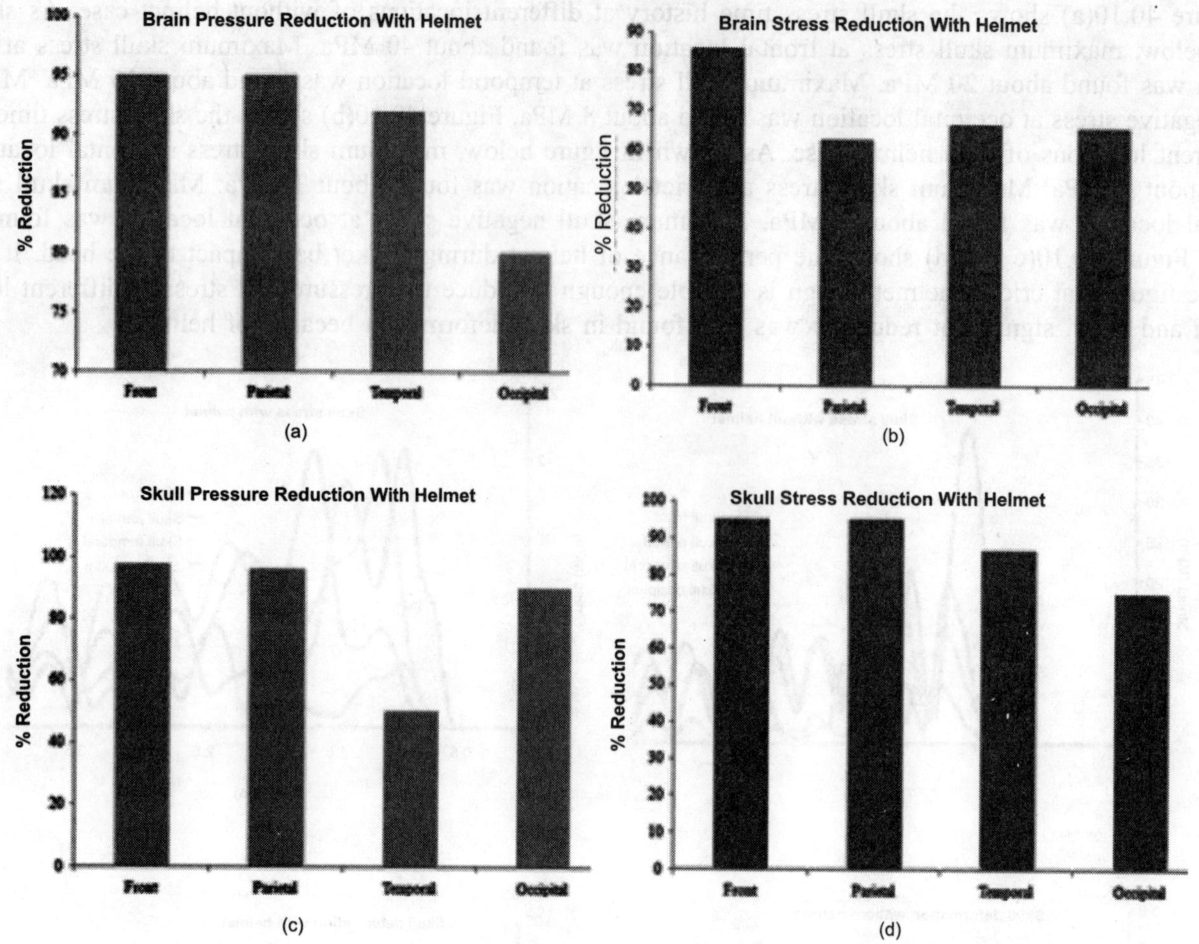

Fig. 40.11 Performance of helmet

The same effects can also be seen in cricket ball impact injuries. As shown in figure below at the same time brain experienced positive (Coup) and at the same time negative pressure at occipital called Countercoup pressure.

REFERENCES

1. http://en.wikipedia.org/wiki/Human_skull
2. http://www.webmd.com/brain/brain-damage-symptoms-causes-treatments
3. http://www.sciencedaily.com/articles/b/brain_damage.htm
4. http://www.medicinenet.com/brain_damage_symptoms_causes_treatment/article.htm
5. http://serendip.brynmawr.edu/bb/kinser/Structure1.html
6. http://en.wikipedia.org/wiki/Finite_element_method
7. http://math.nist.gov/mcsd/savg/tutorial/ansys/FEM/
8. http://www.altairhyperworks.com/Product,7,HyperMesh.aspx?AspxAutoDetectCookieSupport=1
9. https://www.google.co.in/search?q=cricket+ball+dimensions&biw=1358&bih=563&tbm=isch&
10. http://en.wikipedia.org/wiki/Cricket_ball
11. Zhang L, Yang KH and King, AI. (2001) Comparison of brain responses between frontal and lateral impacts by finite element modeling, J. Neurotrauma, 18(1), pp. 21-30

41

Verification of the Hagen-Poiseuille Flow in Cardiovascular System of Humans: A Numerical Analysis

S. P. Rout, B. Pattanayak and K. Dash

Department of Mechanical Engineering
Institute of Technical Education & Research
Siksha O Anusandhan University, Bhubaneswar, India

Email: satyarout@soaunivetrsity.ac.in

ABSTRACT

Blood is a suspension of cellular elements–red blood cells (erythrocytes), white cells (leukocytes), and platelets in an aqueous electrolyte solution, the plasma. Red blood cells (RBC) are shaped as a biconcave saucer with typical dimensions of 2 × 8 μm. Erythrocytes are slightly heavier than the plasma (1.10 g/cm³ against 1.03 g/cm³); thus they can be separated by centrifugation from the plasma. Blood flow in the circulatory system is generally unsteady. In most regions it is pulsatile due to systolic and diastolic pumping. In pulsatile flow, the flow has a periodic behaviour and a net directional motion over the cycle. Pressure and velocity profiles vary periodically with time, over the duration of a cardiac cycle. In the simplest model, blood flow in a vessel is modelled as a laminar, steady, incompressible, fully developed flow of a Newtonian fluid through a straight, rigid, cylindrical, horizontal tube of constant circular cross section. Such a flow is called as the Hagen-Poiseuille flow. The purpose of the present work is to study, how valid is the Hagen-Poiseuille model for a Rheological fluid.

Keywords: Hagen-poiseuille flow, rheological fluid, Reynolds number, blood.

41.1 INTRODUCTION

Elaborate study of pulsatile blood flow through the cardiovascular system is quite complex. The difficulty is due to the complicated nature of blood as a suspension of red and white blood cells, platelets and lipid globules dispersed in a colloidal solution of proteins. So far many experimental relationships have been presented for non-Newtonian behaviour of blood. In normal body, blood flow in vessels is generally laminar. However, at very high flow rates, particularly in ascending aorta, the flow may become turbulent at or near to peak systole. Disturbed flow may occur during deceleration phase of the cardiac cycle (Chandran et al., 2007)). Turbulent flow may also occur in large branching arteries. However, under normal conditions, the critical Reynolds number, R_e for blood flow in long straight smooth blood vessels is relatively high, and the flow remains laminar. The maximum Reynolds number may be as high as 9000. The average value of R_e in the vena cava is about 3000. Arteries have varying sizes and the maximum R_e is about 1000. In the case of blood flow, laminar flow conditions generally prevail at a high Reynolds number of 10000 (Mazumdar, 2004). The steady flow assumption is inapplicable until the flow has reached smaller muscular arteries described by several authors (Ku, 1997 and Pedley, 2000). The Newtonian fluid assumption is inapplicable at low shear rates such as those that would occur in arterioles and capillaries (Pedley, 2000).

$$\frac{\partial V}{\partial t} + V.\nabla V = -\frac{1}{\rho}\nabla P + V.\nabla V \qquad ...(1)$$

41.1.1 The Hagen-Poiseuille Flow Model

The vector from of Navier-Stokes equation and continuity equation are

$$\nabla . V = 0 \qquad ...(2)$$

We will be regarded the flow in long, straight, constant diameter sections of a pipe as a fully developed laminar flow. The gravitational effects (mass force) will be neglect. The velocity profile is the same at any cross-section of the pipe. Although most flows are turbulent rather than laminar, and many pipes are not long enough to allow the attainment of fully developed flow, a theoretical treatment and full understanding of fully developed laminar flow is of considerable importance. First it represents one of the few theoretical viscous analysis that can be carried out exactly without using other ad hoc assumptions or approximations.

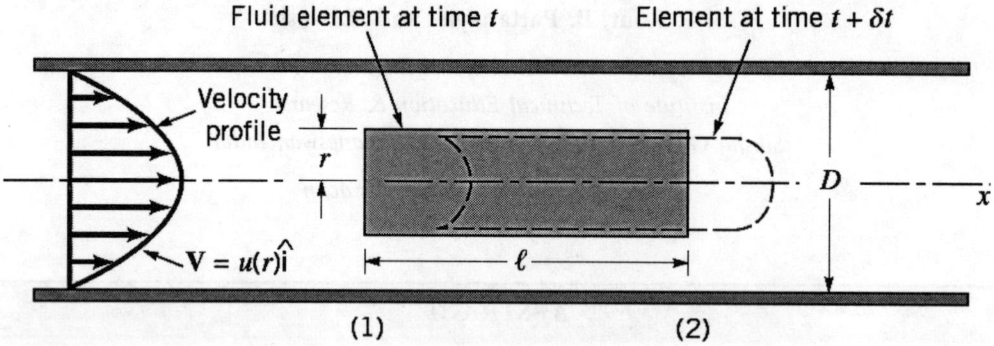

Fig. 41.1 Motion of cylindrical fluid element within a pipe

Solution of Navier-Stokes equation in a cylindrical coordinate (r, θ, x). Axial flow velocity, $u = u(r)$, in a pipe of radius 'a'

$$u = \frac{r^2 - a^2}{4\mu}\left(\frac{dp}{dx}\right) \qquad ...(3)$$

Pressure drop: in a fully developed flow, the pressure gradient, (dp/dx). is constant, and, it may be expressed in terms of the pressure gradient along the entire tube:

$$\left(\frac{dp}{dx}\right) = -\frac{\Delta p}{L} = -\frac{(p_1 - p_2)}{L} \qquad ...(4)$$

Where dp is the imposed pressure difference, subscripts 1 and 2 denote inlet and exit ends, respectively, and L is the length of the entire tube. The equation becomes

$$u = \left[\frac{\Delta pa^2}{4\mu L}\right]\left[1 - \left(\frac{r}{a}\right)^2\right] \qquad ...(5)$$

The maximum velocity y occurs at the center of the tube, $r = 0$, and is given by

$$u_{\max} = \left[\frac{\Delta pa^2}{4\mu L}\right] \qquad ...(6)$$

The volume flow rate is:

$$Q = \int_0^a u\, 2\pi r\, dr = -\frac{\pi a^4}{8\mu}\left(\frac{dp}{dx}\right) = \frac{\pi a^4}{8\mu}\frac{(p_1 - p_2)}{L} = \frac{\pi a^4}{8\mu}\frac{\Delta p}{L} = \frac{u_{\max}}{2}\pi a^2 \qquad ...(7)$$

The shear stress at tube wall is:

$$\tau_{xr}\Big|_{r=a} = \tau = -\mu\left(\frac{du}{dr}\right)\Big|_{r=a} = -\frac{a}{z}\left(\frac{dp}{dx}\right) = \frac{a}{z}\frac{\Delta p}{L} \quad ...(8)$$

Total peripheral resistance is evaluated to be

$$R = \frac{\Delta P}{Q} = \frac{8\,\mu l}{\pi\, a^4} \quad ...(9)$$

41.2 GEOMETRY & COMPUTATIONAL DOMAIN

Many researchers have attempted to find out whether there is a critical size at which both laminar and steady flow is experienced which is conducive for the Hagen-Poiseuille model. It was found out that smaller muscular arteries and arterioles in the circulatory system may exhibit both the behaviour. In the present study we consider palmar digital artery for validation of the Hagen-Poiseuille model. Due to the axis-symmetric nature of the geometry we have considered a two-dimensional model for our current numerical simulation which will not hamper the results at the same time saving computational time. The numerical solution is carried out in ANSYS Fluid Flow Fluent (FFF) 14.5.

Table 41.1 Physiological data of the artery for thermo-fluid model

Segment name	*Proper palmar digital artery*
Length (meter)	0.20
Proximal radius (meter)	0.00055
Distal Radius (meter)	0.00045
Velocity of blood flow (m/s)	0.2

(Through courtesy: Elad David and Einav Shamuel (1980) McGraw Hill)

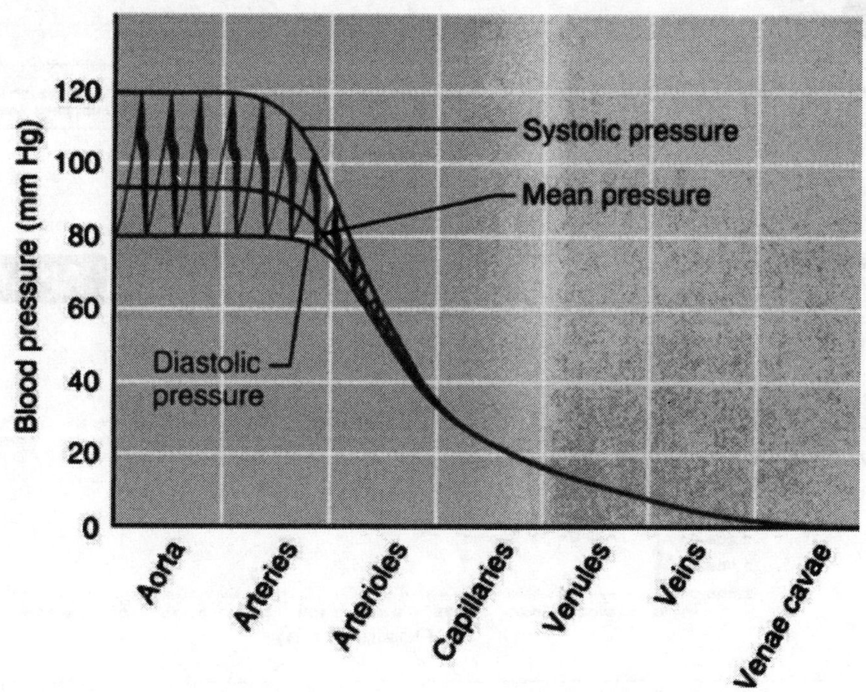

Fig. 41.2 Blood pressure in various vessels

41.3 BOUNDARY CONDITIONS

For blood flow we have applied the following boundary conditions: At the inlet of artery velocity boundary condition is applied whereas the outlet is pressure boundary condition. At the top of the artery wall boundary condition is used there by all velocity components are prescribed as zero (no slip condition). At centre line axi-symmetric boundary condition is used.

41.4 RESULTS & VALIDIATION

The residuals fall below the convergence criterion of 1e-6 at approximately 45 no. of iterations.

41.4.1 Velocity Distribution Along Symmetry Plane

The velocity vectors provide a clear picture downstream of the artery, flow near the wall retarded by viscous friction and fully developed flow is experienced after an entry length. The Hagen-Poiseuille model is experienced.

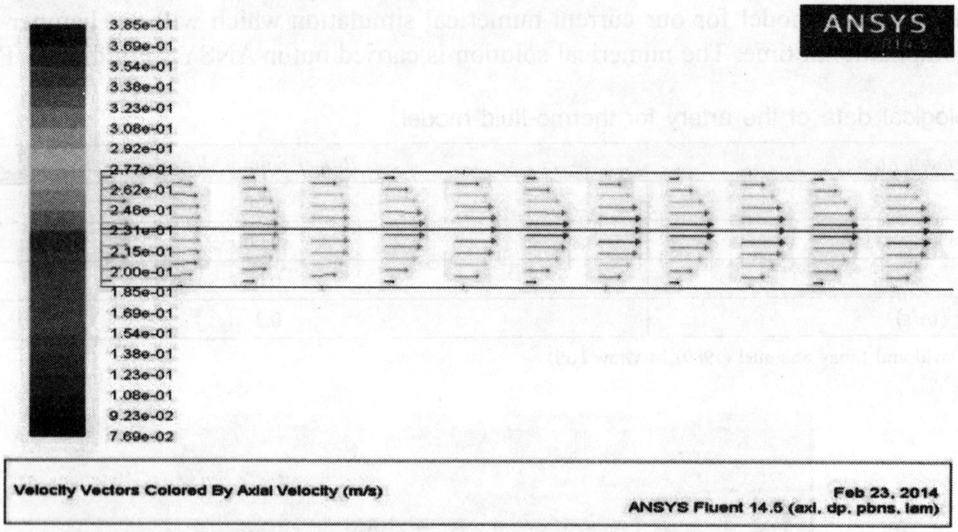

Fig. 41.3 Velocity vectors throughout artery

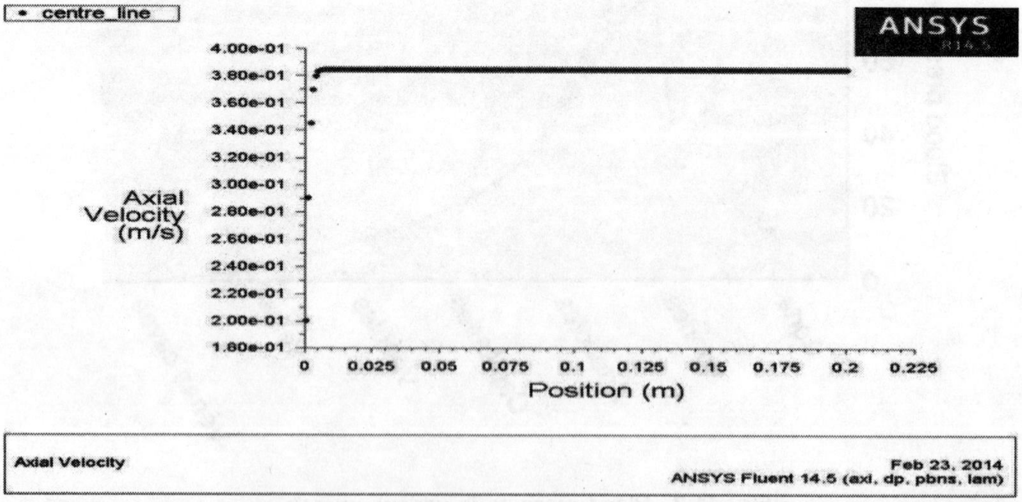

Fig. 41.4 Axial velocity along centre line of artery

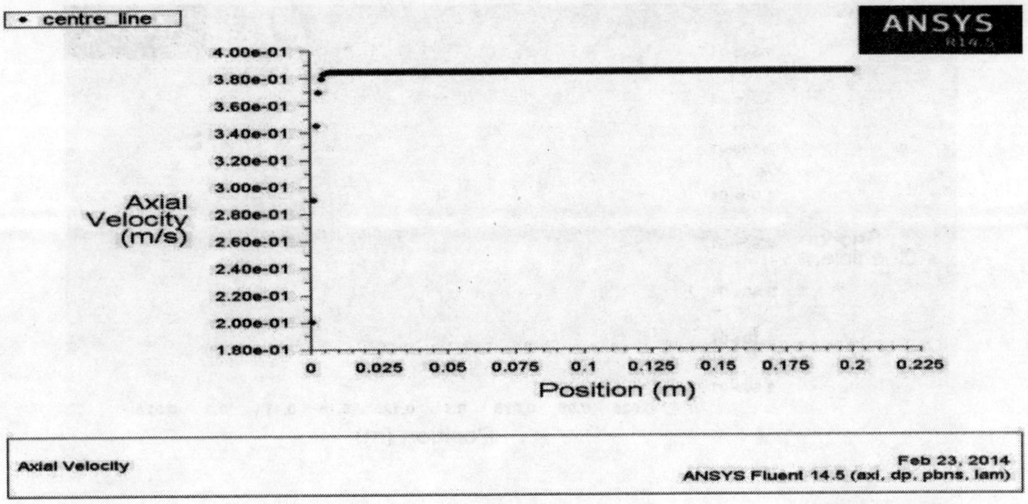

Fig. 41.5 Comparison of axial velocity along centre line of artery

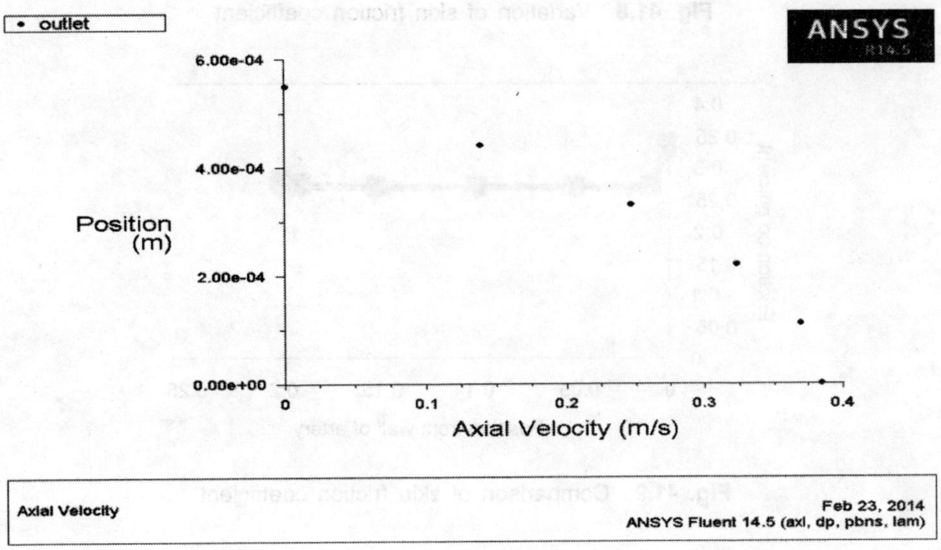

Fig. 41.6 Variation of axial velocity along the radius

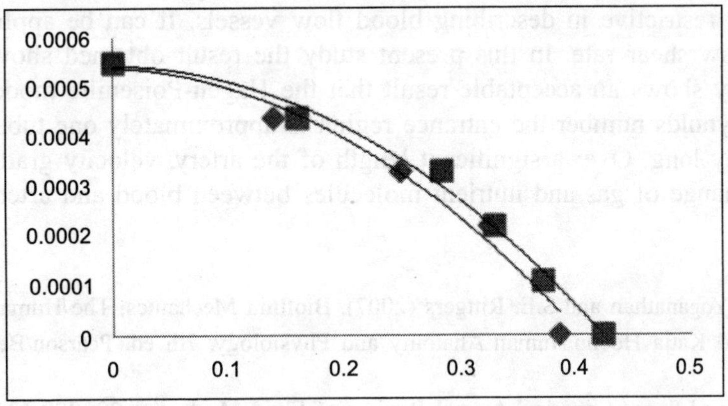

Fig. 41.7 Comparison of axial velocity along the radius

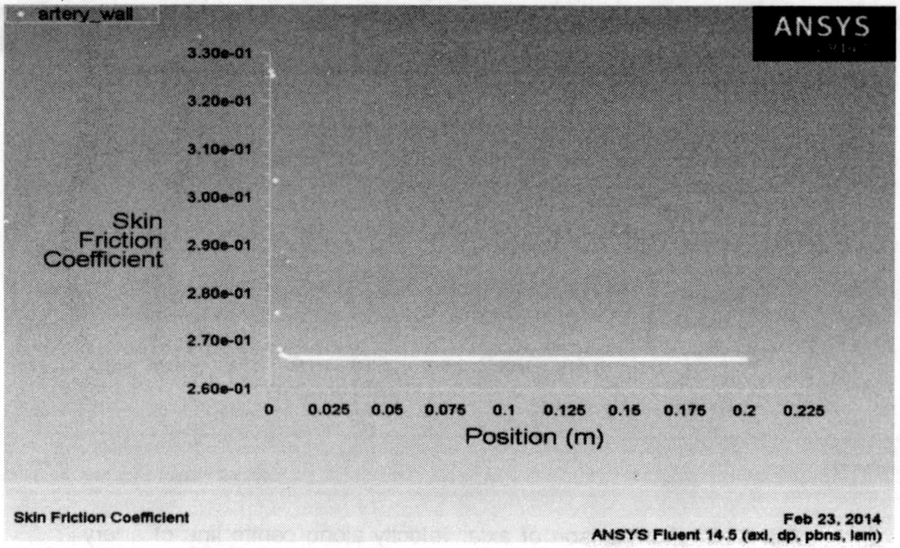

Fig. 41.8 Variation of skin friction coefficient

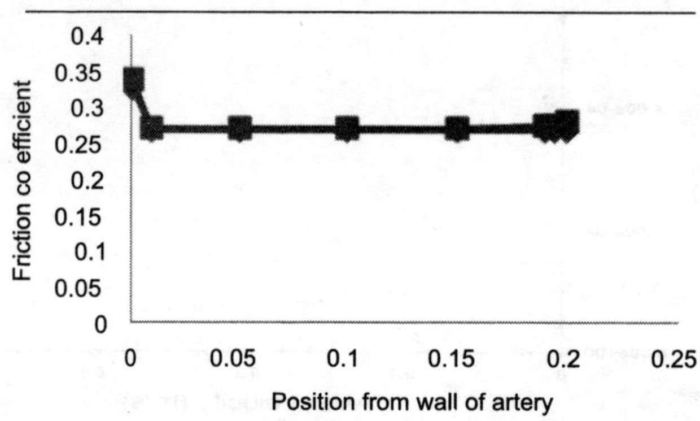

Fig. 41.9 Comparison of skin friction coefficient

41.5 CONCLUSION

For a very complex flow of Rheological fluid like blood, a far simplistic model like the Hagen-Poiseuille flow model can be applied but, restrictive in describing blood flow vessels. It can be applied to micro vessels but not to the capillaries due to low shear rate. In this present study the result obtained shows that the flow of blood in proper palmar digital artery shows an acceptable result that the Hagen-Poiseuille model is reasonable in this case. For steady flow at low Reynolds number the entrance region is approximately one tube radius. In large arteries the entrance region is relatively long. Over a significant length of the artery, velocity gradients are high near the wall. This affects the mass exchange of gas and nutrient molecules between blood and artery wall.

REFERENCES

1. Chandran, K.B, A.P. Yoganathan and S.E. Rittgers (2007). Biofluid Mechanics, The Human Circulation: Taylor & Francis.
2. Elaine N. Marieb and Katja Hoehn,Human Anatomy and Physiology, 7th ed. Pearson/Benjamin Cummings, San Francis Co, 2007, p. 725.
3. Ku, D.N. (1997). 'Blood flow in arteries' Annual Review of Fluid Mechanics 29: 399-434.
4. Majumdar, J.N. (1999). 'An Introduction to Mathematical Physiology and Biology' Cambridge University Press.
5. Pedley, T.J. (1980). 'The Fluid Mechanics of Large Blood Vessels' Cambridge University Press.
6. Elad David and Einav Shamuel (1980). 'Data book on thermo physical properties of blood' McGraw-Hill.

42

Study of the Mechanical Behaviour of Tibia Subjected to Complex Loading

A. K. Dash, D. K. Agarwalla, B. Tripathy, S. K. Bhuyan and P. S. K. Nayak

Department. of Mechanical Engineering, Institute of Technical Education & Research

Shiksha O Anusandhan University, Odisha, India

Email: das.deb1106@gmail.com

ABSTRACT

The tibia connects the knee with ankle bones. Although the tibia is a common site of stress fractures in runners, the loading of the tibia during running is not well understood. It supports most of the body weight and is an important part of both the knee joint and ankle joint. When the tibia is subjected to different complex loading conditions, the bone resista stress in the mid portion up to a certain limit. When the load exceeds the maximum limit the bone experiences a pain or a crack. In the current research computational analysis has been carried out to identify the maximum stress zone in the tibia when subjected to compressive force, bending moment and combination of both. Vibration modal analysis has also been performed to calculate the maximum stress zone and maximum deformation zone. From the analysis of the results it can be concluded that mid portion of the tibia will experience maximum stress and deformation when subjected to complex loading.

Keywords: Tibia, complex loading conditions, bone resista.

42.1 INTRODUCTION

The tibia is the more massive medial bone of the leg skeleton. It is the second longest bone of the body, only exceeded in length by the femur. It is located on the medial side of the lower leg, articulating with the FIBULA laterally, the TALUS distally, and the FEMUR proximally. Its strong shaft, consisting of thick walls of compact bone, is triangular in cross section. Tibia arises from the Latin term meaning a pipe or flute. In later Latin it directly translated as the shin-bone or the shank bone. The functions of the tibia include supporting movement of the extremities, creating insertion points for muscles, producing blood cells in bone marrow, and storing some minerals, such as calcium and phosphorus.

Tibia bone was subjected to different dynamic loads and the damages in the bone were recorded. During normal walking conditions subjected to loads how the tibia bone reacts to the situation, these all were observed. Mishra and Samanta [1] have come up with an idea that how we can study about frequency and damping coefficient of a vibrating tibia by designing tibia as a smaller conical structure. This involves mechanical properties of bonny tissues and material damping. The application of design has been provided through numerical computation. Taylor et al. [2] have investigated about the determination of orthotropic bone elastic constants using FEA and modal analysis. A cadaveric bone is CT scanner and natural frequencies were measured and a finite element mesh was created by comparing it with FE bone model. The mass was made equal, maximum orthotropic elastic constants were established giving maximum error 7.89. Finally, elastic constants of the bone were compared with those measured using ultrasound techniques. This produce the difference between maximum density and Young's modulus by >17. Hyun-Jun Kin et al. [3] have concluded that the flexible fixation method was known to provide beneficial conditions for healing bone fractures. They investigated the healing efficiency of flexible composite bone plates applied to a tibia using finite element analysis (FEA). The finite element model was constructed by the estimation of time dependent properties of living tissue using healing rates. The calculation results showed that composite bone plates gives high healing rates. Walter Rabl et al. [4] have investigated on the fracture behaviour and morphology of the human tibia. They have given the information that the load limits were 2475 N to 12206 N and direct fractures occurred in 46% of the specimens after ventral loading and in 80%, after medial or lateral loading and in 100% after dorsal loading. Thus, they have concluded that ventral medial loading produced direct wedge fractures of the messerer type and dorsal loading resulted in transverse fractures with longitudinal fissures at the impact site of loading stamp. G Lowet et al. [5] have experimented about the vibrational behaviour of a healing tibia using Finite Element Modeling. They developed two finite element models of fractured tibia with healing callus. They have calculated the static tortional stiffness as well as resonant frequencies and made shapes of the first four vibration modes from these two models. The resonant frequencies were found to increase with increasing stiffness of the callus. Mode shapes were similar to those for the intact tibia when the stiffness of the callus was 5% of the stiffness of the intact bone. If the Young's modulus of the callus was equal to or greater than 5% of the Young's modulus of the intact bone then the relation between resonant frequencies & torsional stiffness was valid. James R. Funk et al. [6] have discussed about the results on the line of action in the tibia during axial compression of the leg. Compression of the leg induces bending in the tibia, which can lead to tensile failure of the bone in the mid shaft. They have found that when the knee was fixed, the effective centre of pressure was generally isolated to a small area in the posterior portion of the medical tibia codyle. Finally, they have concluded from the orientation of the compressive load vector in the human tibia that the line of action through the tibia appeared to follow the external load path to the extent allowed by the available joint contact surfaces. Fuhao Mo et al. [7] have investigated about the injury tolerance of tibia for the car pedestrian impact and indicated that tibia fracture tolerance could vary largely due to the impact locations. Previous tibia bending tolerances for pedestrian safety only but this work aim to give result about the injury tolerance of tibia fracture with combined experimental data and numerical simulation. In the previous literature eleven new reported quasi static bending tests of tibia mid shaft and additional eleven dynamic mid shaft bending test result were taken. For the present investigation they have implemented finite element simulations with lower limb model according to three point bending and pedestrian impact conditions. The result was obtained from the simulations on the different impact locations. Amel-farzad et al. [8] have discussed about the stainless steel orthopedic implant. They examined inside a patient thigh with no issue up to two years. Different investigation were made and was observed that the plate got apparently fractured in few months along with different damages were observed. The main failure was determined to be corrosion fatigue. Sudhakar [9] have discussed about a fractured vitallium plate for determination of the basic mechanism responsible for premature failure. Several tests were carried out for the cause of the failure. From the surface analysis and information provided on the implant it was found that the mechanism of corrosion fatigue at the interface of the plate and the screw used were the main failure of the implant. Sudhakar [10] have discussed to determine the root cause for the fracture of an orthopedic implant (nail for shinbone) that failed prematurely. Detailed scanning was carried out and on basis of surface analysis as well as the information provided on the implant, the stainless steel nail failed due to ductile fracture by the presence of non-metallic inclusions. Rieger et al. [11] have discussed on loosened total hip endoprotheses using vibrational technique. Hip models were constructed and 4 different states of implant were simulated to access the vibration. The observations were recorded of different states and finally could determine the implant loosening and localization of the loosened component by vibrational technique. Asgari et al. [12] have come up with about stress, strain reasoning and finite element modeling for intact bone, stem less implant and stemmed one. Stem less implant reconstruction is compared to traditional stemmed implant and intact bone as control solution. The models were studied along two paths of medial and lateral aspect. From those it concluded that stem less implant had less large deviation from intact bone. However, the fixation of this type of implant and its effect must be carefully considered for designing final product. Luca and Marco [13] have investigated on a extensive experimental validation of the mechanical behaviour of the whole bone composite model for different loading conditions. They have measured the stiffness of the tibia with a torsional load applied along the long axis and with a bending load applied both in the latero-medial and in the antero-posterior direction. Composite synthetic models of the human tibia are now used as substitutes for cadaveric specimen. The bending stiffness of the composite tibias matched well with cadaveric specimens. Due to the structure of the reinforcement material, composite tibias were much stiffer than the cadaveric specimens. Thus, they have concluded that composite tibias are suitable to replace cadaveric specimens for certain types of test.

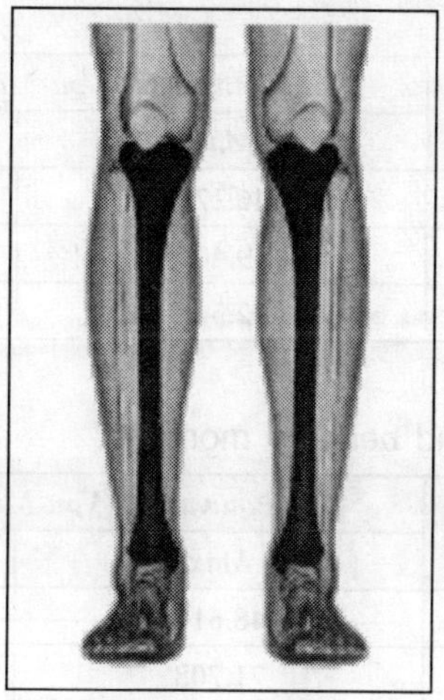

Fig. 42.1 Anatomical structure of the tibia bone

42.2 MATERIAL AND METHODS

42.2.1 Material Properties for FE Model

In this study, the material of the model was taken as isotropic properties. The isotropic properties having Young's modulus, Poisson's ratio and density were assumed as 16000 MPa, 0.3 and 1750 kg/m^3 respectively.

42.2.2 Designing of the Specimen

The designing of the tibia bone was done through the computational modeling software. The dimensions were recorded from the prosthesis of the tibia bone and modeling was done through the surface modeling process using the above software. The dimension lengths *X*, length *Y*, length *Z* are 80.596 mm, 353.35 and 64.542 respectively.

42.2.3 Finite Element Analysis

FEA consists of a computer model of a material or design that is stressed and analyzed for specific results. It is used in new product design, and existing product refinement. FEA uses a complex system of points called nodes which make a grid called a mesh Fig. 42.1. This mesh is programmed to contain the material and structural properties which define how the structure will react to certain loading conditions. Nodes are assigned at a certain density throughout the material depending on the anticipated stress levels of a particular area. Regions which will receive large amounts of stress usually have a higher node density than those which experience little or no stress. The model created in computational modeling software was imported to the computational analysis software, selecting the required material as stated above [section 42.2.1]. The compressive force, bending moment, both compressive force and bending moment and vibration modal analysis was applied to the designed model of the tibia bone with a medium meshed quality having hexahedral structure.

42.3 RESULTS

Table 42.1: Analysis of compressive force

Particulars	*Force (N)*	*Equivalent (Von-Mises) Stress (MPa)*		*Total Deformation (mm)*
		Max	*Min*	
Compression	400	20.318	0.012413	0.09375
	500	25.398	0.015516	0.1172
	600	30.478	0.018627	0.14064

Table 42.2: Analysis of bending moment

Particulars	*Force (N) (N-m)*	*Equivalent (Von-Mises) Stress (MPa)*		*Total Deformation (mm)*
		Max	*Min*	
Bending Moment	10	46.271	0.0088788	0.089731
	15	69.407	0.131318	0.1346
	20	92.543	0.017758	0.17946

Table 42.3: Analysis of compressive force and bending moment

Particulars	*Force (N) (N-m)*	*Equivalent (Von-Mises) Stress (MPa)*		*Total Deformation (mm)*
		Max	*Min*	
Compressive Force (N) and Bending Moment	(400,10)	48.611	0.020605	0.12479
	(500,15)	71.708	0.025093	0.1712
	(600,20)	94.923	0.030197	0.21861

Table 42.4: Analysis of vibration modal analysis

Vibration model analysis (Hz)	119.64	148.17	621.34	743.79	1435.9	1604
Maximum Deformation (mm)	0.55522	0.53468	0.50451	0.46159	0.98709	0.51903

Fig. 42.2 The finite element model

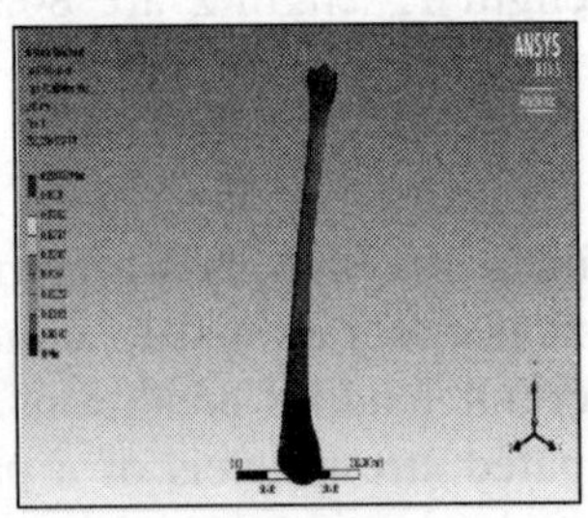

Fig. 42.3 Total deformation in compressive

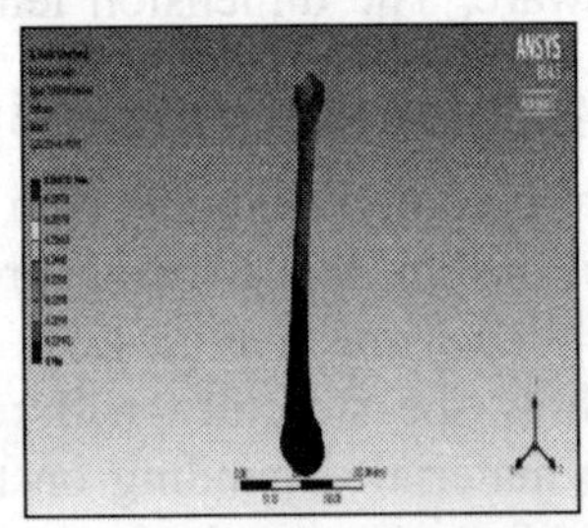

Fig. 42.4 Total deformation in bending moment 10 n-m

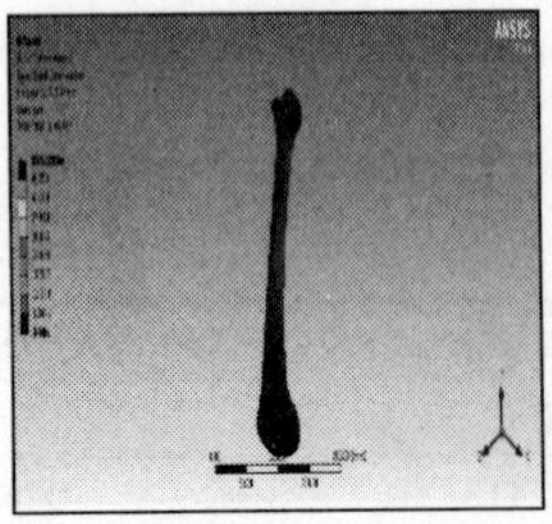

Fig. 42.5 Total deformation in vibration model

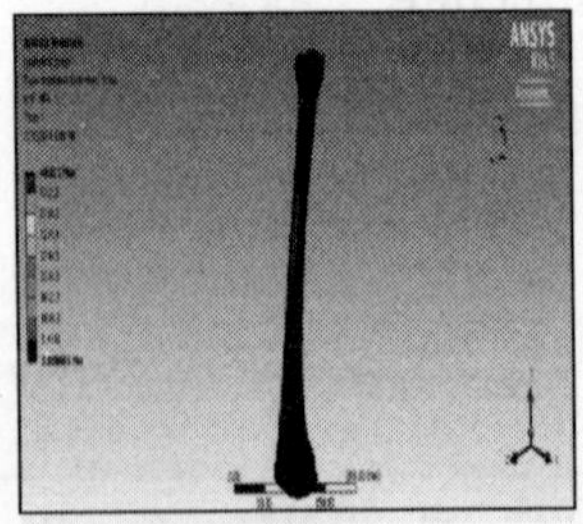

Fig. 42.6 Stress analysis in combined compressive force analysis 119.64 Hz. and bending moment (400 N, 10 N-m)

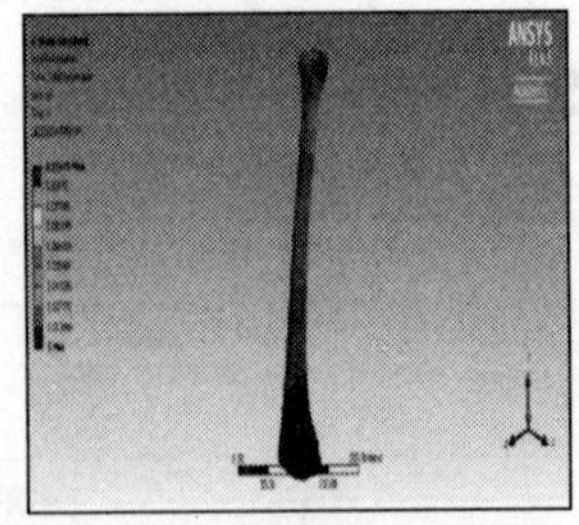

Fig. 42.7 Total deformation in combined compressive

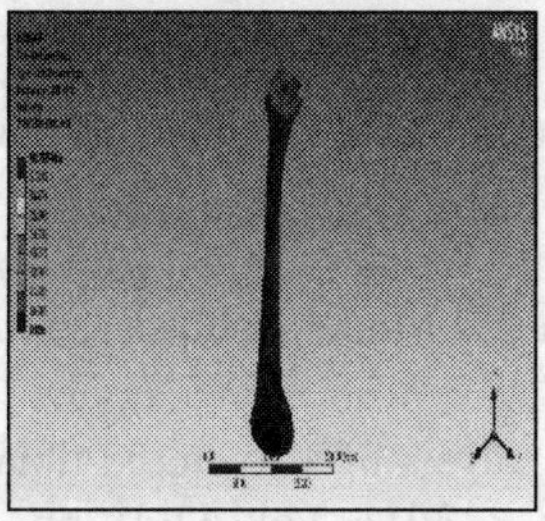

Fig. 42.8 Total deformation in vibration model

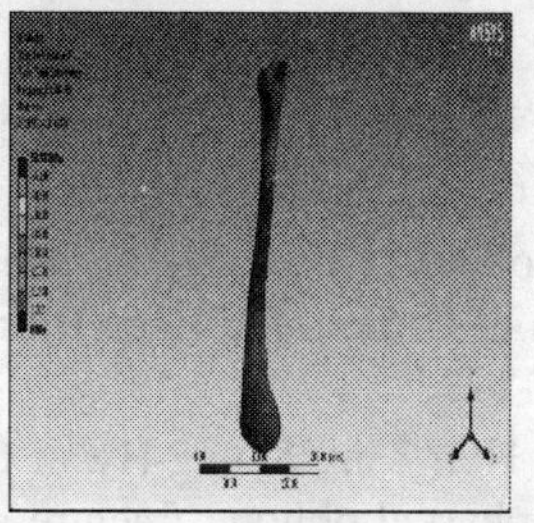

Fig. 42.9 Total deformation in and force bending moment (400 n, 10 n-m) analysis 1435.9 Hz vibration model analysis 1604 Hz

42.4 DISCUSSION

Using the computational analysis software the total deformation in compressive force, bending moment and vibration modal analysis were recorded and illustrated in the above Tables 42.1, 42.2, 42.3 and 42.4 and Figs. 42.2, 42.3, 42.4, 42.5, 42.6, 42.7, 42.8 and 42.9. Showing Analysis of compressive force ranging from 400 N to 500 N, analysis of bending moment ranging from 10 N-m to 20 N-m, analysis of combined compressive force and bending moment and vibration modal analysis with different frequencies. Also shown in Fig. 42.2 in The finite element model that total deformation in compressive load 400 N, total deformation in bending moment 10 N-m, total deformation in vibration modal analysis 119.64 Hz, stress analysis in combined compressive force and bending moment (400 N, 10 N-m), total deformation in combined compressive force and bending moment (400 N, 10 N-m), total deformation in vibration modal analysis 1435.9 Hz and total deformation in vibration modal analysis 1604 Hz. It is found that:

- When the 400 N, 500 N and 600 N of compressive force were applied to the model the maximum deformation and maximum equivalent stresswere found to be 0.14064 mm and 30.478 MPa respectively when tibia is subjected to 600 N.
- When the 10 N-m, 15 N-m and 20 N-m of bending moment were applied to the tibia bone model the maximum equivalent stress and maximum deformation were found to be 92.543 MPa and 0.17946 mm respectively when tibia is implemented to 20 N-m.
- When both compressive force and bending moment were applied to the same model of same magnitude, then the maximum deformation and maximum equivalent stress were found to be 94.923 MPa and 0.21861mmrespectively when tibia is applied to 600 N, 20 N-m .
- Finally, the vibration modal analysis was tested with different frequencies andthe maximum deformation was found to be 0.98709 mm in 1435.9 Hz from Table 42.4 but in frequency-1604 Hz the total maximum deformation found to be 0.51903 mm which is less than 1435.9 Hz.

42.5 CONCLUSION

The section describes conclusion being drawn from the above analysis.

1. During the analysis when the tibia is subjected to compressive load of increasing magnitude from 400 N to 600 N the maximum deformation and maximum stress has been measured on the top portion of the tibia.
2. In the above analysis, the bending moment of increasing magnitude is subjected to tibia ranging from 10 N-m to 20 N-m, the maximum deformation and maximum stress has been measured on the top portion of the tibia.
3. When the tibia is implemented with combined compressive force and bending moment ranging from 400 N to 600 N and 10 N-m to 20 N-m respectively the maximum stress and total deformation has been evaluated on the upper portion of the tibia.
4. During the analysis when tibia is subjected to vibration modal analysis with different frequencies the maximum deformation was measured.

From the analysis it is observed that in all cases the maximum stress in the tibia was found to be in the mid portion of the structure.

REFERENCES

1. J. C. Mishra and S. Samanta,"A Mathematical Analysis of the Vibrattion Characterstics of Human Tibia" *Comput. Math. Applic,* 1988, Vol. 16 (12), pp. 1017-1026.
2. W.R. Taylor, E. Roland, H. Ploeg, D. Hertig, R. Klabunde, M.D. Warner, M.C. Hobatho, L. Rakotomanana, S.E. Clift,"Determination of Orthotropic Bone Elastic Constants Using FEA and Modal Analysis", *Journal of Biomechanics*, 2002, Vol. 35 (6), pp. 767-773.
3. Hyun-Jun Kim, Suk-Hun Kim, Seung-Hwan Chang," Biomechanical analysis of a fractured tibia with composite bone plates according to the diaphyseal oblique fracture angle", *Composites: Part B*, 2011, Vol. 42(4), pp. 666-674.
4. Walter Rabl, Christian Haidb, Martin Krismerc ," Biomechanical properties of the human tibia: fracture behavior and morphology", *Forensic Science International*, 1996, Vol. 83(1), pp. 39-49.
5. G. Lowet, X. Dayuan and G. Van der Perre," Study of the Vibrational behaviour of a healing tibia using finite element modelling" *Journal of Biomechanics*, 1996, Vol. 29(8), pp. 1003-1010.
6. James R. Funk, Rodney W. Rudd, Jason R. Kerrigan, Jeff R. Crandall," The line of action in the tibia during axial compression of the leg", " *Journal of Biomechanics*, 2007, Vol. 40(8), pp. 2277-2282.
7. Fuhao Mo, Pierre Jean Arnoux, Jean Jaques Jureb, Catherine Masson," Injury tolerance of tibia for the car–pedestrian impact"*Accident Analysis and Prevention*, 2012, Vol. 46: pp. 18-25.
8. H. Amel-Farzad, M.T. Peivandi, S.M.R. Yusof-Sani, "In-body corrosion fatigue failure of a stainless steel orthopaedic implant with a rare collection of different damage mechanisms", *Engineering Failure Analysis*, 2007, Vol. 14(7), pp. 1205-1217.
9. K.V. Sudhakar, "Investigation of failure mechanism in vitallium 2000 implant", *Engineering Failure Analysis*, 2005, Vol. 12(2), pp. 257-262.
10. K.V. Sudhakar, "Metallurgical investigation of a failure in 316L stainless steel orthopaedic implant", *Engineering Failure Analysis*, 2005, Vol. 12(2), pp. 249-256.
11. Johannes Sebastian Rieger, Sebastian Jaeger, Christian Schuld, Jan Philippe Kretzer, Rudi Georg Bitsch, "A vibrational technique for diagnosing loosened total hip endoprostheses: An experimental sawbone study", *Medical Engineering & Physics*, 2013, Vol. 35(3), pp. 329-337.
12. S.A. Asgari, A.M.S. Hamouda, S.B. Mansor, H. Singh, E. Mahdi, R. Wirz, B. Prakash,"Finite Element Modeling of a Generic Stemless Hip Implant Design in Comparison with Conventional Hip Implants", *Finite Elements in Analysis and Design*, 2004, Vol. 40 (15), pp. 2027-2047.
13. Luca Cristofolini, Marco Viceconti," Mechanical validation of whole bone composite tibia models", *Journal of Biomechanics*, 2000, Vol 33(3), pp. 279-288.

43

A Review on Stump-Socket Interface Pressure Distribution

Chitresh Nayak, Amit Singh and Himanshu Chaudhary

Department of Mechanical Engineering, Malaviya National Institute of Technology, Jaipur, India

Email: chitresh2001@gmail.com

ABSTRACT

The concept of measuring pressure at the interface between the stump and the prosthetic socket could provide valuable information in the process of prosthetic socket fabrication, modification, and fit. Recent developments in prosthetic socket design have created renewed interest in monitoring the stress distribution at the socket-residual limb interface. X sensor, (FSA), tekscan, novel, Fuji film and many more biomedical engineering research studies involve measuring contact areas, forces, and pressures, have been commercially designed for in situ interface pressure measurement over the past decade. The measurement of low interface pressure between the skin, support surfaces and pressure garments are may be effective for subject. The socket of the prosthesis forms a critical interface between the amputee and their replacement (prosthetic) limb. Current socket design principles emphasize the importance of redistributing the normal stresses (pressures), present at the interface, to areas of the residual limb which can tolerant them.

Keywords: Prosthetic socket, pressure, stump, residual limb, sensors.

43.1 BACKGROUND

Each year, hundreds of thousands of people undergo a limb amputation. After China, India is home to the second highest number of people living with diabetes in the world (65 million, 2013) and is expected to reach over 109 million, 2035. In India and other developing countries, the amputation rate is about 45% for diabetic foot problems. In India, an estimated 50,000 amputations occur per year due to diabetes related foot problems. A vast majority (85%) of these diabetes-related amputations are preceded by foot ulcers [1]. The mortality rate is 50% within 3 years after Above Knee Amputation and 5 years after Below Knee Amputation.

Throughout our daily lives our bodies or some parts of them are in contact with a supporting structure. Whether it is the back of a sleeping man on a bed or the toe of a pirouetting ballerina on the stage, the soft tissues beneath the skeleton experience forces due to the ever present gravity. Sustaining the forces is something we are normally not conscious of until pain reminds us to relieve these tissues of stress. Measurements of forces at the body surface are very problematical, and nowhere more so than when the tissue is enclosed as in a shoe or prosthesis. In these situations knowledge of the dynamics of the forces occurring at the body/support interface is a major component in designing efficient and safe load transfer from skeleton to the support surface.

Studies quantifying the stresses at the prosthetic interface have been conducted for over 40 years. Computational modeling using finite element analysis is a common and simple method for predicting stresses and strains at the interface in that estimates of pressures at any region of the limb can be made, and elaborate experimental setups are not required. However, while convenient, computational models do not account for all potential factors at the interface, specifically the biological tissue reactions to loading [2]. Thus, computational modeling may simply serve well as a means of verification of experimental data.

Table 43.1: Limb amputation in diabetes: Differences-western vs Indian experience

Limb Amputation	*Western*	*Indian*
Mean age at amputation	75 years	61.25 years
Morality at 2 years	50%	14.28%
Contralateral limb amputation at 2 years	30-50%	8.92%
AK-BK amputation	1 : 2	1 : 6
Indication neuropathic	10%	76.78%

Experimental methods involving the use of sensors to quantify the stresses at the prosthetic interface are more desirable in that measurements obtained from actual trials incorporate all factors contributing to discomfort at the interface. However, devising an accurate measurement system remains a challenge. Many factors, such as the size of the sensor, the type of force transducer, and differences in pressure dynamics between types of movement, often compromise the accuracy of stress measurements [2]. For example, too large of a sensor provides an inaccurate average of too large an area of the limb, while sensors that are too small produce undesirable edge effects [2]. Thus, careful consideration of these factors is necessary in experimental procedures.

43.2 INTRODUCTION

Prosthesis is often used to restore appearance and functional mobility to individuals following lower limb amputation. Coupling between the residual limb and the prosthesis is typically achieved by a socket, which surrounds the residual limb, and to which the remaining components of the prosthesis are attached. In current prosthetic practice, the socket is a critical element of a successful prosthesis, as it is the sole means of load transfer between the prosthesis and the residual limb.

The soft tissues of the residual limb are not well-suited for load bearing, and the load tolerance of the residual limb depends on the biological and physiological structure of these tissues. Whenever tissues are exposed to excessive or prolonged loading, there is a risk of tissue trauma due to local circulatory deficits and/or abrasion. Thus, for persons with lower limb amputation, where large loads must be borne by the soft tissues, great care is taken in the design of the prosthetic socket to minimize discomfort and possible tissue trauma. Nevertheless, although a review paper on interface stress has relevance to many areas, this paper will focus on commercially available interface pressure measuring systems and Interface Stress Measurements Methods.

It is this desire for quantitative description of prosthetic interface stress distribution that has motivated the experimental and numerical investigations summarized in this paper. The objective of this paper is to summaries a commercially interface pressure measuring systems, to discuss the working principle, sensor size, range, quoted, and to speculate on work made possible through advances in pressure/force transducer.

The pressures at the contact interface between the skin of the residual limb and the internal surface of the socket are vital indicators for proper fit, comfort, and load bearing capability of the prosthetic limb. Since socket fit is an important characteristic for patient satisfaction [3], studies have evaluated socket fit by measuring pressures in lower limb prosthetic sockets [4]. Achieving a good limb-socket interface by ensuring an optimal distribution of the interface loads is critical to successful prosthetic fitting and rehabilitation of lower limb amputees.

The accuracy of interface pressure measurements relies upon sensor characteristics and measurement procedures which may give rise to a variety of errors [5]. Specifications for the "ideal" sensor to measure the inter-face stresses associated with the skin of the human body were proposed by Ferguson-Pell [6] and Grant [7]. Sensors should be:

(a) Small and thin, e.g., 1 mm in thickness, 10 mm in diameter, and highly flexible;
(b) Have a continuous output;
(c) Be able to measure shear as well as normal forces;
(d) Be free from error of measurement on curved surfaces and from the effects of temperature and moisture;
(e) Low cost.

43.3 TRANSTIBIAL PROSTHESES

Over 50,000 transtibial, or below-knee, amputations are performed annually in the India. While prosthetic limbs help restore mobility and normal appearance to the amputee, the skin and underlying tissue at the residual limb do not react well in load-bearing activities. The prosthetic socket plays an essential role in assisting with load transfer and providing stability at the prosthetic interface. Thus, it is important that the socket is designed to optimally distribute the load to the residual limb to maximize comfort to the patient and ensure proper movement. A typical prosthetic contains a socket, foot, and liner as shown in Fig. 43.1.

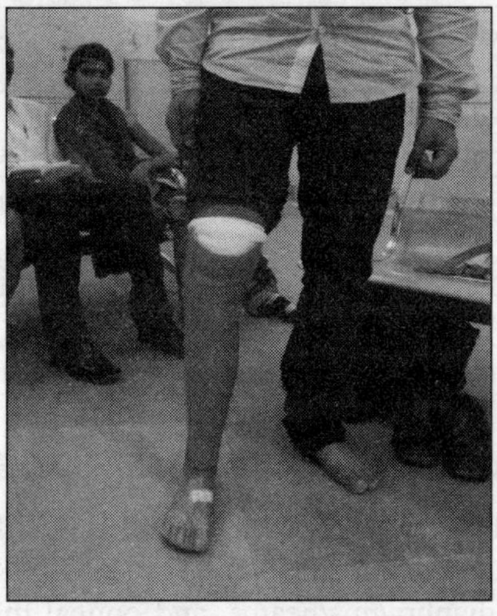

Fig. 43.1 A prosthetic system consists of a foot, liner, and socket

43.4 FUNCTIONAL OUTCOMES AND QUALITY OF LIFE

A major problem for people with transtibial prostheses is failure to accept and use their prosthesis because of discomfort experienced at the prosthetic socket interface [8-10]. Consequently, the prosthesis supplied can have a major impact on the patient's quality of life. To give the best prospect of continued use, the prosthesis must be comfortable and functional for the user. Although evidence-based practice is paramount to the provision of optimum care and enhancing quality of life for the amputee, a limited amount of evidence is available to prescribing clinicians that provides a clear understanding of what constitutes a "good socket fit."

Advanced age, the presence of many illnesses, and above knee amputation are known to retard functional recovery after amputation [11-13].

In an earlier study of male veterans (median age 62), with amputations due to either critical limb ischemia or diabetes who underwent rehabilitation at 17 months follow-up, 29% were able to ambulate outdoors, 25% ambulated indoors only, and 46% were non-ambulatory. Only 42% were using prosthetic limbs. Patients with major limb loss following rehabilitation often remain independent despite infrequent prosthetic usage. This highlights the importance of training for independence at the wheelchair level particularly among those who for various reasons are not considered to be prosthetic candidates [6]. In an exploratory analysis, subjective quality of life was rated high among amputees following rehabilitation for prosthetic fitting except in the area of physical functioning which continued to be seen as problematic through the rehabilitative and 3-month follow-up periods. Satisfaction with the prosthesis was strongly related to pain and body image [14].

43.5 PRESSURE MEASUREMENTS

The pressure distribution at the interface between the residual limb and the prosthetic socket is a critical consideration in socket design and fit. Pressure measurements within prosthetic sockets have been conducted for about 50 years. The information obtained has been used either to increase the understanding of socket load transfer, to assess the socket design, or to validate the computational modeling. Interfacial pressure measurements require a proper measurement technique, including the use of transducers, their placement at the prosthetic interface, as well as the associated data acquisition and conditioning approach. An ideal system should be able to continually monitor real interfacial stresses; both pressure and shear, without significant interference to the original interface conditions. A variety of transducers have been developed for socket pressure measurements. They can be classified, based on their operation principle, sensor size, range, number of sensors, output device

Table 43.2: Comparison of commercially available interface pressure measuring systems

	X Sensor	*FSA*	*Tekscan*	*Novel*	*Fuji Film*
Principle of operation	Capacitive	Piezoresistive	Resistive	Capacitive	Piezoresistive
Sensor size (mm)	Seat 12 × 12	Bead 19 × 50 Foot 9 × 16	Foot 5 × 5	2.7 × 2.7	430 × 430 (maximum sensing area)
Range (mmHg)	0 0-220	Bed 0-220 Foot 0-1500	Seat 0-220 Foot 0-7500	Bed 0-200 Foot 0-1800	Bed 0-220 Foot 0- 2375
No. of sensor	Seat 2304 Bed 10,240	Up to 32 × 32	Over 2000	Up to 2304	Up to 64
Quoted accuracy	10% or 10 mmHg	10%	Clinically ± 3% Laboratory ± 1%	Typically ± 5 %	±10%
Output device	Computer	Computer	Computer	Computer	Computer

43.6 INTERFACE STRESS MEASUREMENTS METHODS

From earlier this century a variety of methods have been employed to qualify or quantify the residual limb/prosthetic socket interface stresses. These have included dye loaded stump socks, fluid or pneumatic membranous cells, variable capacitance stress sensors, strain gauged diaphragms/beams, and optical methods. The devices employed have been designed to be inserted between the limb and socket of a standard prosthesis, recessed into the wall of an adapted socket, or mounted on the supports of isolated sections of adapted prostheses or integral to special sockets. The following briefly reviews the previous methods employed to measure interface stresses.

Similar devices and other methods have been used to investigate stresses at other body/support interfaces. Most notably: with reference to stresses under the normal and pathological foot [9, 10] and for evaluating seating and bedsore problem [15]. Significantly, all but two of these devices [16, 17] have been designed to investigate only the normal stress at the body/support interface.

The techniques for placement of transducers at the residual limb and socket interface can be divided into two categories. They are either inserted between the skin and the liner/socket, or positioned within or through the socket and/or the liner. Only thin sensors, such as the diaphragm deflection strain-gauge sensors [18-21], the fluid-filled transducers [22], the pneumatic transducers [23, 24], and the printed circuit sheet sensors [25-30], are suitable for insertion between the skin and socket. Mounting is relatively easy and it is not necessary to damage the prosthesis. However, for many of these sensors, interference is unavoidable from their protrusions into the socket volume, because of their finite thickness [31, 32]. The diameter of each sensing element is another important consideration. Too big a sensing element can measure only an average pressure over the area, while too small a sensor may be affected by its edge effects, especially for a stiff sensor. Positioning the transducers within or through the socket with the sensing surface being flush with the skin would make the thickness of the transducer become less critical.

Table 43.3: Sites of interface stress measurement

Name	*Description of location*
Anterior medial proximal	At the level of the tibial tubercle, medial side
Anterior medial midlimb-proximal	Mid-proximal, anterior tibial border, medial side
Anterior medial midlimb-distal	Mid-distal, anterior tibial border, medial side
Anterior medial distal	Distal residual limb, anterior tibial border, medial side
Anterior lateral proximal	At the level of the tibial tubercle, lateral side
Anterior lateral midlimb-proximal	Mid-proximal, anterior tibial border, lateral side

Contd...

Anterior lateral midlimb-distal	Mid-distal, anterior tibial border, lateral side
Anterior lateral distal	Distal residual limb, anterior tibial border, lateral side
Lateral proximal	Lateral distal residual limb
Lateral distal	Lateral femoral epicondyle
Posterior proximal	Center of the popliteal fossa, on posterior midline
Posterior midlimb	Mid-calf, on posterior midline
Posterior distal	Distal calf, on posterior midline

43.7 CONCLUSION

The methodologies used in this study successfully measured pressure magnitude and distribution patterns at the limb-socket interface of the upper limb prosthesis using hundreds of pressure sensors when tested on human transradial, transhumeral, and shoulder amputees. Preliminary results indicate a complex but clinically relevant relationship between interface pressures, anatomic features of the residual limb, arm position, carrying weight, and discomfort levels due to interface pressure.

REFERENCES

1. Shobha Shukla, Managing Editor of Citizen News Service (CNS), New Interventions To Treat Diabetic Foot Infections, accessed June 2013.
2. Mak, A., Zhang, M., & Boone, D. (2001). State-of-the-art research in lower-limb prosthetic biomechanics-socket interface: A Review. Journal of Rehabilitation Research and Development, 38(2), 161-174.
3. Dillingham, T.R., Pezzin, l.E., Mackenzie, E.J. & Burgess, A.R. (2001). Use and satisfaction with prosthetic devices among persons with trauma-related amputations: a long-term outcome study. Am J. Phys Med Rehabil, 80, 563-71.
4. Portnoy, S., Yarnitzky, G., Yizhar, Z., Kristal, A., Oppenheim, U., Siev- Ner, I. & Gefen, A. (2007). Real-time patient-specific finite element analysis of internal stresses in the soft tissues of a residual limb: a new tool for prosthetic fitting. Ann Biomed Eng, 35, 120-35.
5. Beebe D. Accuracy of pressure and shear measurement. In: Webster J.G., Editor. Prevention of Pressure Sores. Bristol: Adams Hilger, 1991: 155-74.
6. Ferguson-Pell M.W. Design criteria for the measurement of pressure at body/support interfaces. Eng Med 1980; 9(4):209–14.
7. Grant I.J. Interface pressure measurement between a patient and a support surface. Care Br J Rehab Tissue Viability 1985; 1(1): 7-9
8. Mccoll I. Department of Health and Social Security (DHSS). Review of artificial limb and appliance centre services: the report of an independent working party. London (UK): DHSS; 1986.
9. Chadderton H.C. Prostheses, pain and sequelae of amputation, as seen by the amputee. Prosthet Orthot Int 1978; 2(1): 12–14. [pmid: 724411].
10. Nielsen C.C., Psonak R.A., K.A. Lter TL. Factors affecting the use of prosthetic services. J. Prosthet Orthot. 1989; 1(4): 242-49.
11. Mackenzie E.J., Bosse M.J., Castillo R.C., Smith D.G., Webb L.X., Kellam J.F., et al., 2004. Functional outcomes following trauma-related lower-extremity amputation. Journal of Bone and Joint Surgery AM 86-a(8): 1636-1645.
12. Nehler M.R., Coll J.R., Hiatt W.R., Regensteiner J.G., Schnickel G.T., Klenke W.A., et al. 2003. Functional outcome in a contemporary series of major lower extremity amputations. Journal of Vascular Surgery 38(1): 7-14.
13. Weiss G.N., Gorton T.A., Read R.C., Neal L.A. 1990. Outcomes of lower extremity amputations. Journal of the American Geriatrics Society 38(8) 877-883.
14. Zidarov D., Swaine B, Gauthier-Gagnon C. 2009. Quality of life of persons with lower-limb amputation during rehabilitation and at 3-month follow-up. Archives of physical medicine and rehabilitation 90(4): 634-645.
15. Ferguson-Pell. M.W, Bell. F., Evans. J.H., (1976), interface pressure sensors: existing devices their suitability and limitations, in bedsore biomechanics, Kenedi. R.M, Cowden. J.M., Scales. J.T (Eds), Macmillan Press, London.
16. Appoldt. F.A., Bennett. L., Contini. R., (1970), tangential pressure measurements in above-knee sockets, Bull. Prosthet. Res., 10 (13), 70-86
17. Tappin. J.W, Pollard J., Beckett. E.A., (1980), Method of measuring 'shearing' forces on the sole of the foot, Clin. Phys. Physiol. Meas., 1(1), 83-85

18. Rae J.W., Cockrell J.L. Interface pressure and stress distribution in prosthetic fitting. Bull Prosth Res 1971; 10-16: 64-111.
19. Pearson J.R., Holmgren G., March L, Oberg K. Pressure in critical regions of the below-knee patellar-tendon-bearing prosthesis. Bull Prosth Res 1973; 10-19: 52-76.
20. Burgess E.M., Moor A.J. A study of interface pressure in the below-knee prosthesis (physiological suspension: an interim report). Bull Prosthet Res 1977; Fall: 58-70.
21. Winaski D, Pearson Jr. Least-squares matrix correlations between stump stresses and prosthesis loads for below-knee amputees. J. Biomech Eng 1987; 109: 238-46.
22. Van Pijkeren T.V., Naeff M., Kwee H.H. A new method for measurement of normal pressure between amputation residual limb and socket. Bull Prosthet Res 1980; 17(1): 31-4.
23. Mueller S.J., Hettinger T.L. Measuring the pressure distribution in the socket of the prostheses. Orthpadie-technik Heft 1954; 9: 222-5.
24. Kroupskop T.A., Brown J., Goode B., Winningham D. Interface pressures in above-knee sockets. Arch Phys Med Rehabil 1987; 68: 713-4.
25. Engsberg J.R., Springer M.J.N., Harder JA. Quantifying interface pressure in below-knee-amputee socket. J. Assoc Children's Prosthetic-orthotic Clinics 1992; 27(3): 81-8.
26. Houston V.L., Mason C.P., Lablanc K.P., Beatties A.C., Garbarini M.A., Lorenze E.J. Preliminary results with the dva-tekscan bk prosthetic socket/residual limb stress measurement system. Proceedings of 20th Annual Meeting American Academy of Orthotist and Prosthetist, Nashville, TN, 1994; p. 8-9.
27. Covery P., Buis AWP. Conventional patellar-tendon-bearing (PTB) socket/stump interface dynamic pressure distribution recorded during the prosthetic stance phase of gait of a trans-tibial amputee. Prosthet Orthot Int 1998; 22: 193-8.
28. Covery P., Buis AWP. Socket/stump interface dynamic pressure distribution recorded during the prosthetic stance phase of gait of a trans-tibial amputee wearing a hydrocast socket. Prosthet Orthot Int 1999; 23: 107-12.
29. Zhang M, Turner-Smith AR, Tanner A, Roberts V.C. Frictional action at residual limb/prosthetic socket interface. Med Eng Phys 1996; 18(3): 207-14.
30. Zhang M, Mak Aft, Chung AKI. Dynamic pressure maps over areas of prosthetic sockets. Proceedings of 9th World Congress of ISOP, Amsterdam, 1998; p. 709-11.
31. Appoldt F.A., Bennett L., A preliminary report on dynamic socket pressure. Bull Prosthet Res 1967; Fall: 20-55.
32. Appoldt F.A., Bennett L., Contini R. Socket pressure as a function of pressure transducer protrusion. Bull Prosthet Res 1969; Spring: 236-49.

44

Comparison Between Bowling Workload, Strength Ratios and Quality of Life in Fast Bowlers with and without Shoulder Injuries

Ajit Dabholkar, Minal Dalvi and Sujata Yardi

Department of Physiotherapy

Padmashri Dr. D.Y. Patil University, Nerul, Navi Mumbai–400706

Email: ajitdabholkar78@yahoo.co.in

ABSTRACT

Bowlers exceeding bowling workload guidelines experienced the greatest number of recurring injuries, aggravated by continued bowling though injured. The bowlers are continually repeating an explosive action that places great strain on shoulder musculature influencing their performance and quality of life. The Strength of Shoulder Internal & External rotators as well as Scapular Protractors & Retractors is not statistically significant when compared between the fast bowlers with and without shoulder injuries. Right External rotator/Internal rotator ratio was found to be statistically significant between the fast bowlers with and without shoulder injuries. The Bowling Workload between the fast bowlers with and without shoulder injuries was not found statistically significant. Quality of life was found to be affected in the fast bowlers with shoulder injuries.

Keywords: Bowling workload, strength ratios, scapula stability, musculoskeletal system.

44.1 INTRODUCTION

Cricket bowlers like other throwing sports involve repeated forceful ballistic arm actions which will put a great deal of eccentric load on the shoulder rotator cuff muscles predisposing them to injuries (**Stretch, 2001**). Scapular instability is found in as many as 68% of rotator cuff problems and 100% glenohumeral problems. As with first class fast bowlers increased bowling frequency is significantly associated with increased injury risk. A repeatedly faulty technique is more likely to play a role in injury incidence when coupled with an increased workload and poor physical preparation.[1] Due to this their performance gets affected which in turns affected their quality of life. Thus, the aim of this study is to compare the strength ratios, bowling workload and quality of life between the fast bowlers with and without shoulder injuries.

44.2 MATERIALS

Baseline Push- Pull Dynamometer®, Universal Goniometer, Soft Measuring tape, Logbook for documenting bowling workload, Western Ontario Rotator Cuff Index, Weighing Scale

44.2.1 Study Design

Cross-sectional study

44.2.2 Study Setting

Padmashree Dr. D.Y. Patil Sports Academy.

44.2.3 Inclusion Criteria

Fast bowlers with rotator cuff injuries. For asymptomatic group, fast bowlers without any shoulder injuries.

44.2.4 Exclusion Criteria

Spin bowlers, presence of fracture around shoulder joint, presence of rotator cuff tears, presence of subluxation/dislocation of glenohumeral joint.

44.2.5 Procedure

Approval for the study was gained from Institutional Ethics Committee. All subjects were explained the purpose of the study and consent was taken from all of them prior assessment. In total 30 subjects were recruited; 15 in each group.

44.2.6 Outcome Measures

Muscle strength of external rotators, internal rotators, protractors and retractors[7] were taken with BASELINE Push-Pull dynamometer®.The 3 recordings of the strength were assessed with the rest period of 30 secs between each testing and average was taken for the documentation. The strength ratio of the External/Internal rotators and Protractors/Retractors was calculated with the resultant average values of the respective muscles.

Bowling workload: The bowling workload was documented with respect to the following parameters: Frequency (Number of days of net practice in a week), Duration (Number of minutes or hours of net practice in a week), Number of balls bowled in a week.

Assessment of modified lateral scapular slide test (MLSST): The patient sits or stands with the arm resting at the side. The examiner measures the distance in 3 positions (Kibler et al.)

Assessment of quality of life: The quality of life is documented with the help of Western Ontario Rotator Cuff Index amongst fast bowlers with shoulder injuries.

44.3 RESULTS AND OBSERVATIONS

Table 44.1: Strength of internal rotator

	Mean	*Standard Deviation*	*Confidence Interval*		*p value*
			Upper 95%	*Lower 95%*	
FB RT	10.533	3.502	12.473	8.594	
FBSI RT	9.867	3.067	11.566	8.168	0.45*
FBS LT	10.267	3.453	12.179	8.354	
FBSI LT	10	2.507	11.389	8.611	0.9826*

Table 44.2: Strength of external rotator

	Mean	*Standard Deviation*	*Confidence Interval*		*p value*
			Upper 95%	*Lower 95%*	
FB RT	9.867	2.774	11.403	8.33	
FBSI RT	8.267	2.815	9.826	6.708	0.1333*
FBS LT	9.733	3.195	11.503	7.964	
FBSI LT	10.8	3.278	12.615	8.985	0.284*

Table 44.3: Ratio of external & internal rotator

	Mean	*Standard Deviation*	*Confidence Interval*		*p value*
			Upper 95%	*Lower 95%*	
FB RT	0.9657	0.1974	1.075	0.8563	
FBSI RT	0.8313	0.1741	0.9277	0.7349	0.0198**
FBS LT	0.9707	0.2066	1.085	0.8562	
FBSI LT	1.095	0.2678	1.244	0.9471	0.1565*

Table 44.4: Strength of protractors

	Mean	*Standard Deviation*	*Confidence Interval*		*p value*
			Upper 95%	*Lower 95%*	
FB RT	13.333	0.7475	14.937	11.73	
FBSI RT	13.333	0.7475	14.937	11.73	0.9662*
FBS LT	13.333	3.436	15.237	11.43	
FBSI LT	13.467	2.973	15.113	11.82	0.9829*

Table 44.5: Strength of retractors

	Mean	*Standard Deviation*	*Confidence Interval*		*p value*
			Upper 95%	*Lower 95%*	
FB RT	12.067	2.314	13.348	10.785	
FBSI RT	13.733	2.604	15.176	12.291	0.1338*
FBS LT	12.667	2.469	14.034	11.299	
FBSI LT	14	3.024	15.675	12.325	0.2607*

Table 44.6: Strength ratio of protractor and retractor

	Mean	*Standard Deviation*	*Confidence Interval*		*p value*
			Upper 95%	*Lower 95%*	
FB RT	1.139	0.39	1.355	0.9233	
FBSI RT	0.9807	0.2396	1.113	0.848	0.2101*
FBS LT	1.101	0.3761	1.309	0.8927	
FBSI LT	0.961	0.1448	1.041	0.8808	0.2447*

Table 44.7: Bowling workload

	Mean	*Standard Deviation*	*Confidence Interval*		*p value*
			Upper 95%	*Lower 95%*	
FB	350.13	129.43	421.82	278.45	
FBSI	367.8	152.49	452.26	283.34	0.5612*

Table 44.8: Quality of life (WORCI)

	Sum of Ranks	*Mean of Ranks*	*P value*
Physical Symptoms	557.00	37.133	
Sports/Recreation	726.50	48.433	
Work	610.00	40.667	0.0124**
Lifestyle	323.50	21.567	
Emotions	633.00	42.200	

Table 44.9: Scapular stability

	Mean	*Standard Deviation*	*Confidence Interval*		*p value*
			Upper 95%	*Lower 95%*	
SUP ANG FB	0.28	0.2981	0.4451	0.1149	
SUP ANG FBSI	0.4667	0.4419	0.7114	0.2219	0.2431*
INF ANG FB	0.52	0.4296	0.7579	0.2821	
INF ANG FBSI	0.6667	0.5563	0.9748	0.3585	0.5145*
ARM 90 FB	0.68	0.8866	1.171	0.189	
ARM 90 FBSI	1.053	0.7425	1.465	0.6421	0.0994*
ARM 120 FB	0.4	0.5412	0.6997	0.1003	
ARM 120 FBSI	0.92	0.763	1.342	0.4976	0.0286**

44.4 DISCUSSION

In our study according to Tables 44.1 & 44.2 we found that there is decreased in mean strengths of external rotators than the internal rotators between the fast bowlers with and without shoulder injuries, but there is no statistical significant difference between the strengths of internal rotators and external rotators between the fast bowlers with and without shoulder injuries. This may be attributed to imbalance between agonist and antagonist muscles, though both the groups had posterior soft tissue tightness which restricts the internal rotation of the shoulder. According to Table 44.3, there was statistical significant difference found in the Right External rotators/Internal rotators ratio found in our study. Balasubramanian Sundaram et al. found that there is significant differences observed only for right handed fast bowlers (n = 31) and spin bowlers (n = 28) in both External Rotation Passive Range of Motion and Internal Rotation Passive Range of Motion between their dominant and non-dominant Glenohumeral Joints may be because only a few left handed fast bowlers (n = 4) and spin bowlers (n = 3) participated in this study[2] as most fast bowlers are right handed dominant which is similar to our study. According to Tables 44.4 and 44.5, we found that there is decreased in the mean of protractors (serratus anterior) strength and retractors (rhomboids) strength in injured group compared to uninjured group but there is no statistical significant difference in protractors (serratus anterior) and retractors (rhomboids) strength between the 2 groups. The mean of protractors (serratus anterior)/ etractors (rhomboids) ratio is decreased in injured compared to uninjured group but has no statistical significant difference was found in their ratios as shown in Table 44.6. There was no imbalance between the scapular muscles found. According to Table 44.7, we found that there is no statistical significant difference in the bowling workload between the fast bowlers with & without shoulder injuries.

P L Gregory et al. found that there was no difference between the workloads of injured and uninjured English fast bowlers. This may be simply because the injured bowlers could not bowl for a significant period of time after injury.[3] He concluded that the expected increased incidence of bowling injury in young fast bowlers who bowl most was not observed, although more than 12% exceeded the recommended limit.[4,]

Roxanne Davies et al. found that observed that the most serious injuries occurred in bowlers with the highest bowling workloads. These bowlers compounded matters by continuing to bowl while injured, thus causing further injury. According to Table 44.9 an assessment of scapular stabilization with Kibler's LSST test was done where we found that there is only statistical significance difference between fast bowlers with & without shoulder injuries with arm abducted at 120° this is due to overactivity of lower trapezius and the weakness of serratus anterior which pulls the scapula away from the spine as compared to normal side leading to asymmetric scapular movement[6,8,9]. According to Table 44.8, we found that quality of life of fast bowlers with shoulder injuries was affected which was assessed by WORC index, in which, the domains affected were in the following order with sports and recreation being the most affected, followed by emotional factors, work, physical symptoms and the least affected being lifestyle.

44.5 CONCLUSION

Right External rotator/Internal rotator ratio was found to be statistically significant between the fast bowlers with and without shoulder injuries. Quality of life was found to be affected in the fast bowlers with shoulder injuries. Scapula dyskinesia was identified in bowlers with rotator cuff injury.

44.6 CLINICAL IMPLICATIONS

The scapular muscles strength and strength ratios should be assessed in bowlers with rotator cuff injuries and the strengthening for the same should be done. Bowling workload should be documentated to prevent injury. Quality of life should be assessed in fast bowlers with rotator cuff injuries.

REFERENCES

1. Roxanne Davies The Nature and incidence of fast-bowling injuries at an elite, junior level and associated risk factors. South American Journal of Sports Medicine, 2008; Vol. 20 No. 4.
2. Balasubramanian Sundaram et al. Glenohumeral Rotational Range of Motion Differences between Fast Bowlers and Spin Bowlers in Elite Cricketers. International Journal of Sports Physical Therapy, 2012 December; 7: 576-585.
3. R.J. Dennis, C.F. Finch. et al. Is bowling workload a risk factor for injury to australian junior cricket fast bowlers [British Journal of Sports Medicine 2005; 39:843-846]
4. P.L. Gregory et al. Is risk of fast bowling injury in cricketers greatest in those who bowl most? A cohort of young English fast bowlers. British Journal of Sports Medicine 2004; 38:125-128.
5. R.J. Dennis, C.F. Finch, et al. Is bowling workload a risk factor for injury to Australian junior cricket fast bowlers? [British Journal of Sports Medicine 2005; 39:843-8
6. Kibler WB. Role of the scapula in the overhead throwing motion. Contemp Orthopedic. 1991; 22:525-532.
7. Nichole Turner, Kristen Ferguson, et al. Establishing Normative Data on Scapulothoracic Musculature Using Handheld Dynamometry [Journal of Sport Rehabilitation, 2009; 18, 502-520 © 2009 Human Kinetics, Inc.]
8. Ann M. Cools et al. Isokinetic Scapular Muscle Performance in Overhead Athletes with and without Impingement Symptoms [Journal of Athletic Training. 2005 Apr-Jun; 40(2): 104-110.]
9. Kevin E. Wilk, et al. Shoulder Injuries in the Overhead Athlete [Journal of Orthopaedic & Sports Physical Therapy February 2009; Vol. 39 No. 2]
10. Rene Cailliet, Sports Injuries to the shoulder, F.A. Davis Company, Shoulder Pain, Edition 3, New Delhi, Jitendar P Vij, 2005, pp. 152-154
11. David J. Magee, Shoulder, Orthopedic Physical Assessment, 5th edition, New Delhi Elsevier, 2008, pp. 302
12. Maria Zuluaga. The Shoulder Complex, Pearson Professional Pvt Ltd., Sports Physiotherapy (Applied Science & Practice), Pam Jonas, Melbourne, Judy Waters, 1995, p. 361-364.

45

Study of Stress Distribution and Deformation of Tooth and Dental Implant Under Compressive Loading: A 3D Finite Element Analysis

A. K. Dash, A. Satpathy, D. K. Agarwalla, S. Satapathy, P. Sahoo, Surabhi and A. Nayak

Centre of Biomechanical Science, Institute of Technical Education and Research, Bhubaneswar

Dept. of Periodontics & Oral Implantology, Institute of Dental Sciences, Bhubaneswar

Email: dips50k@gmail.com

ABSTRACT

Dental implants have become a significant aspect of tooth replacement in prosthodontic treatment. Despite of high success rates, complications and failures still occur. The long-term success of an implant depends upon the stress distribution on the dental implant and its capacity to withstand the induced stresses due to the loads. This study evaluates the stress distribution on the dental implant and mandibular first molar tooth. The purpose of this study has been to develop a three-dimensional model of osseointegrated dental prosthesis and natural molar tooth and to analyse the stress distribution and the total deformation under various loading conditions. The basic purpose of the bioengineering which analyse biomechanical principles in vitro studies has been to extrapolate the findings relevant to the risk factors instead of experiencing them empirically in clinical application. The 3D modelling of tooth and the dental implant has been formulated in solid works software. Total equivalent stress and total deformation has been found by meshing in Finite Element Analysis (FEA) software. All materials used in the models have been considered to be isotropic, homogeneous and linearly elastic. Different loading conditions of 300 N, 400 N and 500 N have been considered to be applied on the tooth for the present study. Results from the analysis indicated that the total equivalent stresses have been found to be maximum in the apical areas of the dental implant and minimum on the lateral surfaces, on the other hand, stresses have been found to be maximum on the inner surface of the tooth. Similar has been the case found for deformation. From the analysis, it has been found that, the stress distribution on the tooth has been largely different from the stress distribution in the dental implant and has been more even and symmetric in case of a dental implant.

Keywords: Stress, dental implant, tooth, finite element analysis, compressive load, deformation.

45.1 INTRODUCTION

Placement of endosseous dental implants has become an increasingly common practice. Ideally implants should be placed parallel to each other, parallel to adjacent teeth and be vertically aligned with axial forces. The relation between implant design and load distribution at implant bone interface is an important issue to understand. Many factors affect load transfer at the bone implant interface such as the type of loading, material properties of the implant and prosthesis, implant geometry, surface structure, implant design quality and quantity of surrounding bone and nature of bone implant interface. Endosseous implants are widely used in current clinical practice. As the name suggests, they are surgically inserted directly into the jaw bone. They are of several types including screw threaded, cylinder/smooth and bladed. Dental implants have now considered the first choice for replacement of missing teeth. In spite of high success rates, complications and failure still occur. Failures of dental implant are due to infection, loose implants, improper placement, insufficient jaw bone analysis, poor quality implant and underlying health issues. Kayabas et al. [1] investigated static, dynamic, and fatigue behavior of the dental implants and found that implants can be designed and studied in a simulated environment before it is placed on the patient. This will save time for the design and prevent any permanent damage caused by improper placement of implant. Baggi et al. [2] evaluated stress based performances of five commercially available dental implants and demonstrated that endosseous implants are strongly affected by a number of shape parameters as well as by anatomy and mechanical properties of the site of placement. Ali et al. [3] compared the mechanical behavior of a dental prosthesis under two loading types (dynamic and dynamic with overload) in order to clarify causality of overload as related to implant failure and bone loss. Quaresma et al. [4] evaluated the influence of 2 commercially available dental implant systems on stress distribution in the prosthesis, abutment, implant and supporting alveolar bone and concluded that conical implant connected to a solid, internal, conical abutment furnishes lower stresses on the alveolar bone and prosthesis and greater stresses on the abutment. Merdji et al. [5] evaluated stress distributions in the bone and the dental prosthesis under combined dynamic load and developed a 3-D model of an osseointegrated molar dental prosthesis. Tian et al. [6] studied to determine whether angled abutments could result in decreased stress on surrounding bone of single-unit dental implants and found that angled abutments resulted in decreases stresses when implants have been not placed in ideal axis position. Baris et al. [7] evaluated the effects of different inter-implant distances on stress distribution in the bone around the endosseous titanium implants and found high compressive stress at lingual surface of the cervical region. Ruthera et al. [8] investigated the transmission of an internally generated combined acoustic and vibration signal to detect implant loosening and found a significant correlation between the measured frequency and pull out strength for different level of fixation. Wang et al. [9] have used fringe technique to map the in-plane strain distribution in slices from human tooth crowns under compression and found that strain inside enamel is much less than in dentin. Perez et al. [10] aimed to apply a probabilistic computational approach to predict the fatigue life of three different commercial dental implants and concluded the highest failure probability has been located at the upper screw threads. Lee et al. [11] investigated the effects of thread shape and taper on initial stability of the implant bone structure and found that the site with the greatest stress in the surrounding jaw bone has been consistently concentrated at the root radii. Sinan et al. [12] studied the effects of implant diameter, collar length and slope, body length on the interfacial stress distribution and found that among the six contour parameters, the slope and the length of the implant collar, and the implant diameter influences the interfacial stress levels the most. Djebbar et al. [13] used the finite element method to compute the distribution of stresses in dental prosthesis. The induced stress on the dental prosthesis elements is high and can be fatal for the structure and the patient. Genna et al. [14] numerically analyzed using finite element analysis some issues related to the low-cycle fatigue failure of dental implant subjected to cyclic and transverse loads and concluded that screw loosen after a number of loading cycles. Liang Kong et al. [15] numerically analyzed finite element analysis effects of the implant thread pitch on the maximum Von-Misses stresses in jaw bones and implant-abutment complex and found that thread pitch exceeding 0.8 mm is optimal biomechanical property and over-sized pitch should be avoided. This paper presents a study of stress distribution and deformation of dental implant subjected to compressive loading in a simulated environment.

45.2 METHODOLOGY

Finite Element Analysis (FEA) of a computer generated model of a material or design that is stressed and analysed for specific results is used in new product design and existing product refinement. In case of structural failure, FEA may be used to help determine the design modifications to meet the new conditions. There are generally two types of analysis that are used in the industry, i.e. 2-D modelling and 3-D modelling. Nodes are assigned at a certain density throughout the material depending on the anticipated stress levels of a particular area regions which will receive large amounts of stress usually have a higher node density than those which experience little or no stress. A three-dimensional finite element model of a commercially available threaded endosseous implant made of commercially pure titanium of 4.2 diameter and 10 mm length made by Osstem and a mandibular first molar was created based on the physical properties. A porcelain jacket crown was considered along with the dental implant for simulation purposes.

Procedural sequence:

(1) Profile projection of dental implant was done.

(2) Data obtained from the projection was used in solid modelling for the implant with porcelain crown using SOLIDWORKS SOFTWARE.

(3) Modelling of mandibular first molar was done based of the measurements made on an extracted mandibular first molar using SOLIDWORKS SOFTWARE.
(4) Three-dimensional solid models were then imported to ANSYS SOFTWARE version 13.
(5) A mesh model was generated for tooth and dental implant consisting of nodes and elements having values 31397 and 19160 respectively. Boundary conditions were set.
(6) Root of the tooth and cylindrical implant has been constrained to be fixed. Other surfaces were treated as free surfaces.
(7) Models were subjected to compressive loads of 300 N, 400 N and 500 N applied on the crown of the tooth and dental implant at 90 degrees directed apically.
(8) Values of total deformation and equivalent stresses were calculated on the various parts of the models.

45.2.1 Material Properties

The material properties have been specified using the mechanical properties of Young's modulus, Poisson's ratio of the dental implant and the dentin. All materials have been assumed to be isotropic and homogeneous.

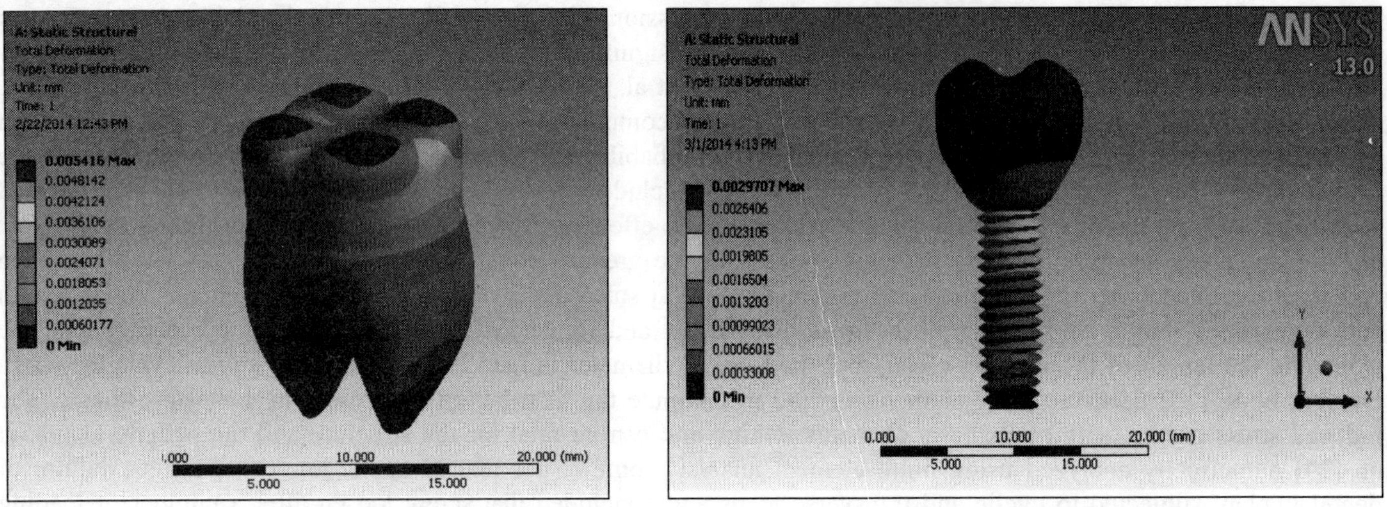

Fig. 45.1 (a) Three-dimensional model of tooth (b) Three-dimensional model dental implant

Table 45.1: Number of nodes and elements in analysed models

Parameters	*No. of nodes*	*No. of elements*
Model of tooth	36042	21180
Model of implant	31397	19160

Table 45.2: Properties of materials used in stress analysis

Material	*Density (ρ) (kg/m³)*	*Young's Modulus (E) (MPa)*	*Poisson's Ratio (v)*
Titanium	4500	110	0.35
Porcelain	2400	80	0.3
Tooth	1750	18600	0.31

Table 45.3: Equivalent stress and total deformation values of the tooth subjected to various loading conditions

Load (in N)	*Equivalent Stress (σ) (MPa)*		*Deformation (Δ) (mm)*	
	Maximum Value	*Minimum Value*	*Maximum Value*	*Minimum Value*
300	41.565	1.576×10^{-3}	3.24×10^{-3}	0
400	55.42	2.102×10^{-3}	4.332×10^{-3}	0
500	69.275	2.620×10^{-3}	5.416×10^{-3}	0

Table 45.4: Equivalent stress and total deformation values of the dental implant subjected to various loading conditions

Load (N)	*Equivalent Stress (σ) (MPa)*		*Deformation (Δ) (mm)*	
	Maximum Value	*Minimum Value*	*Maximum Value*	*Minimum Value*
300	103.54	3.814×10^{-2}	4.131×10^{-3}	0
400	134.01	4.186×10^{-2}	5.941×10^{-3}	0
500	159.23	6.427×10^{-2}	6.817×10^{-3}	0

45.3 RESULTS AND DISCUSSION

Finite Element Method in the mechanical analysis of dental implant is important in terms of accuracy and reliability as the degree of measure of such parameters is very high. In this biomechanical analysis, FEM can be used to simulate and analyse stress distribution dividing the three-dimensional problematic geometry into a collection of very small elements. In such a model using FEM analysis, it consists of elements, nodes and predefined boundary conditions. Finite Element Analysis is always a better option for examination rather than Analytical method because; division into a number of elements and nodes can be effectively achieved in FEM, which is not possible in Analytical model. Results obtained through FEA can be very close to the actual situation observed in clinical studies. Such has been observed from the results of the study of the stress analysis. Table 45.1 depicts the stress and total deformation values of the dental implant subjected to various loading conditions. Tooth when subjected to compressive loads of 300 N, 400 N and 500 N, the maximum stress induced under the loading conditions of 500 N is 69.275 MPa. The minimum equivalent stresses for the same cases have been found to be 1.576×10^{-3} at 300 N. Maximum total deformation values have been found to be 5.416×10^{-3} at 500 N. The minimum total deformation in all cases has been found to be zero.

Table 45.2 depicts stress and total deformation values of the dental implant subjected to various loading conditions. Maximum stress induced under the loading conditions of 500 N has been 159.23 MPa. The minimum equivalent stresses for the same cases have been found to be 3.814×10^{-2} at 300 N. For the same, the maximum total deformation values have been found to be 6.817×10^{-3}. The minimum total deformation in all cases has been found to be zero.

Figures 45.1 (a) and (b) represent the solid models of the tooth and dental implant. Figures 45.2(a) and (b) indicate the mesh model of the tooth and dental implant respectively. In Fig. 45.2(a) the number of nodes has been assigned to be 36042 and the number of elements to be 21180. In Fig. 45.2(b) the number of nodes are 31397 and the number of elements 19160. The machine used has been tetrahedral mesh.

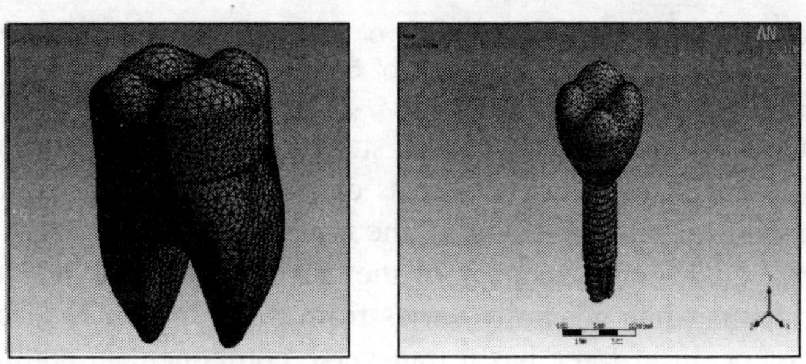

(a) (b)

Fig. 45.2 (a) Finite element model of tooth (b) Finite element model of tooth with implant

All the analysis data has been for the subjected load of 500 N. As represented in the Fig. 45.3(a), the values of equivalent stresses for a tooth vary from 7.69 MPa at the top of the tooth to 0.002 – 30.79 MPa at the root. The values at the outer and the inner surface are 0.0026 and 50 MPa respectively. But, as represented in the Fig. 45.3(b), when the dental implant has been used the values at the top, root, inner and outer surfaces are 8, 47.816, 35.872 and 0.039 MPa respectively.

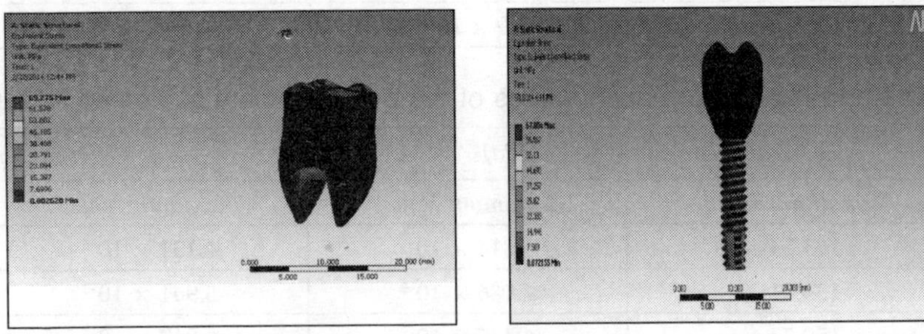

(a) (b)

Fig. 45.3 (a) Equivalent stress distribution in subjected to a load of 500 N (b) Equivalent stress distribution in a tooth dental implant subjected to a load of 500 N

Under similar conditions of the load of 500 N, as represented in Fig. 45.4(a), for tooth, the value of total deformation from root, top, outer to inner surface of the tooth varies from 0, 0.00541, 0.0006-0.0054 to 0.0018 mm respectively. But in as in the manuscript Fig. 45.4(b), for the dental implant model, the values at the top, root, inner and outer surface varied from 0.00415, 0, 0.002 to 0.001–0.004 respectively.

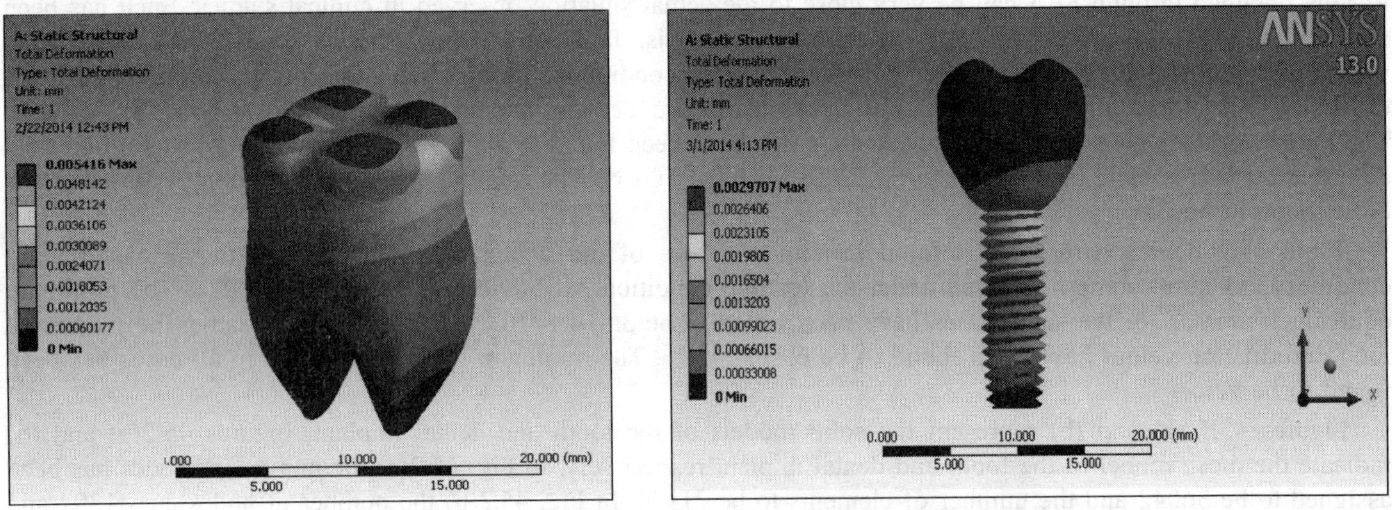

(a) (b)

Fig. 45.4 (a) Total deformation of tooth subjected to a load of 500 N (b) Total deformation of dental implant subjected to a load of 500 N

FEA has been effectively used in this case for proper study of the stress distribution generated in the implant subjected to various values of compressive load. The stress distribution around osseointegrated dental implants are affected by a number of biomechanical factors including the type of loading, material properties of the implant and the prosthesis, implant geometry, quality and quantity of the surrounding bone and the nature of the bone-implant interface. The biting force of a human bite generally varies from 80 N to 200 N Stress has been generated through the vertical force only. In reality, occlusal force has a transverse component in addition to a vertical component. It has been apparent that the presented model has been only an approximation of the clinical situation. A quantitative and qualitative analysis have been performed, based on a progressive visual colour scale, predefined by the software used, ranging from dark blue to red. But from a factual point of view the findings should be substantiated by clinical research. As we have seen from the diagrams and figures the amount of stress that has been induced on the tooth has been found to be higher than in case when the implant has been used, much of the stresses have been transferred to the implants we can see from the cross-sectional view of the diagram indicated by Fig. 45.5(a).

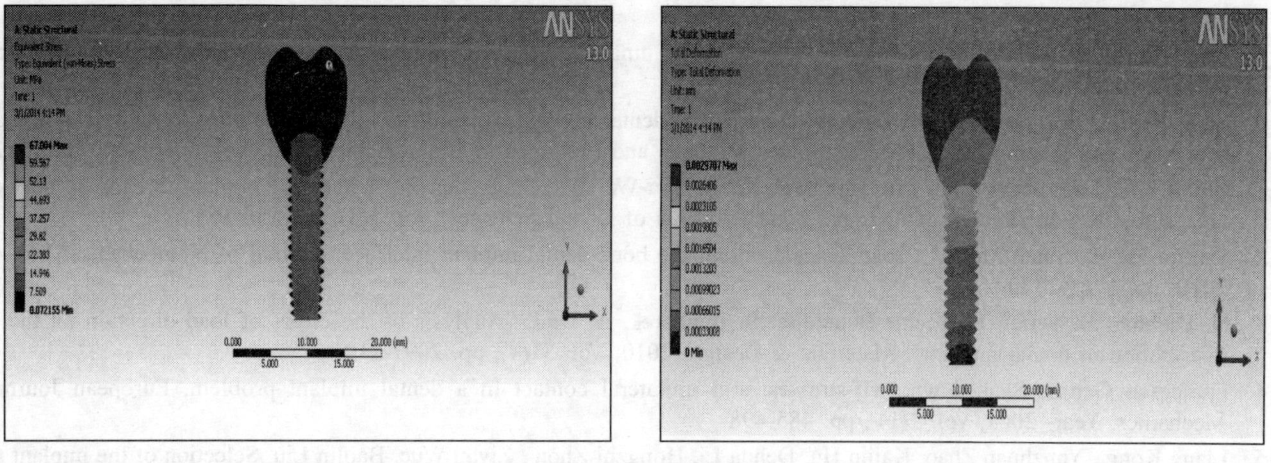

(a) (b)

Fig. 45.5 (a) Cross-sectional view of the dental implant showing the stress distribution (b) Cross-sectional of the deformation in the dental implant

45.4 CONCLUSION

For nearly all dental prosthesis systems, the major objective is to increase the life expectancy of the dental implant. As an overall conclusion, from the three-dimensional finite element numerical analysis, the following assumptions are obtained from this study by the biomechanical consideration. It is visible that the mechanical stress has been propagating in the inner part of the tooth and decrease in magnitude towards the outer region. The mechanical stress reaches the highest value at the junction of the tooth and the supporting bone, which is then transferred to the implant when the dental implant has been used. The maximum total equivalent stress occurs inside the dental implant itself. In this study various compressive loading conditions of 300 N, 400 N and 500 N have been established. For the loading condition stated, the maximum stress values had not reached the yield strength of the tooth or the titanium dental implant. It has been seen that the implant is durable in all conditions of compressive loading. Implant has been found to be durable under the given conditions before being tested in the clinical situation. Thus, further time waste in any misinterpretation of the analysed dental implant data and stress distribution has been minimised.

REFERENCES

1. Og˘uz Kayabas, Emir Yu¨zbasıog˘lu , Fehmi Erzincanl, "Static, dynamic and fatigue behaviors of dental implant using finite element method", Advances in Engineering Software, 2006, Vol. 37(10), pp. 649-658.
2. Luigi Baggi, Ilaria Cappelloni, Franco Maceri, Giuseppe Vairo, "Stress-based performance evaluation of osseointegrated dental implants by finite-element simulation", Simulation Modelling Practice and Theory, 2008, Vol. 16(8), pp. 971-987.
3. Benaissa Ali , El Bahri Ould Chikh , Hadj Miloud Meddah , Ali Merdji , Bel abbes Bachir Bouiadjra, "Effects of overloading in mastication on the mechanical behavior of dental implants" , Materials & Design, 2013, Vol. 47, pp. 210-217.
4. Sergio E. T. Quaresma, MSc; Patricia R. Cury, PhD; Wilson R. Sendyk, PhD; Claudio Sendyk, PhD; A Finite Element Analysis of two different Dental Implant: stress distribution in the prosthesis, abutment, implant, and supporting bone; Journal of Oral Implantology; Vol. 34(1); 2008; pp. 1-6.
5. A. Merdji , B. Bachir Bouiadjra , B. Ould Chikh , R. Mootanah , L. Aminallah , B. Serier , I. M. Muslih, "Stress distribution in dental prosthesis under an occlusal combined dynamic loading " Materials & Design, 2012, Vol. 36, pp. 705-713.
6. Kebin Tian, Jiang Chen, Lilin Han, Jin Yang, Wenxiu Huang, Dong Wu, "Angled abutments result in increased or decreased stress on surrounding bone of single-unit dental implants: A finite element analysis" Medical Engineering & Physics, 2012, Vol. 34(10), pp. 1526-1531.

7. Barıs, S,ims,ek , Erkan Erkmen, Dervis, Yilmaz, Atılım Eser; Effects of different inter-implant distances on the stress distribution around endosseous implants in posterior mandible: A 3D finite element analysis; Medical Engineering and Physics; Vol. 28(3); 2006; pp. 199-213.
8. Cathérine Ruthera, Hannes Nierath, Hartmut Ewald, James L. Cunningham,Wolfram Mittelmeier, Rainer Bader, Daniel Kluess, Investigation of an acoustic-mechanical method to detect implant loosening, Journal of Medical Engineering, Year- 2013, Vol. 19(3), pp. 1-7.
9. R.Z. Wang, S. Weiner,"Strainstructure relations in human teeth using Moir'e fringes", Journal of Biomechanics 31, pp. 135-141.
10. M.A. Pérez, "Life prediction of different commercial dental implants as influence by uncertainties in their fatigue material properties and loading conditions" Computer Methods and Programs in Biomedicine, 2012, Vol. 108(3), pp. 1277-1286.
11. Chia-Ching Lee, Shang-Chih Lin, Ming-Jen Kang, Shu-Wei Wu, Ping-Yuen Fu,"Effects of implant threads on the contact area and stress distribution of marginal bone", Journal of Dental Sciences, Vol. 5(3), pp. 156-165.
12. Samira Faegh, Sinan Müftü*; Load Transfer along the bone-dental implant interface; Journal of biomechanics; Vol. 43(9); 2010; pp. 1761-1770.
13. N. Djebbar, B. Serier, B. Bachir Bouiadjra, S. Benbarek, A. Drai, "Analysis of the effect of load direction on the stress distribution in dental implant", Materials & Design, 2010, Vol. 31(4), pp. 2097-2101.
14. Francesco Genna, Shakedown, self-stresses, and unilateral contact In a dental implant problem, European Journals of Mechanics, Year- 2004, Vol. 1(1) , pp. 485-498.
15. Liang Kong , Yunzhuan Zhao, Kaijin Hu, Dehua Li, Hongzhi Zhou , Ziyan Wuc, Baolin Liu, Selection of the implant thread pitch for optimal biomechanical properties: A three-dimensional finite element analysis, Journal of Advances in Engineering Software, 2008, Vol. 40(7), pp. 474-478.
16. Grzegorz Milewski, Numerical and experimental analysis of effort of human tooth hard tissues in terms of proper occlusal loading, Acta of Bioengineering and Biomechanics, 2005, Vol. 7(1), pp. 47-59.

46

'pTwisle': A Pneumatic Twisting Actuator

Sadanand Shriniwas Deshpande and G. Agarwal

Department of Mechanical Engineering, Govt. Polytechnic, India

Dept. of Mechanical Engineering

Malviya National Institute of Technology, Jaipur, India

Email: ssa.deshpande@gmail.com

ABSTRACT

Scientists and researchers faced challenges in using robots in non-industrial environments, especially in medicinal applications. These robots were indeed a requirement for surgical operations, physiotherapy, exoskeletons, prosthetic treatments, etc. The traditional actuators in robots based on hydraulics, pneumatics and electrical engineering were found very 'stiff' and difficult to manipulate and control. 'Soft' and 'Compliant' actuators, Pneumatic Artificial Muscles (PAMs) were tried in between 1950 and 1980 and were found difficult to control and manipulate. In the mean time, precise electronic sensors and control systems were evolved. These muscles are made of rubber tube and are enclosed in nylon braided sleeve. The two ends of tubular structure are closed by end-caps. One of the caps is fitted with air inlet port to admit air into it and other is sealed, so that no escape of air is there. There are many variants of PAMs and are widely being preferred in rehabilitation devices and many robotic applications. Predominant features of this actuator are compliant in nature and higher power to weight ratio. The actuators are generally applied in agonistic and antagonistic as observed in skeleton human muscles. However, the other forms of motions in skeleton have been less attempted by research community, one of it is, 'Pronation and Supination of human arm'. After trying various ideas for twisting motion of PAM, a mechanism was designed and a prototype was developed. In order to know the behaviour of the prototype, its static modelling was done. To verify the model the static characterisation was verified by experimentation.

Keywords: Pneumatic artificial muscle, twisting actuators, compliant actuators, soft actuators, static modelling, pneumatic Twisle.

46.1 INTRODUCTION

In our society the care takers for senior citizens are becoming less and expensive. In future, robots will take over these agile caring tasks. In conventional robots, the actuators are stiff and make it unsafe for human handling operations. For such serving applications for senior citizens and physically challenged persons, the stiffness is to be controlled precisely without failure in all working conditions or we have to design perfect compliant actuators.

The Pneumatic Artificial Muscles (PAMs) are latest compliant (spongy) actuators. These are in tubular structure. These are made of rubber tube and are enclosed in nylon braided sleeve. The two ends of tubular structure are fitted with end-caps. One of the caps is fitted with air inlet port to admit air into it and other is sealed, so that no escape of air is there. When high pressure air is allowed to enter into the port, it exerts pressure on the walls of rubber tube. This bulges the tube in radial direction. In order to compensate increased volume due to rise in diameter, the tube shortens in length. The material being soft, it causes the shortening effect. The nylon braided sleeve structure covering the rubber tube actually takes major share to cause contractile force in axial direction. At the beginning of the contraction, the maximum force is generated, while it drops to zero as the stroke reaches to its maximum contraction position and an equilibrium state is reached. The following features have been highlighted in literature.

- These are light in weight actuators.
- The force-velocity response curve is similar to skeleton muscles (Klute et al., 2002).
- These actuators possess high output to weight ratio than other conventional actuators.
- The rubber tube being an elastomeric material and under pneumatic actuation, they are compliant within nature which ensures safety to humans in the vicinity.
- Gives non stick-slip operation as there are no sliding parts.
- These PAMs can generate up to 4000 N contractile force.
- Generally, the maximum contraction at 6 bar is 30% of its nominal length (Chou and Hannaford, 1994, 1996, Saga et al., 2007).

Comprehensive review of the PAMs can be found in articles by Tondu and Lopez, 2000 and Daerden 2002.

These actuators are also called Pneumatic Rubber Muscle Actuators, Pneumatic Artificial Rubber Muscles (PARMs), Pneumatic Rubber Artificial Muscles (pRAMs), and Fluidic Muscles (Festo 2009), Air Muscles (Shadow 2009). Though there are slight variations in construction, the functioning remains same.

While designing of humanoids, scientists have achieved the motions of arms by replacing the skeletons by suitable mechanisms. Most of the skeleton muscles are arranged in agonistic-antagonistic set as it has been observed in the motions of forearms, fingers, palms, etc. One of the promising motions in the human arm is 'pronation and supination'. In this article, the parametric results of newly developed twisting motion actuator are presented. The application for Indian Patent is pending.

46.2 PRESENT ARTICLES FROM LITERATURE

A report (Schindling, 2006) describes "manhood invented the wheel and further the rotary motor which produces powerful and fast motion but also stiffness, high energy costs and a lack in flexibility, hence lays great constraints in robotic potentials. Strangely rotary joint doesn't exist in nature. Evolution over millions of years decided to base biological locomotion on linear motors-muscles. Biological muscles primary function is to contract and exert force controlled by electric stimulation. Further, they operate as brakes, shock absorbers, springs and struts." The natural muscles contract and do not twist. This message inspired to create a new actuator-Twisting Muscle.

The type of actuators are categorised by actuation energy–electric, hydraulic, pneumatic. The class of pneumatic actuators can produce linear motion (by expansion or by contraction), bending or rotary. Industrial robots demand rigid actuators in order to maintain accuracy level, but in service robots, the stiffness is most unwanted aspect. The stiff robot actuators may harm the delicate objects and living beings. In order to have better application in exoskeletons, deployment of compliant or flexible nature of the actuators, many authors have presented their arts for rotary motions apart from linear motions as well. Wearable elbow exoskeleton was presented by Maruta et al., 2002 and Zhang et al., 2008, in which the PAM was mounted over grooved pulley. The flexible pneumatic actuators were demonstrated like elephant trunks, in which the bending of actuator takes place (Trivedi et al., 2008). Similarly, 'air tube actuators' for strengthening of muscles are illustrated in Tsukagoshi et al., 2002. Development of rotary actuators (Yamamoto et al., 2002) using pressure cuffs of a sphygmomanometer sandwiched between plates. The thin plates spread like fan by supplying air to these cuffs. These actuators were successfully mounted on exoskeleton for nurse to lift patients. It produced 30 Nm torque. Another design of rotary actuator with folded lamellas to articulate elbow motion (Kargov et al., 2007) was put forth. This actuator produce 2700 rotary motion and exert 20 Nm torque is a success. A pneumatic pleated REC (Rotary Expandable Chamber) actuator (Ivlev 2009) is a soft one and generates 50 Nm torque at 3 bar for 600. A comprehensive literature review on flexible actuators can be found in De Greef et al., 2008. It hardly mentions twisting type actuators. The above references are mostly related to rotary actuators. However, the actuators for twisting are very limited. In Hirai et al., 2000, simulation studies about the cylindrical actuators with actual actuation of two prototypes were presented. Though the actuator can twist at 900, it has some translation motion. For rotary twisting type actuators, the simulation studies were presented also by Yee and Coghill, 2002. The art, 'inner skeleton robot for forearm motion assist' was presented by Kiguchi and Kundu, 2006. The 'switched reluctance actuators' were used in the assist device, as those can be fitted into body. The motions pronation and supination were achieved in this construction. The upper arm rehabilitation exoskeleton using PAMs in 'agonistic-antagonistic action' was demonstrated by Tsgarakis and Caldwell, 2003. It has seven degrees of freedom with training control architecture. In most of the humanoids, the forearm pronation-supination motions are prefixed at wrist level. The accessibility of many points in space by the arms being achieved using all rotational joints. Very less attempts have been made towards compliant actuators for twisting type of motion. Though electric motors can easily produce the required twisting motion, those are stiff and difficult control for particular rotation. No doubt there have been innumerable attempts for developing bending, axially contracting, expanding and rotary compliant actuators, till date some compliant actuator exists for 'axial cum twisting' action but no compliant actuator exists in pure 'twisting' action. In this study a PAM is used as a basic force generator to imitate torque cum twisting motion in pronation-supination.

46.3 PNEUMATIC TWISLE -'PTWISLE'– A NEW PROTOTYPE – DRAWINGS CONSTRUCTION AND WORKING

Please visit link **ipindia.nic.in >status of patents> type application number 1519/DEL/2010** to see the detailed drawings, construction and working of the 'Pneumatic Twisle'. Figure 46.8 shows the device. This work presents current generation twisting muscle actuated by compressed air. The device consists of a PAM and two pipes. The two pipes are coaxial over a PAM. These can slide and rotate over each other. When ends move towards each other, the pipes do slide over each other. The smaller diameter pipe has a helical slot. The length of this pipe is more than other pipe. The bigger diameter pipe is fitted with a roller-screw assembly, which moves inside the helical slot. Whenever the contraction force is generated inside the muscle, the ends try to come nearer. This action is constrained by helical slot and roller mechanism and cause twisting. When pressure is released, the air is forced out of muscle due to spring-back action of muscle tube, thus no bias torque is needed to bring back the 'Pneumatic Twisle' to its original position. The two guide pipes can be observed are roto-scopic in nature, i.e. able to 'slide and rotate' over each other. The relative motion between the end-caps and pipes is constrained fully.

46.4 MODEL FOR POWER 'PTWISLE'

In this article, to estimate the power of muscle, we have taken biomechanical approach and named the actuator as 'Pneumatic Twisle'. The suggested model as the characteristic model for Pneumatic Twisle is evolved in present study as:

$$P_m = \{d\,(F_m + a)\,(V_m + b) + (T_m{}^{*}\,\omega + e)\} \qquad ...(1)$$

where, F_m is the contraction force developed by Pneumatic Twisle,
a is a constant used for biological muscle having force units,
V_m is the contraction velocity of biological muscle,
b is constant used for biological muscle having unit of velocity,
d is a constant-dimensionless,
T_m is the torque developed in Pneumatic Twisle,
ω is the angular velocity,
e is a variable having units of power and
P_m is the power of muscle activity.

The constants depend on the physical dimensions of the PAM and arrangement to convert force into torque. Henceforth, this model will be stated as **'SSDGA'** model. This model is based on the classical model given by A.V. Hill in 1938 under Isotonic Contraction. In isotonic contraction, testing of muscles are carried out under constant load and allowed to contract. The model is stated in the form

$$P_m = (F_m + a)\,(V_m + b) \qquad ...(2)$$

As per this study, constant d is 0.85, i.e., 0.15% power of linear contracting muscle can be converted into torque. If we use rubber hoses instead of plastic parts, less torque will be generated and can be used in low torque applications in addition more compliance can be achieved which is deserved in service robots.

46.5 EXPERIMENTATION WITH 'PTWISLE'

In this article pTwisle and Pneumatic Twisle are synonymous. An experimental setup was prepared in 1/10th cost of sophisticated setup. High pressure air admission to actuator is controlled by 3/2 D.C. solenoid valve ('3'- number of ports, '2' number of states 'ON or OFF', 'D' Direction, 'C' Control). The valve is controlled through circuit board having 5 channel 8-bit A/D converter microcontroller (PIC16F73), 4 relay control with serial port. The range of A/D converter is 0-5 V. The voltage regulator chip is 7805 which regulates voltage in ±0.05 V. The time duration of switching ON state of the DC valve and sensor data can be accessed through HyperTerminal software. The sampling time is 0.01 s. After 0.5 s, 3/2 DC solenoid valve is switched off. The deflation time is monitored for 0.5 s.

The angle sensor employed is a resistance pot of 'ELcon' make B type 100 K model 503. A 'Sharp' make contactless distance sensor GP2D12 was used to measure the contraction. It is used with 'Kodak' make R-27 card (part number KK9760). A 'Suraksha' make torque sensor is used for torque measurement. Its range is 0.0 to 10.0 Nm in output voltage range is 0-5 V. The length of the muscle is 350 mm and outer diameter of the muscle tube is 20 mm. The Pneumatic Twisle is tested under no-load conditions and in open-loop mode.

The pressure is adjusted by using pressure regulator for four pressure levels at 2, 3, 4 & 5 bar and the actuator is allowed to twist. The coaxially fitted angle sensor is twisted. The readings are acquired for actuation time of 1.0 s. After releasing the air, the Pneumatic Twisle takes its original position. The response time and data in both inflation and deflation are recorded. The high pressure air is allowed to enter the muscle as soon as the 3/2 DC solenoid valve is switched ON by using HyperTerminal software. The sampling time is 0.01 s. It is observed that the inflation takes place within 0.1 s in open-loop mode without axial or torsional load condition.

In the first set of experimentation, contraction and twist angle are measured. Thereafter, torque sensor is attached replacing the angle sensor. In the second set of experimentation, torque and contraction values are obtained.

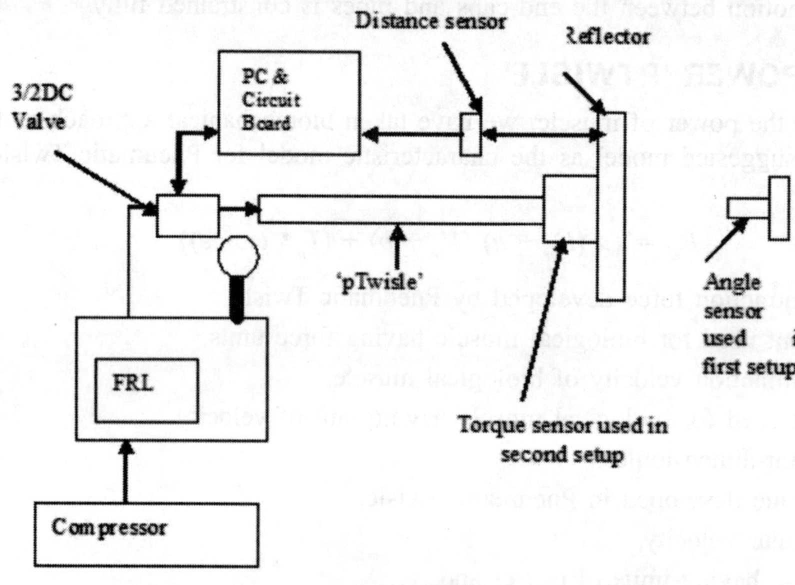

Fig. 46.1 Experimental setup

46.6 RESULTS OF EXPERIMENTS

The experimentation with *Pneumatic Twisle* was carried out at air pressure of 2, 3, 4 and 5 bar. The air is admitted in open loop mode. The results of the experiments are presented in ensuing section. In the first set of experimentation, contraction and twist angle are measured. Thereafter, torque sensor is attached removing angle sensor. In the second set of experimentation, contraction and torque values are obtained. The contraction of the device is measured twice, with angle sensor and with torque sensor attached.

In Fig. 46.2, twisting angle β vs. time response is shown. Observations for more time are not shown as equilibrium condition is reached. It is observed that the angular velocity generated in *Pneumatic Twisle* using 20 mm diameter PAM is quiet in agreement with fastest angular velocities of pronation-supination. This confirms we can achieve slowest angular velocity in closed loop operation.

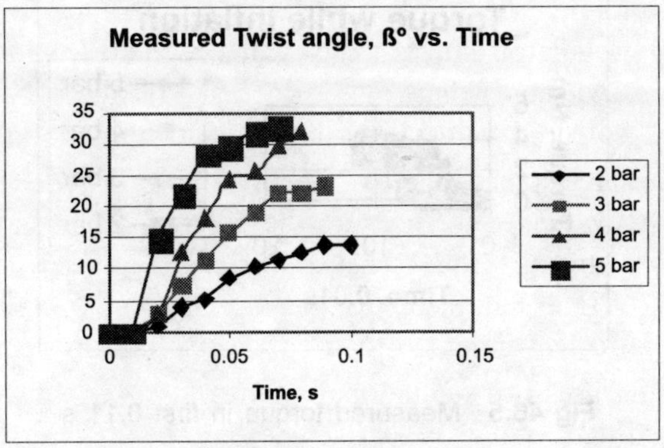

Fig. 46.2 Measured twist angle β° vs. time

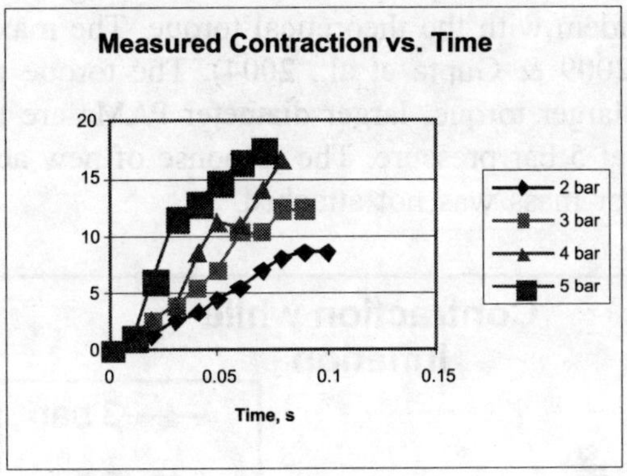

Fig. 46.3 Measured contraction vs. time

Measured contraction vs. *time* while inflating the muscle is shown in Fig. 46.3. More contraction is observed at higher pressures.

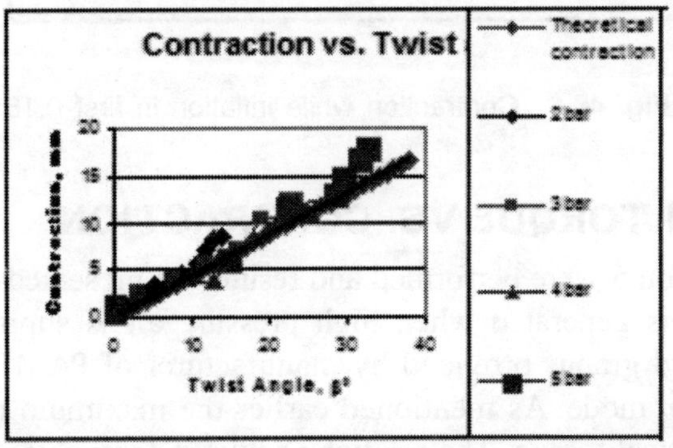

Fig. 46.4 Measured contraction vs. twist angle, β°

The *theoretical* and *actual contractions* against *twist angle* β are depicted in Fig. 46.4. The errors in actual and measure values are within 5% except for 5 bar. The errors are by virtue of wide tolerances in plastic parts and can be sorted out.

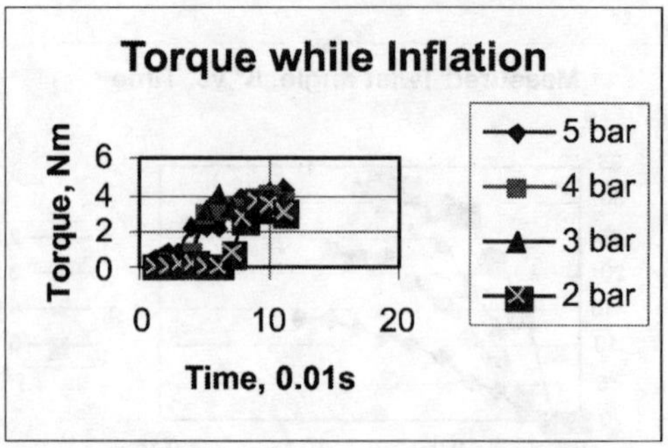

Fig 46.5 Measured torque in first 0.11 s

Torque sensors with angle measurement facility in static conditions are very costly. Thereby, in the second set of experimentation only torque and contraction were measured by replacing angle sensor. *Actual torque* measured at the free end of *Pneumatic Twisle* in open-loop mode is shown in Fig. 46.5. The maximum torque is 4.25 Nm at 5 bar. These values are also in tandem with the theoretical torque. The maximum values of pronation-supination torque are 9.1 Nm (Ramsay et al., 2009 & Gupta et al., 2004). The torque produced by the *Pneumatic Twisle* is less than biomechanical values. For larger torque, larger diameter PAMs are to be used. As observed in Fig. 46.6, the maximum contraction is 27 mm at 5 bar pressure. The response of new actuator to contraction is different than Fig. 46.3 as in that case torque sensor mass was not attached.

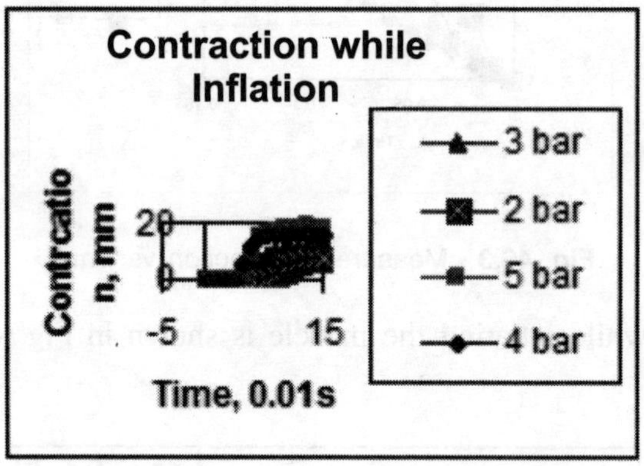

Fig. 46.6 Contraction while inflation in first 0.15

46.7 RELATION BETWEEN TORQUE VS. CONTRACTION

The experiments as narrated in Section 5 were performed and results are presented in subsequent figures. As mentioned in Section 1, the contractile force is generated when high pressure air is supplied to the *Pneumatic Twisle*. This parameter is available from standard graphs provided by manufacturer of PAM and is used to obtain *actual torque* likely to be produced in *closed-loop* mode. As mentioned earlier the maximum measured torque agrees (percentage discrepancy –5.8%) with the theoretical torque. The relations between *torque vs. % contraction* is shown in Fig. 46.7 (a), *torque vs. contraction* (mm) in (b) and *torque vs. twist angle* (c).

Please note that these graphs are to be used when pressure controlled closed-loop operations are adopted.

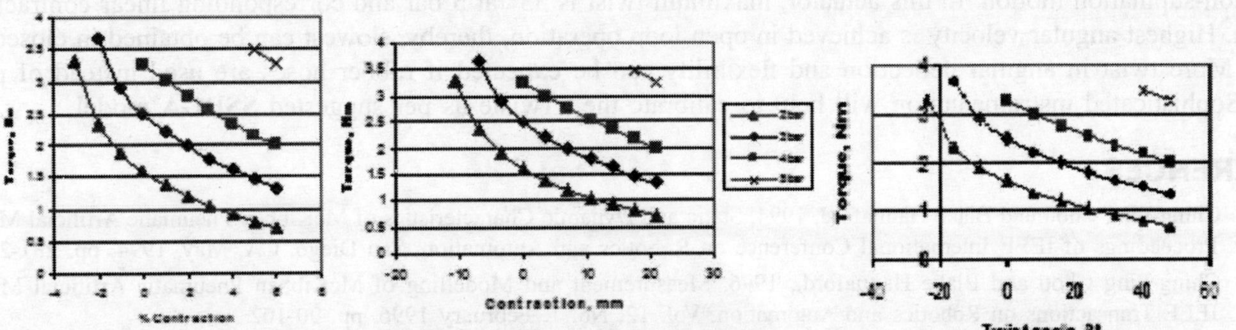

Torque vs. %Contraction (a)

Torque vs. % Contraction for Muscle Length 350 mm (b)

Torque vs. Twist angle, β° for Muscle Length 350 mm (c)

Fig. 46.7 Torque vs. contraction

46.8 COMPARISON WITH EARLIER RESULTS

The literature on 'power in open-loop testing' is not available. Most of the articles are dedicated to closed-loop pressure control and dealt with contractile force only and not with velocities. In this article we dealt with open-loop operation as it delivers maximum velocity at maximum force and at maximum power.

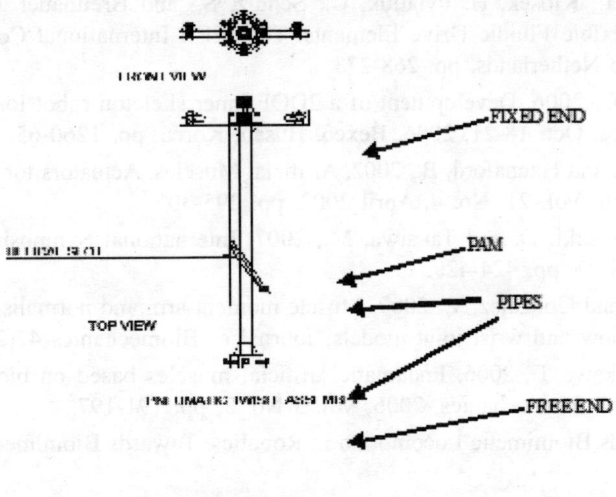

Fig. 46.8 pTwisle/pneumatic twisle assembly

46.9 ERROR ANALYSIS AND FURTHER DEVELOPMENT

The non-linearity in relations and errors in experimentation are observed. Those probably contributed by (a) the flexural properties of the plastic parts used to covert part of contractile force into torque (b) quick admission of air (c) 8-bit A/D converter (d) wide range in voltage regulation (e) wide tolerances on plastic parts. These could be eliminated by using sophisticated data acquisition system and using both modes of experimentation, i.e., closed-loop and open-loop.

46.10 CONCLUSION

The maximum twist angle expected in design for 'pronation-supination' motion is 180° without axial contraction, is still a challenge in flexible actuators. We have developed *linear-cum-twisting* actuator assembly inspired from pronation-supination motion. In this actuator, maximum twist is 33° at 5 bar and corresponding linear contraction is 18 mm. Highest angular velocity is achieved in open loop operation, thereby, slowest can be obtained in closed-loop mode. More twist in angular deflection and flexibility can be expected if rubber hoses are used instead of plastic parts. Sophisticated instrumentation will help to calibrate the pTwisle as per suggested **SSDGA** model.

REFERENCES

1. Ching-Ping Chou and Blake Hannaford, 1994, Static and Dynamic Characteristics of McKibben Pneumatic Artificial Muscles, Proceedings of IEEE International Conference on Robotics and Automation, San Diego, CA, May, 1994, pp. 281-286.
2. Ching-Ping Chou and Blake Hannaford., 1996, Measurement and Modelling of McKibben Pneumatic Artificial Muscles, IEEE Transactions on Robotics and Automation, Vol. 12, No. 1, February 1996, pp. 90-102.
3. Daerden F., and Lefeber D., 2002, Pneumatic artificial muscles: actuators for robotics and automation, European Journal of Mechanical and Environmental Engineering, 2002, Vol. 47 (1), pp. 41-50
4. De Greef, A., Lambert, P., and Delchambre A., 2008, Towards flexible medical instruments: Review of flexible fluidic actuators, Precision Engineering (2008) In press.
5. Gupta, A., and O'Malley, M.K., 2004, Design of haptic arm exoskeleton for training and rehabilitation. Proceedings of IMECE'04, ASME International Mechanical Engineering Congress and RD&D Exopo, Anaheim, California, USA, November 13-19, 2004, pp. 1-8.
6. Hirai Shinichi, Cusin Pierre, Tanigawa Hiroki, Masui Tomohiro, Konishi Satoshi and Kawamura Sadao, 2000, Qualitative Synthesis of Deformable Cylindrical Actuators through Constraint Topology, Proceedings of the IEEE/RSJ International Conference on Intelligent Robots and Systems, pp. 197-202.
7. Ivlev, O., 2009, Soft Fluidic Actuators of rotary type for safe physical human machine interaction, 11th International Conference on rehabilitation robotics, Kyoto International Conference Center, Japan, June 23-26, 2009, pp 1-5.
8. Kargov, A., Breiteiwieser, H., Klosek, H. Pylatiuk, C., Schulz, S., and Bretthauer, G. 2007. Design of a Modular Arm Robot System based on Flexible Fluidic Drive Elements. The IEEE International Conference on Rehabilitation Robotics, June 12-15, Noordwijk, The Netherlands, pp. 268-273.
9. Kiguchi, K., and Kundu, S.K., 2006, Development of a 2DOF inner skeleton robot for forearm motion assist, SICE-ICASE International joint conference, Oct. 18-21, 2006, Bexco, Busan, Korea, pp. 1260-65.
10. Klute G.K., Czerniecki, J.M., and Hannaford, B., 2002, Artificial Muscles: Actuators for Biorobotic Systems, The International Journal of Robotics Research, Vol. 21, No. 4, April 2002, pp. 295-309.
11. Maruta, H., Noritsugu, T., Sasaki, D. and Takaiwa, M., 2007, International Symposium on microNano Mechatronics and Humanscience , 2007, MHS'07, pp. 424-429.
12. Ramsay, J.W., Hunter, B.V. and Gonzalez, V., 2009, Muscle moment arm and normalised moment contributions as reference data for musculoskeletal elbow and wrist joint models, Journal of Biomechanics 42(2009), pp. 463-473.
13. Saga N., Nagase J. and Saikawa T., 2006, Pneumatic artificial muscles based on biomechanical characteristics of human muscles, Applied Bionics and Biomechanics, 2006, Vol. 3 No. 3, pp. 191-197.
14. Schindling, E., 2006, Towards Biomimetic Locomotion in Robotics, Towards Biomimetic Locomotion in Robotics Goteborg, Sweden, 2006, 18 pages.
15. Shadow 2009, Shadow Robot Company: 2009.
16. Tondu, B., and Lopez P., 2000, Modeling and Control of McKibben Artificial Muscle Robot Actuators, IEEE Control Systems Magazine, April 2000, pp. 15-39.
17. Trivedi, D., Dienno, D., and Rahn, C.D., 2008, Optimal Model-based design of soft robotic manipulators, Journal of Mechanical Design, September 2008, Vol. 130, pp. 091402-1 to 9.
18. Tsagarakis, N.G., and Caldwell, D.G., 2003, Development and control of a soft actuated exoskeleton for use in physiotherapy and training, Autonomous robots 15, pp. 21-33.
19. Tsukagoshi, H., Kitagawa, A., and Kamata, Y., 2002, Wearable fluid power composed of transformed flat tube actuators, Proceedings of the 2002 IEEE/RSJ International Conference on intelligent robotics and systems EPFL, Lausanne, Switzerland, October 2002, pp. 1178-1183.
20. Yamamoto, K., Hyodo, K., Ishii, M., and Matsuo,T., 2002, Development of power assisting suit for assisting nurse labor, JSME International Journal Series C, Vol. 45, No. 3, pp. 923-930.
21. Yee, N., and Coghill, G., 2002, Modelling of a novel rotary pneumatic muscle, Proceedings of Australian Conference of Robotics and Automation, pp. 186-190.
22. Zhang, J.-F., Yang, C.-J., Chen, Y., Zhang, Y., and Dong, Y.-M., 2008, Modeling and Control of a curved pneumatic muscle actuator for wearable elbow exoskeleton, Mechatronics, 18(2008), pp. 448-457.

Index